T0136318

Accessing the E-book edition

Using the VitalSource® ebook

Access to the VitalBook™ ebook accompanying this book is via VitalSource® Bookshelf — an ebook reader which allows you to make and share notes and highlights on your ebooks and search across all of the ebooks that you hold on your VitalSource Bookshelf. You can access the ebook online or offline on your smartphone, tablet or PC/Mac and your notes and highlights will automatically stay in sync no matter where you make them.

1. **Create a VitalSource Bookshelf account at** *https://online.vitalsource.com/user/new* or log into your existing account if you already have one.

2. **Redeem the code provided in the panel below to get online access to the ebook.** Log in to Bookshelf and click the **Account** menu at the top right of the screen. Select **Redeem** and enter the redemption code shown on the scratch-off panel below in the **Code To Redeem** box. Press **Redeem**. Once the code has been redeemed your ebook will download and appear in your library.

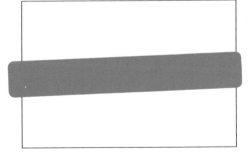

DOWNLOAD AND READ OFFLINE

To use your ebook offline, download BookShelf to your PC, Mac, iOS device, Android device or Kindle Fire, and log in to your Bookshelf account to access your ebook:

On your PC/Mac

Go to *http://bookshelf.vitalsource.com/* and follow the instructions to download the free **VitalSource Bookshelf** app to your PC or Mac and log into your Bookshelf account.

On your iPhone/iPod Touch/iPad

Download the free **VitalSource Bookshelf** App available via the iTunes App Store and log into your Bookshelf account. You can find more information at *https://support.vitalsource.com/hc/en-us/categories/200134217-Bookshelf-for-iOS*

On your Android™ smartphone or tablet

Download the free **VitalSource Bookshelf** App available via Google Play and log into your Bookshelf account. You can find more information at *https://support.vitalsource.com/hc/en-us/categories/200139976-Bookshelf-for-Android-and-Kindle-Fire*

On your Kindle Fire

Download the free **VitalSource Bookshelf** App available from Amazon and log into your Bookshelf account. You can find more information at *https://support.vitalsource.com/hc/en-us/categories/200139976-Bookshelf-for-Android-and-Kindle-Fire*

N.B. The code in the scratch-off panel can only be used once. When you have created a Bookshelf account and redeemed the code you will be able to access the ebook online or offline on your smartphone, tablet or PC/Mac.

SUPPORT

If you have any questions about downloading Bookshelf, creating your account, or accessing and using your ebook edition, please visit *http://support.vitalsource.com/*

SYSTEMS ENGINEERING

*Design Principles
and Models*

SYSTEMS ENGINEERING

Design Principles and Models

Dahai Liu

CRC Press
Taylor & Francis Group
Boca Raton London New York

CRC Press is an imprint of the
Taylor & Francis Group, an **informa** business

CRC Press
Taylor & Francis Group
6000 Broken Sound Parkway NW, Suite 300
Boca Raton, FL 33487-2742

Printed on acid-free paper
Version Date: 20150410

International Standard Book Number-13: 978-1-4665-0683-1 (Pack - Book and Ebook)

Library of Congress Cataloging-in-Publication Data

Liu, Dahai.
    Systems engineering : system design principles and methods / Dahai Liu.
        pages cm
    Includes bibliographical references and index.
    ISBN 978-1-4665-0683-1 (hardback)
    1. Systems engineering. I. Title.

    TA168.L476 2015
    620.001'171--dc23   2015003480

To my wife, Juan, and my children, Meryl and Jenny,

for their love, patience, and support.

Contents

Section II Systems Methods, Models, and Analytical Techniques

Section III   Systems Management and Control Methods

Preface

With more complex systems being designed and utilized, understanding concepts of systems and mastering systems engineering methods are of utmost importance for systems designers. Complexity exists in all levels of systems and their subsystems, even components, which makes the design challenging. Guided by systems science, systems engineering applies scientific theories and models and incorporates the factors of system elements, including humans, machines, and environments, into unified models to analyze the effects of various system elements on system behaviors, which, in turn, seek the optimal course of actions to design these complex systems. Systems engineering is multidisciplinary in nature, integrating concepts, models, and techniques from a number of disciplines, including natural science, social science, and engineering, at both basic research and applied research levels. For the past several decades, systems engineering has grown rapidly in its scope and application and shown significant benefits for the design of large, complex systems.

I have been educating systems engineers for over ten years, teaching systems engineering at both undergraduate and graduate levels. My students are different from most systems engineering majors, as these undergraduate and graduate students have backgrounds in a large variety of areas, including psychology, physiology, and some engineering fields. With my experience teaching students with a variety of backgrounds, I feel that a textbook in systems engineering targeted at nontraditional engineers is in demand. The current systems engineering textbooks are either too technical or at a high conceptual level, giving these students very limited choices. Students need a good book that not only gives them exposure to the concepts of systems and systems engineering, but also enough technical expertise for them to use and apply what they learn on the job. That is the rationale for developing this book.

This book is written primarily for students with diverse backgrounds to learn about systems and systems engineering, and, more specifically, to be able to use and apply the models and methods in the systems engineering field. The materials included in the book have been taught for many years in the Human Factors and Systems Department at Embry-Riddle Aeronautical University; it has integrated feedback from students and colleagues and is written at a level appropriate for those groups of students to learn systems engineering, especially those nonengineering students who have no prior exposure to this subject. Engineering students, on the other hand, may also find this book useful and handy, as it provides a comprehensive overview of the subjects as well as the relevant analysis models and techniques. This book should serve well as a reference book for professional systems engineers.

This book has ten chapters, organized in three parts, including systems and systems engineering concepts (Chapters 1 and 2), systems methods, models, and analytical techniques (Chapters 3 through 9), and systems management and control methods (Chapter 10). The approach leans toward process-oriented and model-based systems engineering, with necessary topics covered in different chapters. Chapter 2 describes the system life cycle and systems engineering design process. Systems design starts with requirements; beginning with a good set of requirements is critical for the success of the design, and developing and analyzing requirements are not usually covered by a systems engineering book; that is why Chapter 3 is dedicated to this subject. It is believed that students will have a solid understanding of system requirements after having read this detailed information; a software package (CORE) is also used to introduce the requirement management process. Chapter 4 describes the design process in greater detail, especially the functional models. Chapters 5 through 9 explain the necessary technical measures and models that most systems designs will involve, including reliability, maintainability, supportability, usability (human factors), decision-making models under risks and uncertainty, optimization models, process models (queuing models and simulation with Arena software), and cost analysis using engineering economy models. The last chapter (Chapter 10) gives a comprehensive overview of systems management based on systems engineering management plans (SEMP) and systems control based on critical path method (CPM) and program evaluation review technique (PERT) models. In terms of the mathematics involved, it is believed that no specific prerequisite requirements are needed, except for fundamental algebra and basic probability and statistics theory; Appendix I is provided at the end of the book for readers to review these materials. This book can be used as a textbook at either undergraduate upper-level or graduate starting-level courses; exercise problems are provided at the end of each chapter for students to gain hands-on experience with the concepts.

I thank each individual who has encouraged me and assisted me in the process of writing this book. Specifically, I would like to thank Meaghan Hart, Julian Archer, and Hemali Virani for their assistance in preparing this book. Special thanks go to the senior editor Cindy Carelli, project coordinator Laurie Oknowsky, project editor Richard Tressider, and Michele Smith of CRC Press for their assistance, advice, and support throughout the process. I would also like to thank Katie Thacker and Bethany Maddox of Vitech Corporation and Jon D. Santavy of Rockwell Automation for their cooperation with the project. And finally, I want to express my appreciation to my family; their love and support are the inspiration for my career.

Dahai Liu

Author

Dahai Liu is currently a professor in the Department of Graduate Studies in the College of Aviation at Embry-Riddle Aeronautical University, Daytona Beach, Florida. He has been teaching systems engineering since 2002. His research areas are system modeling and simulation, human factors engineering in unmanned systems, human/machine learning, and artificial intelligence. Professor Liu has over 70 publications in professional journals and proceedings of conferences, technical reports, and a number of book chapters. He has conducted numerous research and consulting projects funded by the government and industrial companies such as the National Aeronautics and Space Administration, the Office of Naval Research, the Federal Aviation Administration, and Honeywell. Professor Liu received his Ph.D. in industrial and management systems engineering from the University of Nebraska at Lincoln in 2002 and his M.S. in systems engineering from Tsinghua University in 1996.

Section I

Systems and Systems Engineering Concepts

1

Introduction: Systems and Systems Engineering

Systems engineering and systems science has become one of the most important, comprehensive, and fundamental fields, and has wide application in almost every area of our society; from the macromanagement of government to the daily production of manufacturing facilities, from the development of spacecraft to the design of small consumer products, systems engineering is being applied at all times, at different levels. As one of the engineering disciplines, systems engineering studies systems; in this first chapter of the book, we will review the basic scientific concepts of systems, understand the background for systems science, and examine the evolvement of systems and systems engineering from a historical perspective. This will help readers to understand the origin of systems science and the need for systems engineering. In this chapter, we will first define systems and systems engineering, describe the unique characteristics of systems engineering, and give a brief historic introduction to systems engineering development.

1.1 Systems

The term *system* is no stranger to us; we have been exposed to different kinds of systems since our early school years. According to the *Oxford English Dictionary*, a system is "a set of connected things or parts forming a complex whole," or "a set of principles or procedures according to which something is done; an organized scheme or method." Although the basic definition does not change, the term means slightly different things in different disciplines; for example, in physiology, human systems are sets of organs working together to serve common human physiological functions, such as the human nervous system, which serves as the body control center and communications network for the human body; or the human digestive system, which turns food into the energy humans need to survive and process food residue and waste for disposal. For human factor professionals, a system is the combination of human, machine, and

3

the surrounding environment. For systems engineering, we must define "system" more specifically:

> *A system can be broadly defined as a set of integrated components that interact with each other and depend upon each other, to achieve a complex function together. A system can be decomposed into smaller subsystems or components and a system may be one of the components for a larger system.*

The following characteristics are present in all systems:

- *A system has a main function or a meaningful purpose.* To perform a function, a system takes input and generates output through the mechanism defined in the system structure. Systems' functions, functional models, and analyses are illustrated in greater detail in Chapters 3 and 4. In this chapter, we briefly introduce the function's concepts. For example, an automobile system's function is to transport humans/goods to destinations. To perform this function, the system receives inputs and energy that allow the driver to start, control and drive the system get to the destination location; its inputs include the driver's control and energy provided by the gasoline or electricity, and the output of the system is the velocity change (momentum).

- *A system has a hierarchical structure.* As seen in the definition, a system usually consists of many subsystems that can also be smaller-scale systems; meanwhile, a system is usually a subsystem of a higher-level system. For example, our planet is a large biological system, consisting of many smaller systems, such as the ocean and continents. Planet Earth is simultaneously a subsystem of the larger solar system.

- *A system has interacting subsystems and components that interact with each other.* Components' interactions are necessary for achieving the system's functions. A system is not simply the sum of all components, but rather it is an integrated whole package of the subsystems and components working in a well-defined hierarchical structure.

- *A system has a life cycle.* The system life cycle is the time span that starts from the concept of the system, continuing through the exploration of the concept, design and development, system operation, and maintenance, until the system is retired and discontinued.

- *A system has reliability.* No system is 100% reliable and error free. As Murphy's law states, "Anything that can go wrong will go wrong" at some point. One of the most important design objectives is to keep Murphy's law in mind and make the system more reliable.

More formally presented, systems have the following elements:

1. Subsystems and components: These are the fundamental constructs/ functional units for systems. If desired, systems can be decomposed almost infinitely, all the way to the microworld level, such as electrons and atoms. For most design purposes, it is not necessary to decompose a system to this level of depth. As a rule of thumb, the decomposition usually stops at the assembly level, that is, at the level at which the commercial off-the-shelf (COTS) items can be obtained externally. The lowest level of assembly components and the other levels of the system are called subsystems. Subsystems are described through a hierarchical numbering system such as 2.0, 2.1, 2.1.1; by definition, 2.1 is a subsystem of 2.0, and 2.1.1 is a subsystem of 2.1. The hierarchical modeling approach will be illustrated in greater detail in Chapter 4. It is important to note that the subsystem and components may sometimes not be tangible hardware items. Depending on the nature of the system, some components can also be software, human, or even information.

2. Systems components have attributes. These attributes, often called design-dependent parameters (DDP), define and specify the systems components. For example, the physical dimensions of a component, mean time between failures (MTBF), power input and output, and so on. The purpose of systems design, to some extent, is to derive these quantitative and qualitative attributes from the systems requirements, so that specific system components can be built or obtained.

3. Systems components and subsystems interact and regulate system behaviors through different relationships. A system starts with user/ customer requirements; *requirements* are the *basis* for systems *functions*; higher-level *requirements* can be *refined* by lower-level *requirements*; higher-level *functions* are *decomposed by* lower-level *functions*. Each *function* is *performed by* one or more *components*. These relationships are essential for successfully translating the systems requirements into component attributes, providing system rationale, and providing traceability for systems design activities. Systems design and modeling, including the specification of the system components and their relationships, is discussed in greater detail in Chapters 3 and 4.

Let us illustrate the systems concepts above with the example of a fixed-wing airplane (such as a Cessna 172). The function of a fixed-wing airplane is flight. In order to achieve this function, an airplane needs to have the necessary components; these components are used to construct the airplane, and they work together and interact, for the airplane to be able to take off, cruise, and land safely. Meanwhile, the components must be able to communicate with other airplanes and air traffic controllers. Among fixed-wing airplanes, major components are the airframe and propulsion systems. The airframe subsystem can be further decomposed into smaller subsystems, such as fuse-lage, empennage, wing, landing gear and so on. The empennage subsystem consists of stabilizers, rudders, tabs, and elevators. Propulsion systems consist of components such as the power-plant subsystem (the engine) and electrical control systems (avionics). These smaller components are themselves a system on a smaller scale, which can also be further decomposed into smaller components, if desired. All the components work and interact with each other to provide some kind of function, in order to serve the bigger system's functions.

1.2 Systems Classification

Depending on the perspectives from which a system is studied, it can be classified into different categories. Understanding the system categories can help us to narrow the scope of the systems and derive common system characteristics. In categorizing the system, one has to keep in mind that none of the classifications makes a clear cut, and furthermore, any subsystem might belong to several different categories; for example, a man-made system could be dynamic and closed-loop controlled. Generally speaking, a system can be classified into one or more of the following categories: natural or man-made system, static or dynamic system, conceptual or physical system, and open or closed system.

1.2.1 Natural Systems versus Man-Made Systems

A natural system is a self-organized system that nature formed after millions of millions of years' selection and development. Examples of natural systems are the planet, oceans, and natural lakes. A natural system sustains itself by self-organizing to a state of equilibrium, for example, the food chain in a natural lake. Any disturbance to this equilibrium can be devastating for the natural system. A man-made system, on the other hand, is made by humans. Man-made systems, such as computers or automobiles, cannot be obtained from nature, but only through the creative efforts of humans. Systems engineering studies man-made systems as objects, with less concern for natural systems. One has to keep in mind, however, that there is no absolute isolation between natural systems and man-made systems. As a matter of fact, man-made and natural systems constantly interact with each other and sometimes make a huge impact on each other. Man-made systems often need inputs from nature (i.e., the automobile needs gasoline, which is made from crude oil from the natural world) and rely on nature to process the waste generated (e.g., the greenhouse gases from the automobile). With more advances in our technological development, man-made systems are becoming more complex and powerful, which requires more resources from Mother Nature. Meanwhile, more waste returns back to nature, often polluting and causing the natural environment to deteriorate. Industrial pollution is becoming a more serious problem now, and with more awareness from humans, environmentally friendly systems designs have become part of system competitiveness and advantages.

1.2.2 Static Systems versus Dynamic Systems

Systems can also be classified as *static* or *dynamic*. Static systems are those structural systems that do not change their state within a specified system life cycle, such as a bridge, a building, or a highway. A dynamic system is one where its state, or the state of its components, changes over time, either in a continuous or discrete manner. A dynamic system's state can be considered a function of time, its change taking place at either a more deterministic rate or a more stochastic rate (think of a large service center's change of state as customers arrive in a random pattern). Similarly to natural versus man-made systems, the distinction between static and dynamic systems is relative, not absolute. Depending on the different perspectives from which a system is being analyzed, a system may be considered static or dynamic. For example, from Earth, within a short period of time, the Great Wall of China is considered a static system; however, if one observes it from outer space, the Great Wall is certainly moving, although at a very slow speed. While there may be no absolutely static systems, within the context of systems engineering, we treat some fixtures or structures as static. By doing so, we can simplify the problem, as we need to focus only on their dynamic components. For example, when we study operations within a production facility, the facility building is considered static; we only need to concentrate on the dynamic aspects of the system, such as material flows and human activities to investigate the effectiveness of production management.

1.2.3 Conceptual Systems versus Physical Systems

Our world comprises physical systems, and physical systems consist of objects that can be seen, touched, and felt. Natural systems such as animals, bacteria, lakes, and humans are all physical systems; physical systems also include man-made systems, such as the computers, appliances, tools, and equipment that we humans use on a daily basis. Conceptual systems are those consisting only of concepts, not real objects, so we cannot visually see or physically touch these systems. Conceptual systems illustrate the relationships among objects and allow us to understand the system and communicate details about the system's structures and mechanism. A simulation model of a factory operation process, a blueprint of the machine assembly, or the information processing model of human cognition and perception would be examples of conceptual systems. Science and mathematics are the fundamentals of conceptual models. A conceptual model can be general for a wide range of objects or it can be specific for a certain physical system. In systems engineering, conceptual modeling serves as a basis for the physical system. Before the actual physical systems are manufactured, parts procured and systems assembled, a conceptual model is usually built first to allow us to analyze the feasibility of such a system and assess the fundamental characteristics of system performance. Data collected from conceptual models helps designers to make the necessary adjustments; it is usually cheaper and quicker to modify a conceptual system than a physical system. Conceptual systems design is a very critical step for systems engineering design, as most of the systems analysis occurs at this stage and most systems specifications will be determined in the conceptual model. From a conceptual model, a physical model can be easily derived. Conceptual design will be covered in Chapter 2.

1.2.4 Open Systems versus Closed Systems

Systems can be classified based on their interaction with the environment. Open systems exchange information, matter, or energy with their surrounding environments, while closed systems do not exchange such things. In a very strict sense, there is no absolutely closed system existing in the universe, while any system can be more or less considered an open system. Generally speaking, systems are thought of as closed systems when the exchange of matter, information, or energy can be ignored. In physics, closed systems are further classified as closed systems or isolated systems. An isolated system has no exchange of anything with the outside environment, but when only energy is exchanged, the system is closed but not isolated. An isolated system is an ideal closed system, which is practically nonexistent, but closed systems can be found in our daily life, such as a sealed container of water or gas. In thermodynamics, the laws of thermodynamics requires the system to be classified exactly as isolated, closed, or open; thus, system energy and entropy (the amount of energy used for work) can be determined. Open or closed systems concepts also apply in engineering systems and social systems, with different components: large numbers of biological objects or groups that interact with each other, rather than physical particles or objects. In systems engineering, a system is considered open—especially when one takes a life cycle perspective—if the system interacts with the environment and its impact on the environment is one of competitive advantage that system design should include in the early development stages.

Besides the four types of classification, systems can also be classified into other different categories. For example, in system simulation, we often distinguish between continuous systems and discrete systems; in the business and management field, there are manufacturing systems and services systems. The classification is really dependent on the specific system we are studying and the objective of the study.

System science and systems engineering have developed along with the development of technology. System concepts and systems thinking have become more critical in industry, especially after the industrial revolution;

manufacturing has been transformed from human labor-based, small-scale workshops to large factories using complex machinery systems. The transition to the machine age enabled many new machines to be developed; machines gradually replaced human labor in many areas, freed humans from repetitive and dangerous work, increased work efficiency, and in turn made more advanced machines to achieve more complex functions and tasks. After many decades of machine age development, technology has served and replaced almost every aspect of our everyday life, including the manufacturing sector. Now, we can build a new machine without worrying about the details or the fundamental parts of the machines, as we can find the parts to assemble them quickly. The focus has now shifted from the machine age to the system age. The new problem is not to make a single device work; it is how to use the available technology to build something more complex, more powerful, more user friendly, and more efficient. When we begin building systems, systems thinking becomes more important than at any time previously. We need to have the big picture first, just as when we draw a large picture, we compose and sketch the system structure first before we put in any details and colors. A good design process will ultimately determine the success of the system, and the philosophy of the big picture, carried out by a structured process, is called systems engineering. In the next section, we will define systems engineering, review its fundamental features as well as its historical perspective, and introduce the profession of systems engineering.

1.3 Systems Engineering

Before we get into systems engineering, let us first talk about engineering. Engineering, according to the American Board for Engineering and Technology (ABET n.d.), is

> the profession in which a knowledge of the mathematical and natural sciences gained by study, experience, and practice is applied with judgment to develop ways to utilize, economically, the materials and forces of nature for the benefit of mankind.

It is the application of what we discover from science in man-made systems. Systems engineering, similarly, applies the knowledge, theories, models, and methods of systems sciences, based on the philosophy of systems thinking, to guide in designing man-made systems. The International Council on Systems Engineering (INCOSE 2012) defines systems engineering as follows:

> Systems engineering is an interdisciplinary approach and means to enable the realization of successful systems. It focuses on defining

customer needs and required functionality early in the development cycle, documenting requirements, then proceeding with design synthesis and system validation while considering the complete problem: operation, performance, test, manufacturing, cost and schedule, training and support and disposal. Systems engineering integrates all the disciplines and specialty groups into a team effort forming a structured development process that proceeds from concept to production to operation. Systems engineering considers both the business and the technical needs of all customers with the goal of providing a quality product that meets the user needs.

In all the systems engineering books available, there are some variations on the definition of systems engineering, but regardless of what definition is being used, systems engineering has the following common characteristics:

1. It is an applied science. Systems engineering applies scientific discoveries and mathematics to make a specific system work. Note we say specific here; that is because systems engineering does not really address the creation of general scientific theory, nor does it work for all systems. Its results only work for a specific system, not every system in general.

2. Systems engineering is concerned with the big picture of the system: it is top-down design processing, as opposed to starting with detailed bottom-up processing. In systems engineering, a system starts with a need that comes from the users/stakeholders of the system. This need is translated into concepts and is expressed in the format of systems *requirements*. Systems engineering design is driven and guided by requirements, which follows a structured iterative design process that constantly involves "analysis-synthesis-evaluation." Through this process, the details of the systems are gradually evolved into tangible, visible physical systems. The systems engineering process is a translation process, transforming users' needs into a working system through guided procedures and proper use of models and analytical methods. In this process, critical thinking and rigorous reasoning (top-down thinking) as well as creative thinking are both important for the success of the system. In turn, systems engineers are required to have a broader scope of knowledge and skills, compared to other traditional engineers; as we will see later, systems engineering is a general design philosophy, which can be applied to all kinds of complex systems design, including aerospace, aeronautics, manufacturing, software, transportation, and so on, to name a few, with special tailoring to the system being designed. A systems engineer not only needs a good understanding of the nature of the system, understanding of the domain knowledge of that specific system, knowledge of how the system works—at least at a high

level—but also have the ability to coordinate with different engineers in the design team. Thus, systems engineers need to have both knowledge of big-picture system thinking and good communication skills. A good systems engineering team is often the core determinant of a system's success.

3. Systems engineering is a multidisciplinary field. As we have discussed the systems engineering process, we have mentioned that systems engineering is multidisciplinary in nature; it is essentially team work. Arthur David Hall III (1969) proposed a three-dimensional morphology of systems engineering: time, logic and profession. The time dimension defines the major design phrases and milestones of the system development processes; the logic dimension defines the logical models and steps through which a system would evolve, similar to the system's life cycle status, including problem definition, design, synthesis, decision making, optimization, planning and actions, and so on. The third dimension, profession, describes the different professions that are involved in systems engineering work. Systems engineering professions include management, designers and engineers, graphical artists, and supporting technicians. Compared to several decades ago, at the present time the extent of this range of professions has expanded a great deal, due the scope and complexity of systems we are designing, which have arrived at a different level. With the advanced development of current technology in computing and the Internet, and the fast-growing trend of globalization and supply chain management, there are remarkable increases in the number and scope of professions. The current disciplines and professions involved can be summarized in four different categories:

   a. The art and science domain: This is the fundamental scientific knowledge pertaining to systems concepts and system analysis methods and models, including applied mathematics (operations research), human-related science (psychology, physiology), graphical design, architecture, environmental science, natural science, biology, and so on. The main mathematical model used in systems engineering is operations research, an applied field in mathematics.

   b. The engineering domain: Engineering teams are a core operational part of any systems design. Depending on different types of systems, the relevant engineering disciplines are included as a part of the systems design team. Moreover, in terms of system-level planning, scheduling and optimization, industrial engineering, human factors engineering, and management science are among the most relevant fields for systems engineering.

c. The management domain: In the current globalized environment, any business faces competition from all over the world. Utilizing the global supply chain is an essential part of core competitiveness for any business, translating to systems designers making management a more important role in the design and distribution process. For an effective design, management must not only provide a framework to manage and control the design team and its efforts, but also, more importantly, build a culture and management style that encourages and motivates employees, reinforcing user loyalty throughout the system life cycle and design process. Almost all management functions play a role in the design, including human resources, accounting, finance, and marketing. These functions provide an external liaison for the systems design teams and make the internal operations more efficient as well.

d. Supporting roles: Besides all the key players in systems design, as mentioned above, there are also personnel that support the design efforts, sometimes indirectly, such as the information technology technicians, legal department specialists, test and equipment technicians, and others. These support functions make sure the daily design activities are carried out smoothly, which is also essential for any design team.

The main professional association for systems engineering is the International Council on Systems Engineering (INCOSE). INCOSE is a non-profit organization dedicated to the field of systems engineering. Founded in 1990, INCOSE has over 9000 members (as of December 2013) all over the world, representing a wide range of expertise and backgrounds from industry and academia. The main mission of INCOSE is to "share, promote and advance the best of systems engineering from across the globe for the benefit of humanity and the planet" (www.incose.org). INCOSE offers professional training and certification in systems engineering; for more information, refer to www.incose.org.

1.4 Brief History of Systems Engineering

Systems engineering is a relatively young field, compared to other engineering disciplines. However, its fundamental concept, system science, can be traced back to ancient literature as early as the eighteenth century, when science knowledge exploded, especially in the natural and physical sciences. The creation of structure in different areas of scientific knowledge set in motion the development of system science. Since then, the fundamental concepts of system science have been presented in many scientific disciplines.

1.4.1 From Reductionism to System Thinking

There are two major milestones in the development of system science that led directly to the growth of the field of systems engineering: the machine age and the system age. They are based on different views of systems. The machine age is largely based on reductionism. Reductionism is a philosophical view of science. It is based on the principle of causality, that is, a system behavior can be explained completely by its fundamental elements or components; it holds the premise that the components of a system are the same when examined separately. Since the early nineteenth century, with the development of new technology, devices and tools, such as the steam engine, the industrial revolution has marked a turning point for modern manufacturing and production. More and more advanced machinery has been developed to replace human power; machines have been developed to achieve more sophisticated functionality. The main principle for designing these machine systems are cause-effect control systems, which are the main characteristic of the machine age. The focus was on designing the control mechanism for the machine; and by integrating different components together, a bigger system can be built. In the machine age, the design of systems largely depends on the reductionism principle, following a bottom-up process. As the technology for individual machines becomes more developed and system development moves toward to a more complex level, reductionism has started to show its limitations when dealing with increasing levels of complexity.

Using reductionism, design methods could fail, especially when applied to complex phenomena such as human society, biological sciences, behavioral sciences, and managerial sciences, as these systems are difficult to examine as isolated entities, and the system is not simply the sum of its elements in a linear manner. The systems approach, however, is more holistic, and follows the concept of expansionism. A system approach argues that even if every part is performing well for its objective, the total system might not be performing well for the system objective if the parts are imperfectly organized (Parnell et al. 2008).

System design moved from the lower level of the machine age to the system age. In the system age, systems approaches look at systems as an integrated whole unit, with the composed components interacting together to serve the systems' purposes, often in a nonlinear manner. The unique characteristic of the system age is the design of systems from a top-down approach; instead of looking at individual components first, system thinking starts with the system as a whole at the beginning, by looking at the big picture of the system, identifying the objectives (requirements) of the system, and having the objective to direct the design of the system. After decades of continuous evolution, systems engineering has become the standard approach for complex system design. In the following section, major historical events area described to briefly show the evolution of systems engineering.

1.4.2 Early Practices

The first recognizable use of the systems approach was in the telephone industry in the 1920s and 1930s. It was Bell Telephone Laboratories who first used the term "systems engineering" in the 1940s. The concepts of systems engineering within Bell Laboratories traces to the early 1900s and describes major applications of systems engineering during World War II. At that time, a group of U.S. and British scientists from various disciplines tried to resolve the problem of achieving the optimal military strategies and actions using limited resources. These practices directly led to the birth of an applied mathematics, known as *operations research*, now a major applied mathematical methodology used in systems engineering. The application of operations research proved substantial through its successful application to military operations. Around the same time, the National Defense Research Committee established a Systems Committee with Bell Laboratories support to guide a project called C-79, the first task of which was to improve the communication system of the Air Warning Service.

In 1946, the Research and Development (RAND) Corporation was founded by the United States Air Force; the systems analysis applied within RAND has been considered an important part of systems engineering development. It was the first to propose and utilize the "system analysis" approach and demonstrate its significance. During this time, the first formal attempt in history to teach systems engineering started at the Massachusetts Institute of Technology (MIT) in 1950 by G.W. Gilman, the director of systems engineering at Bell Laboratories. With the wide application of operations research and control theory, together with the development of digital computers, the scientific foundation for systems engineering has been established.

1.4.3 Government Role

Government has played an important historical role in promoting systems engineering, since most of the early complex systems were requested by government. With more demand for and experience in developing large complex systems, the need to identify and manipulate the properties of a system as a whole—which in complex engineering projects may differ greatly from the sum of the parts' properties—motivated agencies such as the Department of Defense (DoD) and National Aeronautics and Space Administration (NASA) to pay more attention to applying systems engineering discipline. Since the late 1940s and early 1950s, the DoD has applied a systems engineering approach to the development of missiles and missile-defense systems. This undertaking was recognized as the Intercontinental Ballistic Missile (ICBM) Program, later the Teapot Committee, and it served as one of the key historical foundations of systems engineering. The ICBM was an idea given by Bernard Schriever to the Teapot Committee in 1954, and it ultimately changed the organizing principles of managing a systems development contract.

From this point on, "there would be a System Engineering contractor staffed by 'unusually competent' scientists and engineers to direct the technical and management control over all elements of the program" (Hallam 2001). The Teapot Committee helped establish systems engineering as a discipline by "creating an organization dedicated to the scheduling and coordinating of activities for subcontractor R&D, test, integration, assembly, and operations" (Hallam 2001). The coordination required for the system approach to the project was further driven by schedule demands containing concurrent development of designs, subsystems, manufacturing, and so on. It was then, through trial and error, that quantitative tools and methodologies were developed. These very tools formed the basis for the interdisciplinary trade studies and decision aids that enrich the field of systems engineering.

By the 1960s, degree programs in the discipline of systems engineering became widely recognized across U.S. universities and many systems engineering-related techniques were developed within more academic research and industry applications. For example, in 1958 the Program Evaluation and Review Technique (PERT) was developed by the United States Navy Special Projects Office for the Polaris missile system. PERT is a scheduling method designed to plan a manufacturing project by employing a network of interrelated activities, coordinating optimum cost and time criteria. PERT emphasizes the relationship between the time each activity takes, the costs associated with each phase, and the resulting time and cost for the anticipated completion of the entire project. Since existing large-scale planning was inadequate at the time, the Navy teamed up with the Lockheed Aircraft Corporation and the management consulting firm of Booz, Allen, and Hamilton, working in large-scope systems engineering efforts. Traditional techniques such as line of balance, Gantt charts, and other methods were found limited in dealing with variability, and PERT evolved as a means to deal with the varied time periods needed to finish the critical activities of an overall project.

By the mid-1960s, in response to many contractors expressing a need for greater latitude in applying alternative systems engineering techniques, several criteria-oriented military standards were issued, including MIL-STD-499A. Army FM 770-78 and MIL-STD-499A formed the foundation for the current application of systems engineering concepts and requirements in military development programs. It was in the late 1950s and 1960s that the emergence of engineering discipline "specialists" on most development programs occurred.

In 1962, Arthur Hall published his first book on systems engineering: *A Methodology for Systems Engineering*. Hall was an executive at Bell Laboratories and was one of the people who were responsible for the implementation of systems engineering at the company. Hall reasoned that systems engineering was important because products were increasing in complexity, the needs of consumers were expanding, the business markets were growing rapidly, and there was an acute shortage of technically and scientifically trained people handling the complex systems design. Systems engineering was in great demand in this era. According to Hall, systems engineering is a function with five phases: (1) systems studies or program planning; (2) exploratory planning, including problem definition, selecting objectives, synthesis and analysis, selection, and communication; (3) development planning; (4) studies during development, including system components development and integration and testing of components; and (5) current and emerging engineering, that is, current engineering technology and that which will exist while the system is operational (Buede 2009). Hall's book is considered a major milestone for systems engineering education.

In the 1960s, a third defining moment for systems engineering was seen: the birth of NASA's Apollo program, which lasted from 1961 to 1972. In fact, the Apollo program is probably the most classic example of systems engineering in practice to date. The task of sending humans to the moon was daunting and complicated; it involved breaking the underlying goal into multiple sections or manageable parts that participating agencies and companies could work with and comprehend. These various parts then had to be reintegrated into one whole solution, and as a result, careful attention and management involving extensive testing and verification was necessary. The complex nature of these tasks made systems engineering a suitable tool for designing such systems. It was the principles of systems engineering that resulted in the rigorous system solution which contributed to Apollo's overall success.

In 1972, the International Institute for Applied Systems Analysis (IIASA) was founded, with the intention of applying techniques of systems analysis to solve urban, industrial, and environmental problems that transcended international boundaries. Members of the IIASA discussed hallmarks of the IIASA approach to problem solving, which included interdisciplinary emphasis and the maintenance of credibility with both scientist and decision makers (Hughes, n.d.). The IIASA approach is illustrated in the organization's most successful venture: a transboundary air pollution project using the Regional Acidification Information and Simulation (RAINS) model of the impact of acidification in Europe. With efforts such as these, systems engineering has emerged from the government and military fields to almost every area of the industrial and social, and moved to a more comprehensive and advanced stage: the information age.

1.4.4 Information Age

With the development of hardware and software, computer technology made tremendous advances in the 1980s and more in the 1990s, which marked the beginning of the information age, or computer age/digital age. With fast-growing and innovative technologies such as the internet, humans have achieved an era of most effective, rapid technological growth than at any previous time. More and more complex, technology-driven systems are being designed every day, creating a new technological world for a wide range of users. With the shift from traditional industry to a highly computerized economy, the degree of complexity involved in the systems has reached a new high level. The increasing level of complexity has put the practice of systems engineering in greater demand. With the blooming of information technology, limited resources dwindling, and a more competitive market, the world has started a whole new chapter, characterized by globalization and the use of the supply chain in the design process. It is hard to believe that, at the current time, any organization or business is not using some kind of global supply chain for their operations. Having an efficient systems integration process is essential for the success of any complex system design. In the information age, or system age, it is hard to find any complex system which does not follow a systems engineering approach. Nowadays, systems engineering has become a well-accepted standard for practice across government and industries internationally.

In 2009, CNN Money ranked systems engineer as no. 1 in the list of the best jobs in the United States, according to CNN.com:

> demand is soaring for systems engineers, as what was once a niche job in the aerospace and defense industries becomes commonplace among a diverse and expanding universe of employers, from medical device makers to corporations like Xerox and BMW.

After several decades, systems engineering has now grown into a profession; it is a must-have function and department for most of the large corporations, and the major association for systems engineering, INCOSE, has grown significantly since its formation in 1990. Now, INCOSE has more than 9000 members globally, and is still growing every year. Meanwhile, government and industrial organizations are still facing a significant shortage of systems engineers, especially senior-level expert systems engineers; careers in systems engineering will become more abundant in the near future. That is the reason why we are here, to study this subject.

1.5 Summary

Systems engineering has become one of the most important engineering fields in every aspect of our society. It is based on systems concepts and the philosophy of the big picture first, in order to bring complex systems into being from beginning to end in the most efficient manner. This first chapter reviewed the basic concepts of systems, systems science, systems engineering, and the historical evolvement and development of systems engineering. Knowing this basic information about systems will help us understand the origin and need for systems engineering and thus grasp better the ideas and models in systems engineering.

PROBLEMS

1. Define systems. What are the characteristics of systems?

2. Give an example of a system within the systems engineering context. Identify the basic elements involved.

3. What are the system classifications? Briefly discuss these classifications and give examples (other than those examples given in the book).

4. Define science and engineering. What is the difference between science and engineering?

5. What is reductionism and what is expansionism? Briefly discuss the difference between these two terms.

6. Describe the main historical phases of systems engineering development and the main development milestones for each phase.

2

Systems Life Cycle and Design Processes

Let us start the chapter with a "rocks and bucket" story. There is a bucket, some big rocks just enough to fill it, some small stones, some sand, and water. Let us first fill the bucket with big rocks. Do we still have space? Yes, so we put the small stones around the big rocks in the bucket. Now is it full? We still have some room for some sand; after putting the sand in, is it full? No, we can still put water in the bucket: now the bucket is full. Now ask yourself: What can you see from this story? Someone might say that you can always have room for smaller things. While this might be true to some extent, for us systems engineers, the point we get from this story is that for us to fit all the things (big rocks, small stones, sand, and water) into the bucket, we have to follow the order in the story. If we put the small things into the bucket first, in the end we will realize that there is *no way* we can fit all the big rocks into the bucket. This order is analogical to the fundamental philosophy guiding the system engineering process: top-down processing.

The whole idea of systems engineering is the implementation of a top-down process; that is, starting from a problem or a need, it gradually evolves from a design concept to a tangible system that can be used. The key issue in a top-down systems engineering design process is a *big-picture* perspective; that is, designing the system from a life cycle perspective. Nothing lasts forever; everything has a life span. Just like anything else in the world, an engineered system also has a life cycle. Utilizing the big picture means that we have to consider everything involved in the system life cycle from the very beginning of the system design, doing things right in the beginning, even on issues related to the system's retirement. They need to be considered in the design phase, as design decisions in the early stages can have a significant impact on the system's performance later, even on the system's retirement. One of the most important factors in systems design is change or violation of design needs. It has been found that a large percentage of costs incurred in systems design is due to changes in the later design stages. Although, for any system, it is inevitable that change will occur, since unexpected and random events happen all the time, systems design must have the ability to accommodate such an environment; however, controlling and minimizing those changes can significantly reduce the complexity and cost of the whole design process, as the later the change occurs, the more expensive and difficult to implement it becomes. Having a big picture at the beginning of the design process helps us to keep the design changes at a minimum level. The second issue for systems engineering is that it is process oriented; in other words, the process determines the success of the system. Every system starts with a need and it gradually evolves to its final forms through a structured, well-defined process. There is no guarantee that a good process will lead to a good product, since many unexpected issues will occur during the process; however, a structured process will provide the greatest probability of success for bringing the system into being within its cost and time limits. After many decades of practice and implementation, the systems engineering process has been proven to be the most effective and structured approach for designing systems successfully, especially for large complex systems, and it is believed that it will still be the philosophy used for complex system design in the foreseeable future. In this chapter, we will

- Provide an introduction to systems' life cycles and describe the major forms of systems at different stages of system life cycles.
- Describe the elements that are involved in the system design process, including system need, requirements, functions, components, prototypes, models, and their relationships with regard to different phases in systems' life cycles.
- Describe the system engineering process and define the main objectives for each of the phases in the design processes, including the main design activities, major milestone baselines, and product for each of the phases, and the design evaluation approaches applied in each of the phases.
- Review the most common models for the systems design process, including the waterfall model, vee model, and spiral model, and their unique features for different types of systems.

2.1 System Life Cycle

"Life cycle," according to *Webster's Dictionary*, refers to "a series of stages through which something (as an individual, culture, or manufactured product) passes during its lifetime." Almost everything has a life; even the sun on which our planet Earth relies has a life of approximately 14 billion years, starting from a large cloud of dust and gas 5 billion years ago, to become a giant red star in another 5 billion years, and eventually a black dwarf approaching the end of its life.

The life cycle of an engineering system is a sequence of stages/phases in the life of the system. It is the life span or history of the system, from the need to create the system to the point that the system is retired and removed from service. There are some variations among different people regarding the naming of the systems life cycle; for example, Clark et al. (1986) defined systems life cycles as stages of "planning, definition, design, integration acceptance, delivery product"; Blanchard and Fabrycky (2006) used the terms need identification, conceptual design, preliminary design, detailed design, production and construction, utilization and support, and systems phase-out and disposal. Some of the life cycle definitions, such as the ones from Blanchard and Fabrycky, are more like a design process, expressed from the designer's perspective. Here I adopt the one from Chapanis (1996), as it concentrates on the system's status and format at different life cycle stages, rather than focusing solely on the design activities. Before we get into the life cycle description in detail, there are two fundamental characteristics of the systems life cycle that need to be brought to the reader's attention. Firstly, any system starts from a need; this need determines the system's concepts. From the concepts, system functions are derived, decomposed, and allocated to the system's components—hardware or software—all the way to its final configuration: developed, constructed, and delivered. All the different stages of the systems life cycle and different forms of systems are driven by the initial needs. Secondly, from the first characteristics just mentioned, one can easily see that the system evolves from a general form to specific forms. This evolvement is a top-down process, starting with the big picture—the need for the design of the system—with further details added in the various life cycle phases. In the following section, we will describe the major phases in the system life cycle; these life cycle phases summarize a common form for most systems. Despite the fact that different systems have different variations of these forms, the fundamental phases should be similar. Generally speaking, the system life cycle consists of seven major phases: (1) the operational need for the system; (2) system concept; (3) system concept exploration and validation; (4) engineering model development; (5) systems production, deployment and distribution; (6) systems operation and maintenance; and, finally, (7) system phase-out and retirement. We will briefly discuss, in each of the system life cycle phases, what the basic format of the system is, what major questions are being addressed, what information it will include, and what major system outcome is expected for these life cycle phases.

2.1.1 Operational Need

A system starts with a need; a need gives the reason for such a system—that is, *why* does one need such a system and *what* is needed—at a very high level. Needs come from customers. There are two basic types of customers involved in systems engineering (these are not exactly the same as the system users; we will talk about system user classes in greater detail in Section 2.2.3). One type is the government agency; for example, the Federal Aviation Administration (FAA) may want a new flight management system (FMS) to be developed, based on the electric flight bag (EFB) devices to integrate four-dimensional (4-D, spatial plus temporal) flight data into the cockpit. Government agencies such as the FAA and Department of Defense (DoD) play an important role in prompting systems engineering, as most of the systems they need are large and complex, and often have a tight budget and schedule. These government systems usually have more strictly defined structures and specifications; for example, for the DoD, a commonly used structure is Department of Defense Architecture Framework (DoDAF): this is a foundational framework for developing and describing systems architecture that ensures a common denominator for communicating, understanding, comparing, and integrating architectures across different organizations for DoD-related systems. DoDAF establishes data element definitions, terminologies, rules, relationships, and a baseline for consistent development of military systems architectures.

The other type of customers are nongovernment customers, either commercial or nonprofit related. Systems or products needed for this type of customer vary a great deal in terms of scope and complexity, and compared to government customers, their needs are usually less structured. For example, a marketing survey found out that U.S. customers need a new type of hybrid vehicle that uses environmentally friendly fuel, and in addition, need to be able to use household electricity to charge the vehicle, due to the rising price of regular gasoline. In areas that lack electricity, there is a need for a solar-powered portable water purification system, to provide drinking water for a certain number of people daily.

Systems needs originate from different sources; for government systems, the needs are usually generated from national strategic decision-making, and documented and distributed by a request for proposal (RFP). RFP is a formal invitation document, distributed to the prospective firms or academic organizations, to describe the needs for the system to be developed, and inviting the qualified organization to propose a systems development plan. Based on the RFP, potential developers will respond with a structured proposal, describing the system in a more technical and operational form, and propose a solution and plan for the system. The successful bidders will be required to develop a detailed work plan—namely, a statement of work (SOW)—to illustrate the technical and managerial aspects of systems development, including the personnel qualifications, efforts, and the major timeline and budget. The detailed information included in the RFP and SOW will be explained in Chapter 3.

For nongovernment customers, the need for a new system or product originates from organizational strategic planning. Strategic planning for a new system or product emerges from a number of sources, including marketing surveys and customer feedback/complaints, changing environments, new technologies and new resources, depletion of old resources, and so on. Unlike government projects, in which a strict documentation format has to be followed, in nongovernment projects, the operational need is developed internally and does not usually have a unified format which must be adhered to. Regardless of what type of systems are developed, one cannot depend solely on customers' original views as the only source, as these are often vague, incomplete, and too general. That is why one needs to have an operational need. Information from a variety of sources needs to be compiled and translated into one document, describing the system's intended purpose and functions. The following example illustrates typical operational needs:

1. Introduction: This section gives the background and scope of the systems to be designed/requested.
2. Missions: This section defines the highest level of mission that this design tries to accomplish. Missions may be broken down into specific phases and time periods if required.
3. Technical objectives: This section defines the major functions required for the system to be designed. Major function performance parameters are included for these functions. These functions/parameters may be accompanied by technical performance measures.
4. Constraints: This defines the time/cost restrictions on the project and the major milestones and deliverables required.

A sample technical need document can be seen in Table 2.1.

2.1.2 System Concept

After the operational needs are identified, an integrated systems concept will be developed. This concept is illustrated by a conceptual model. A model is the abstract representation of the real-world object; it focuses on the factors most relevant to the system needs, conveying the important relationships between these factors, and ignores nonrelevant factors. For large complex systems, using appropriate models is of the utmost importance, as there are usually hundreds or thousands of factors involved in the design. For the most part, this book is about the models used in systems

TABLE 2.1

Sample Design Needs for a Small Aircraft

Parameter	Description
Range	200 statute miles, with 30 min reserve, day VFR at $\geq$4000 feet MSL over nonmountainous, sparsely populated coastal terrain
Efficiency	$\geq$200 passenger-MPGe energy equivalency
Speed	$\geq$100 mph average on each of two 200 mile flights
Minimum speed	$\leq$55 mph in level flight without stall, power and flaps allowed
Takeoff distance	$\leq$2000 feet from brake release to clear a 50 foot obstacle
Community noise	$\leq$80 dBA at full power takeoff, measured 240 feet sideways to takeoff brake release

engineering. We will explore different types of systems engineering models in Section II. The philosophy of systems engineering can be considered to be model based, as models are an absolute necessity in all aspects of systems engineering design. The conceptual model is the first model that the system evolves from its need. In the conceptual model, systems needs are organized as *requirements*. Requirements are formal documents that define the system's intended purposes. From the requirements, operational concepts will then be developed. These depict a complete picture of the systems operations. Operational concepts are commonly written in narrative format; sometimes they are also called operational scenarios. Scenarios tell a complete story of the system's intended activities, if designed and developed, and how it will be utilized by its operators/users. The following example shows a typical scenario for using an automated teller machine (ATM):

> The customer (including walk-up and drive-through customers; they may also be visually and hearing impaired) requests service by pressing the start key or by inserting their debit card. They receive feedback from the ATM that their request was accepted. The ATM system reads the card and requests the PIN. The customer inputs their PIN and the system processes it; if the PIN is correct, the ATM proceeds to the next selection menu. If the PIN is not correct, the ATM provides feedback and goes back to request the PIN again. If this process is repeated three times, then the ATM blocks the card and provides feedback.

To better explain the system scenarios, some other format of the operational profile can also be developed to accompany them, to give a more comprehensive picture of the system concepts. This profile includes a graphical representation of the systems operations spatial profile, a timeline of the operations temporal profile, a chart of user/operator characteristics, and an interaction diagram to illustrate the exchange of information, materials, and energy flow between users, systems, and the environment. We will describe this in depth in Chapter 3, which is about systems requirement analysis.

As an immediate result of systems concepts development via the methods mentioned above, systems evolve to a conceptual functional model. System *functions* define the purpose of the system and illustrate *what* actions the system would perform to accomplish its purposes. The concepts of functions are developed to give a complete picture of the system's intended actions, hierarchically and operationally. It defines what the system should do to accomplish its requirements and to fulfill customer needs. Functional analysis is one of the most important systems analysis methods, as it serves as the first and most critical bridge between customer needs and the system's technical specification. Functional analysis will be covered in great detail in the systems engineering process sections and in Chapter 4.

2.1.3 Systems Concept Exploration and Validation

Systems concepts, represented by system functional models, need to be further explored in terms of their validity to the system requirements. System concepts are explored and validated by translating the conceptual functional model to the allocated *components* model. A component allocation model is built to verify that, firstly, the conceptual model is feasible to be implemented by systems components, including hardware, software, and human operators. It is a two-way process, in the sense that the conceptual model needs to be adjusted to make the physical allocation feasible, taking into consideration the availability of the current and emerging technology; on the other hand, the functional model serves as the basis for the selection of the components, as there are many candidates for these, with a large variety of suppliers, manufacturers, and different models involved. The decision-making process for the translation from conceptual model to physical models is iterative and requires a careful consideration of all the requirements and their parameters. That brings us to the second verification in this phase, which is that systems requirements are verified by the physical models. You might have seen, on several occasions, that we have used the words *verification* and *validation*; you might wonder whether these two words are the same? If not, what is the difference? In the systems engineering context, "verification" and "validation" are related to each other but do not convey exactly the same meaning. The verification process is to make sure the systems requirements are being met at different phases in the system life cycle; that is, the systems requirements are translated, carried over, implemented, and materialized throughout the system life cycle. Systems requirements are traced through this translation, and the design models are verified against the requirements parameters by using reviews and tests at different stages; sometimes, users are also involved in the tests and reviews. Validation is similar to system demonstration or evaluation: it is to make sure the design outcome is what users want, and that we are designing the correct systems. So, briefly summarizing the two terms, verification means doing things right, following the systems engineering process and structure, while validation is to make sure we are making the right thing, providing the correct outcome for the users' needs. These two activities go hand in hand, and are closely related to each other; without a good verification plan and process, there will be no valid system to be designed.

In the systems exploration and validation phase, systems evolve from the conceptual model to the physical allocation model, represented by the system allocation baseline (Type B system specification). The allocation baseline answers the question of *who* will perform *what* function at what cost. One has to keep in mind that the process is iterative; the allocation models of the systems are reviewed, and usually systems users are involved in this review process.

The allocation model usually includes the following items:

- Systems requirements for subsystems and components, refined and quantitative technical performance measures for functions, including the input, output, and constraining factors for each function
- Functional package and assembly lists, including systems components lists and preliminary configurations for the components
- User systems interaction requirements
- Design data, validation and evaluation documentation, including the faulty tree analysis (FTA) and failure mode and effect criticality analysis (FMECA), human factors tasks analysis, and computer-aided design data

2.1.4 Engineering Model Development

Based on validated system concepts, functional models, and allocation baseline, the actual components are procured and assembled. This is the point at which all the components are integrated together and, for the first time, the system is in its intended form, not only functionally but also in its physical configuration. To obtain a complete and detailed system configuration, the systems specification is translated from *what* (functions) and *who* (allocations) to *how* and *how much*. Details of systems components are specified, including the systems component/assembly specifications, what materials are involved, and how to assemble them together. The system evolves from the functional allocation baseline (Type B specification) to the product baseline (Type C specification), process baseline (Type D specification), and material baseline (Type E specification). There are various types of documents involved in the engineering model, based on different types of systems. Generally speaking, the model usually consists of the part list, material lists, blueprints, computer-aided design data (such as AutoCAD and CATIA design drawings), and specifications for the interfaces between components. Since this is the first time that the all the components are integrated in their actual design forms, the system will be tested iteratively to verify that the derived physical models meet the system requirements. These tests, usually conducted formally with users and stakeholders involved, are intended to identify the last-minute problems before the system is produced, distributed, and deployed. Necessary changes are made to the models if any mismatches between the physical models and system requirements are found, trade-off analyses are carried out to balance out different options, and changes are monitored and strictly controlled in order not to cause any major time delay and cost increase. Minimizing changes is one of the most important objectives for systems engineering. But, due to the complexity of system design, changes are often inevitable. Systems engineering tries to identify and address the changes as early as possible, which is why traceability plays an important role in the life cycle. Different system formats are interconnected, and the evolvement of different models from one version to another is supposed to be requirement driven, to minimize any deviation from the right track. We will talk about change management in the detailed design phase later in this chapter. Here, it is important to point out that managing changes efficiently is an essential aspect of the development of physical system models.

2.1.5 System Production, Distribution, and Deployment

After the physical system model has been built, tested, and finalized, the system will move to the next life cycle phase, which is full-scale production, distribution, and deployment. The system is in its final format, and is transferred to the production assembly line to start formal mass production. At this stage, the system is produced, possibly in multiple copies. All the components are specified, either produced separately if they are specially designed for the system, or procured if using standard commercial off-the-shelf (COTS) items. These components are delivered to the production site, assembled, and then distributed via retail outlets to customers. A final production of the system may involve hundreds or thousands of suppliers, depending on the level of complexity of the system. The supply chain plays an important role at this stage, as greater cost may be incurred in distribution and deployment than in the production/assembly itself. Much of the system's assembly work may be outsourced internationally; for example, to take advantage of cheap labor and materials locally. For instance, the cost of labor and materials of many U.S.-designed systems in third-world countries, such as China, is less than 10% of the total cost of the system. As a matter of fact, the majority of the cost is incurred in system distribution and deployment, in addition to the cost of research and development. The system has now evolved from its models to the final realization of its designed format, together with its supplemental materials including manuals and training services; it is delivered to customers and installed for operations.

2.1.6 System Operations and Maintenance

After the system has been produced, assembled, and delivered to the customer, it will be in fully operational mode for the period of time designed, providing functions to users and operators. Users and operators are usually the same group of people; the difference between these two lies in the complexity of the system itself. If the system is fairly simple in nature, the term *users*, that is, the user–equipment combination, is employed; if the system is very large and complex, the term *operators* is used instead. System users and operators essentially represent the same type of user class; namely, the end customer of the system. There are also other classes of systems user; systems maintainer is one of the classes. Although highly reliable systems are desired, systems do fail sometimes, requiring maintenance activities to be carried out. At this stage of the system life cycle, the emphasis is on customer service and support of the system; since the design of the system has been finalized, no further design changes are possible. However, follow-up tests and evaluations of systems are still necessary to identify any problems within operations and maintenance; these test results, together with feedback from customers, serve as a guide for any engineering changes that may be made for the subsequent version or next generation of the system. Emerging problems are addressed and fixed immediately, especially those pertaining to user safety and hazards. Faulty systems are fixed, recalled, or replaced to minimize the impact from these issues. At this stage of the system life cycle, operation, maintenance, and evaluation efforts are carried out continuously, until the system is ready to be retired.

2.1.7 System Phase-Out and Retirement

There are many reasons for system retirement; the majority of these are incompatibility with emerging technology, discontinuation of supply of materials, changes to legislation and regulations, new trends in customer demand, and so on. At the system retirement stage, the system is often characterized by a reduction in the number of customers, increasing numbers of system problems, high costs, and difficulty of maintenance. System functions are terminated and the system is disassembled to the level of its components; the system is disposed of. The natural resources from which the system is built are limited, so it is unlikely that the system will be completely discarded and wasted; it is desirable to retrieve materials from retired systems as far as possible, to save the cost of the materials, and more importantly, to reduce the amount of waste to the environment. Sustainable development or *green engineering* is one of the most important types of development, depending on the different kinds of components. There may be different end uses for the components: reuse, remanufacturing, or recycling. *Reuse* is the highest level at which a system is preserved—usually a nonfaulty system—to keep it at a degraded function level to prolong its life cycle (one can think of this as a semiretirement phase). For example, older computer systems may not be fast enough for laboratory scientific computation purposes, but since they are still functional, they may be donated to charity for educational purposes in schools and offices. *Remanufacturing* (also known as closed-loop recycling, or sometimes called refurbishing) implies a series of activities to put a retired system back into use in its complete form, by repairing or replacing faulty parts, to become operational again. *Recycling* is a process that retrieves useful raw materials, to reduce waste and the cost of procuring similar fresh new materials. It is believed to benefit the environment by reducing pollution and saving the energy consumption of obtaining new materials. Although a common practice for a long time, it did not catch people's attention until the twentieth century; as an outcome of the industrial revolution, productivity has increased tremendously, as has demand for materials. With more human-made systems, the by-products, waste, and pollution from the production process has influenced our natural environment, which has caused many problems for human health and quality of life. Sustainable development and green engineering of systems have become more and more important for assessing their effectiveness. With more customer awareness and global-ization of system development, green engineering has become an essential part of system competitiveness. The life cycle consideration of system design enables designers to take systems retirement into account in the early phases of the design process—assembling it in a way which will simplify its disas-sembly—to facilitate the system being phased out and retired, to impact the environment at a minimum level, and meanwhile save internal costs and energy.

Where the system is retired only partially, its materials will be recycled or reused for the next generation of the system and some of the concepts will be carried over to the next generation, as part of the requirements for the new system; at that time another life cycle will begin.

2.2 Systems Engineering Processes

2.2.1 Definition of Systems Engineering Process

Parallel to the system life cycle, there are a series of activities that bring systems into being; this series of activities is called systems engineering pro-cesses, or more specifically, system design processes. According to INCOSE. org, the systems engineering process is comprised of seven typical tasks: "State the problem, Investigate alternatives, Model the system, Integrate, Launch the system, Assess performance, and Re-evaluate. These functions can be summarized with the acronym SIMILAR: State, Investigate, Model, Integrate, Launch, Assess and Re-evaluate" or SIMILAR.

Chapanis (1996) also defined systems design as a seven-step process in his process model: "Requirement Analysis, Systems Design, Hardware and Software Design, Hardware and Software Development, Systems Integration, Installation/Transition & Training, and Maintenance & Operation"; the three strands of development hardware, software, and users are woven together into one integrated system. Blanchard and Fabrycky (2006) defined the systems design and life cycle as a process with two major steps, the acquisition phase and the utilization phase; the acquisition phase consists of conceptual design, preliminary design, detailed design and development, and production and construction; the utilization phase consists of product use, phase-out, and disposal. There are many other variations of the processes. Despite the different names for the various phases, they all share the same characteristics, which can be summarized as follows:

1. The system engineering process is requirement driven. Systems design starts with identification of the need; the need is the problem statement that the system is designed to address. Once this need is identified, system requirements are derived and these requirements are tailored throughout the system design processes, from the beginning to the end of the whole process, from concept exploration until the system is retired. 'Requirement driven' is the key element for the system engineering process; requirements are the guide for the design.

2. Systems design follows a general-to-specific evolvement process. The top-down systems' requirement-driven features imply that systems evolve from general requirements to specific functions and components; this process is to ensure the system is designed in an effective and efficient way, by having a big picture first to guide one through the process. Through test and evaluation, the requirements are traced, confirmed, and translated to detailed design specifications.

3. System design activities are iterative in nature; there is no clear cut between design activities and the system design process is by no means a linear manner. In any design process, one can expect the cycle of design-verification-modification to iterate many times until the design outcomes are validated and approved. Chapanis (1996) also calls this a 3D process: define-develop-deploy; Blanchard and Fabrycky (2006) used the circle of analysis-synthesis-evaluation to illustrate this iterative process.

Generally speaking, systems engineering design is an iterative, top-down, and need/requirement-driven process, to gradually translate the high-level general system needs and requirements to specific technical system parameters, so that the system can be developed, assembled, installed, and delivered for operations, in a well-managed, tightly controlled, cost-efficient manner, and within cost limitations. We adopt the terms by Blanchard and Fabrycky (2006) for the major design activities for the process: conceptual design, preliminary design, detailed design, systems development and construction, systems operations, and maintenance.

2.2.2 Basic Concepts and Terminologies for Design Process

Before we get into the details of the system design processes, it is necessary to define some of the terms for readers to understand the material hereafter, as these terms are general in systems engineering and they will be used quite frequently throughout the book. Also, we have listed a comparison of these terms with those being used in some systems engineering management tools; for example, CORE. Since we will be using CORE to illustrate the examples, we are comparing the CORE terminology with general systems terminology definitions, for readers to better understand those examples.

- Requirements: A requirement is a single formal statement containing *shall* to define a need that a system must provide or perform. It has the following syntax: system (or subsystem or component) + shall + verb + noun (i.e., provide a specific function) + (applicable parameters to which this requirement pertains) + (applicable environmental or contextual information for the requirement). For example, "the portable water purification system shall be able to purify at least two gallons of water per minute."
- Functions: A function is a specific action or activity that a system does or provides; it is a meaningful purpose for which the system is developed or designed. Since it is an action, a function usually contains a verb. The common syntax for a function is verb + noun. It is easy to confuse functions with tasks; especially for novice designers, it is a common mistake that user tasks are sometimes defined as functions. For instance, in ATM systems, "deposit cash," "deposit check," and "transfer funds" are all ATM functions; but "insert debit card" and "key in passcode" are user tasks, not functions. One has to keep in mind that tasks support functions; any user task should relate to one or more system functions—in other words, without a function, there are no tasks. That is why functional analysis of the system usually comes before task analysis. Information on functional analysis and tasks analysis will be given in greater detail in Chapter 4.
- Components: Components are what constitute the system; they are the elements of system construction. A system may have different kinds of components. Generally speaking, system components consist of three basic types: physical components—the hardware to build the system; electrical and computer software components—the software of the system—namely, the programs and codes that control and regulate system operations; and lastly, human components, sometimes called livewire. Humans can be important components for some systems; their interaction with hardware and software are essential for the carrying out of the system functions. There are different levels of human involvement in systems; we will be describing the user classes in Section 2.2.3.

- Input/Output: As the dynamic entities of the system, system components need input to perform their functions; and since components are connected to each other, some components may, in the meantime, also generate output to some other components; these inputs/outputs may be materials, energy, information, or actions.

- Baseline: In systems engineering, a baseline is a basis for evaluation; it is a reference point for system design to be evaluated. At certain stages of systems design, to verify that the design meets the system requirements, one needs to conduct tests and evaluations to make sure the design is on the right track; these are conducted on the intermediate results of the design, which are referred to as baselines. For example, after function analysis, the functional baseline (also called Type A specification) is developed, which contains the functional models of the system. When functions are analyzed and allocated to components to perform the functions, the allocation baseline (also called Type B specification) is derived. When the design progresses to more detailed information, the production configuration (product baseline, or Type C specification), the manufacturing and assembly process (process baseline, or Type D specification), and the material baseline (Type E specification) are developed; the system can be built in its final form and verified/validated against the requirements. To some extent, system design is a process of translating user needs to requirements, and then different specifications/baselines of the system can be constructed/developed. The information pertaining to design baselines will be visited again in greater detail in Section 2.3.

Table 2.2 compares some of the basic terms that are used in CORE with general systems engineering terminology. We list the two sets of basic terms side by side so readers can have a clear understanding of the terms utilized in CORE, and use the correct terms in their applications.

TABLE 2.2

Side-by-Side Comparison of Commonly Used CORE Terms with Systems Engineering Concepts

Systems Engineering	CORE	Examples
Document, report	Document	Files that contain requirements and specifications
Requirements	Requirements	Originating requirements, decision decisions
Functions	Functions	System actions
Components	Components	Hardware, software, and people
Input, output	Item	Materials and information

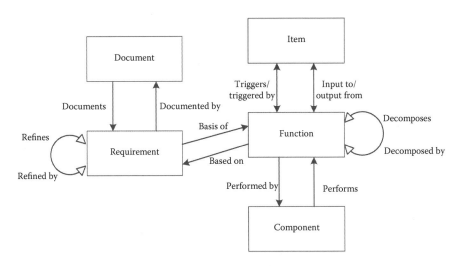

FIGURE 2.1
Illustration of CORE concepts.

Figure 2.1 illustrates the relationship between CORE concepts; these relationships can be easily established within the "relationship" tab in CORE. A more detailed description of the application of CORE in the systems engineering process is included in subsequent chapters (Chapters 3 and 4). We wish to provide an overall view first in this chapter.

2.2.3 Systems User Classes

Not only are humans an inherent part of the system, human components also play a vital role in the success of the system mission. Throughout the system life cycle, there are different kinds of human involvement in the system at different levels. It is difficult to provide a clear-cut classification of the different human roles in system development and operations; especially for a system with high complexity, it is not uncommon to have multiple tiers of human users related to every aspect of system design and utilization. Roughly speaking, there are basically three levels of user involvement within the context of systems engineering:

1. System users/operators: Also sometimes called customers, system users can be further divided into two categories, direct end users and indirect users. The term *end users* is easy to understand; these are the people that interact with the system directly. Chapanis (1996) also argued that different names should be used for this category; *users* refers to small-scale systems, such as a tool, consumer product, or equipment—for example, a micro-computer system user; for large complex systems, such as military aircraft systems or unmanned

aerial vehicle ground control systems, people whose use these type of systems are called *operators*. Besides end users/operators, there are also people who do not directly interact with the system; however, their opinion will impact the end users so much that they cannot be neglected—for example, family members of end users, or managers or supervisors of operators. These indirect users play a significant role in influencing users'/operators' decisions and behaviors; their needs constitute a major part of the system requirements.

2. System maintainers/supporters: Another group of humans involved in the system design and life cycle are those who maintain and support the system. The system has reliability, but sometimes it might fail; this implies that the system needs to be maintained and supported, both from a preventive and a corrective perspective. The system's life cycle cost and operational effectiveness also include maintenance efficiencies. This, in turn, requires that the system design also needs to consider the system maintainers' requirements. As part of the system requirements, the system design should also make maintenance activities easier and more efficient, including indication of the problem area for easy identification of the failure, ease of accessibility for faulty components, standardized tools, equipment and components used in the design to facilitate the maintenance, and so forth. Design requirements for system maintainers' needs should be considered and specified in the early design phase, as they also have a major impact on the system configuration in later stages.

3. System designers: These are the third class of human involved in system design and include the systems engineering team members and related system management personnel/administrators. System designers are those who bring the system into being; the system design's focus for this group of users is the documentation of the design. In other words, how does one manage a systems engineering project so that coordination and communication among team members are smooth? For a system with a high level of complexity, there are easily hundreds, even thousands, of people involved in the design, and communications are often not synchronized, as the team members may be distributed all over the world. Traditional, paper-based documentation may not be sufficient to handle the high volume of data. The design should follow a unified standard. Systems engineering is itself about the structure of the design; it is about following procedures to make sure things are done correctly from the beginning. Current systems engineering projects are usually managed by computerized software to provide real-time coordination, prompt responses, and efficient communication, strictly controlling changes and precisely tracing design efforts. It will also provide some basic analysis modules to analyze the data. For

example, Microsoft Office, CORE (VitechCorp), and DOORS (IBM) are among the most popular commercial systems engineering management software available.

2.3 Systems Engineering Design Processes

The systems engineering design process can be thought of as a translation process or transformation process. It translates system needs into well-structured systems requirements, and through systems modeling, analysis and synthesis, transforms the general requirements to quantitative system specifications and parameters, so that the system evolves from general concepts to a concrete design configuration. The system can then be easily produced or constructed. This process is to make sure the translation of general system needs to specific system configurations are controlled and kept on track, while at an optimum level of design efficiency. The design process for most systems are quite similar, generally speaking; the design process (or system acquisition process) consists of stages of conceptual design, preliminary design, detailed design, and production/construction. The basic methodology involved in the design process can be summarized in three words: analysis, synthesis, and evaluation (Blanchard and Fabrycky 2006). System analysis studies the needs, the environment, system functions/structure, and the proposed solutions to the design problem; and selects the optimum design concept. The concept is continuously evaluated and verified; the validated design concept is implemented by integrating and balancing the other factors from the system environments. This process is iterated at different levels at different design phases. The design progresses as the cycle of analysis-synthesis-evaluation evolves to a more detailed level.

The actual process involved and implementation of design might vary from one system to another, depending on the nature, type, and complexity involved in that specific system, but the fundamental principle underpinning all the designs are essentially the same, which are top down and systems requirements driven. Here, we present a rather general framework of the systems engineering design process, illustrating the purpose of each stage, main design functions achieved, major design activities performed, commonly used design approaches, design products/data outputs, and major design reviews/evaluations conducted.

2.3.1 Conceptual Design

Conceptual design is the first step in the systems engineering design process, yet it is the most important step for the overall process. In the conceptual design phase, the focus is on requirement analysis and elaboration. In the conceptual design stage, the problem statements of the system are defined as requirements. As the sources for requirements, system needs are collected and captured, then analyzed and compiled into one comprehensive requirements document. System operational requirements are defined in the requirements document; this document provides a rationale for the system design, explaining why the design is necessary. For a design to be carried out, not only must it be necessary, but it also needs to be feasible. Based on the systems requirements, systems design concepts are explored, in conjunction with system feasibility analysis. The design concepts are realized by the output of the conceptual design, the functional model of the system. Functional models are evaluated and the first formal design specification, the functional baseline, is generated.

In the conceptual design stage, the major design activities include

- Identify system users and system needs; this could be a new need or one to correct a deficiency of a current system. Translate user needs into a formal definition of system requirements.
- Conduct feasibility analysis to identify the technical, social, environmental, and economic issues for the system design approach; work out a feasible course of action plan for system design processes.
- Develop system operational requirements that describe the system functions and their contextual information. These requirements should include the normal system functions as well as the system maintenance and support concepts for the normal functions. Technical performance measures (TPMs) for these functions are also defined for these functions at the system level.
- Accomplish a system-level functional analysis, defining the hierarchical structure of the functions and the operational relationships between the functions.
- Perform a system-level analysis and trade-off studies, using a number of systems analysis techniques and models.
- Produce the Type A specification of the system, which is the functional baseline, documenting the results for the above activities.
- Conduct the conceptual design review.

Through the above steps, the system needs are translated into a set of operational requirements, containing the information necessary for the designers beginning to explore design concepts, in terms of what functions the system will provide. These requirements also contain quantitative parameters, enabling the design to be verified against the requirements.

2.3.1.1 Identification of Needs

A system starts with a need, a desire, or a want. In the system life cycle section earlier in this chapter, we have introduced the first very life cycle format of the system as the system need. This need could be a new function, a new product or an improvement of an existing product based on current system deficiencies. Usually, system needs come directly from customers, through interaction with them, including customer feedback and market surveys. Commonly used methods for this include questionnaires, interviews, observations, critical incidents, and accident studies. The needs could also result from strategic mission planning for government customers (such as DoD and FAA system acquisition). Depending on the different sources, needs may originate in different formats; some are narrative, some are graphical. It is the designer's responsibility to integrate the different formats of the needs together to compile a unified system need document. This need document serves as a system design mission statement and a starting point for requirement analysis and elaboration, defining the overall mission and objectives for the system to be designed. If necessary, a need analysis is often conducted to define the problem statement thoroughly and completely. There is no general template or approach to follow to perform the need analysis; its goal is to define *what* the system should provide, not *how* it is constructed. According to Blanchard and Fabrycky (2006), a need analysis should provide general answers for the following questions (but not limited to these):

- What is required of the system in functional terms?
- What functions must the system perform?
- What are the "primary" and "secondary" functions?
- What must be done to alleviate the foreseen deficiency?
- When must the functions be performed?
- When must the system be constructed and delivered?
- What is the planned system life cycle?
- What is estimated system life cycle cost?
- Where is the system intended to be used?
- What frequency is the system intended to be used?

Answers to the above questions give us a big overall picture of what needs to be done, not how it is done. We must make sure we do not over-specify the needs by adding details of design solutions. It is not uncommon for designers to have initial personal thoughts on the system needs, especially those designers who are very familiar with similar, existing systems. For example, some designers might have a preference toward a certain type of technology or components; those designers may select such technology or components without exploring other possibilities. One has to keep in mind that having a solution without going through the formal design process is often premature, distorted, and biased by personal experience. Focusing on one preferred solution to the design problem may miss other options that could bring better benefits to the design. Systems engineering is about following the correct process to do things right at the beginning. Of all the philosophies of design, as we have pointed out when looking at the system life cycle, grasping the big picture is the most important one. Having a completely and clearly defined system need is often the first foundation for system success, which has not been emphasized very much historically; rather, people believe they can always "design it now, fix it later," which has been proven to lead to many changes to design in the later stages, causing unnecessary delay and more cost.

2.3.1.2 Feasibility Analysis

Once the system needs are determined, before starting the design process, it is necessary to conduct a feasibility analysis to point out the direction of the general design approach. Feasibility analysis provides an answer to the questions "Is it beneficial to design the system?" and "Can we do it?" As part of the strategic planning for the system design, feasibility analysis provides information on the strengths and weaknesses of the organization. By looking at the potential market and customer groups, considering environmental factors, access to necessary resources, workforce readiness, and current/emerging technology, we can objectively and rationally assess the possibility of success if the design project is carried out. The commonly used criteria in assessing the system design feasibility are the cost involved and the value obtained from the design, which includes the monetary return as well as the sustainable development and growth opportunities for the organization.

Generally speaking, there are three interrelated types of feasibility that need to be taken into account—technical feasibility, economic feasibility, and operational feasibility. Technical feasibility refers to the practicality of a specific technical approach for the proposed system and the availability of technical resources, including the readiness and maturity of current and emerging technology. Since the design takes time to complete, designers are often required to look at the current technology available, and predict the technology that will emerge in the near future. Certainly, there are risks involved, since any future prediction has some degree of uncertainty. No one would have expected that HD-DVD technology for high definition video media would cease to be adopted when it first came on the market. There are occasions when decision makers need to choose the right track for the design approach by specifying what technology would be adopted in the design. Designers should be familiar with the cutting edge technology of the development of similar systems, and grasp the pulse of the technology developed in the field. Compatibility and flexibility are two key factors for technology selection. The story of Wang Laboratories is a good example of and lesson about not conforming to common technology and standards. As another example, as a trend for mobile devices, the 4G network has become more sophisticated and affordable; many designs of mobile devices will consider this technology as a possible solution for certain network systems applications. When computer-aided software engineering (CASE) technology was first introduced in the mid-1980s, many organizations thought it was impractical for them to adopt it, due to the limited availability of the technical expertise to use it. Nowadays, although adopted slowly, object-oriented programming for software systems has become a universal standard. One has to keep in mind that the selection of applicable technology is done in conjunction with system needs; as a matter of fact, the need should drive the technology selection, not vice versa.

Operational feasibility measures how well the proposed system solves the problems, and takes advantage of the opportunities identified during needs definition and how they satisfy the requirements identified in the requirements analysis phase. In other words, operational feasibility tells us how well the system will work in or fit in the current environment, given the current status of the designing organization. Historically, it was found that operational feasibility is often overlooked or taken for granted. Typically, for operational feasibility analysis, if a new system is being proposed, we need to answer the following questions: How does this system work well with the existing operational system? Is there any resistance to the new system from the end users? How does the system fulfill customer needs? Are the operational characteristics of the system compatible with the current existing management style? What are the potential pros and cons of the system's operational efficiency? Do we have the necessary resources to design such a system, including facility support, management support, capacity/capability of the organization, resource readiness, workforce skills/training, and so on? For example, before building a diversion route for an urban freeway system, designers first assess the operational feasibility by conducting capacity analysis at critical intersections of the arterial route. Using simulations at different times of day and traffic conditions, the critical demand volumes that can be handled by the new system are computed, and different diversion strategies are compared to find the optimal solution for the new system structure. Operational feasibility is conducted prior to system design to make sure that the system can be designed in a timely manner with the minimum possibility of failure.

Economic feasibility, also called cost-benefit analysis (CBA), measures the cost-effectiveness of the proposed system. Economic analysis is a systematic process to compare the costs and the benefits of the system over its intended life span, to see if the benefits outweigh the costs. As a basic requirement for the design project to be economically feasible, the cost-benefit analysis is expected to show positive net benefit over cost. Economic feasibility is considered the most important decision factor for the system design project, as the ultimate goal of a business organization is to make a profit; the system design project is considered as a capital investment, and treated in the same way as other financial investments. A general approach involved in economic analysis includes the following steps: identifying the stakeholders; identifying the cost factors; selecting measures for all the cost factors and elements; predicting the monetary outcome of the costs and benefits over the life cycle time period; applying the appropriate model to integrate all the values from different time periods into one common basis of measure; comparing the costs and benefits; calculating the net benefit of the system; and performing sensitivity analysis to make the comparison more robust. This process is rather general; with different types of systems and different levels of complexity, it may vary accordingly. There are two primary types of costs that are almost every system will incur within its life cycle: system development cost and system operation/maintenance cost. The system development cost occurs in the early phases of the system life cycle; it is more of an initial one-time cost. The operation/maintenance cost is an ongoing cost that is incurred throughout the system life cycle until the system is retired. There are further categorizations within each type, depending on the nature of the system, which can be broken down into several levels. Two things need to be pointed out here: Firstly, since this is only the system planning stage, most of the cost factors and estimates are based on the analytical model and empirical data prediction; we cannot expect the figure to be precisely accurate at the beginning. As a matter of fact, it is expected that the cost estimate will be modified and revised when more design data is available, and the estimate will be more realistic in the later design stages; in system economic analysis, a commonly used approach for estimating cost is the activity-based approach. Based on the source of the cost, which is believed to be relevant to certain activities, mapping the amount of effort for each design and operation activity is believed to provide a basis for cost estimation. The cost concept will be further elaborated on in greater detail in Chapter 9. Secondly, as you have probably also realized, most of the cost estimates are in different time periods. As we all know that time puts a value on money (a dollar at the present time is worth more than a dollar in the future—think about the interest paid on the student loan you have), it is imperative to consider time in the formula of the cost-benefit analysis; all the cost and benefit values need to be converted to a common basis—either the current (present) value or a future value, or evenly distributed annually (such as the mortgage payment equivalent amount on a borrowed principal payment). The subject of studying the time value of money and its related models is called *engineering economy*; we will review this subject in detail in Chapter 9. The benefits involved in the system design are primarily tangible benefits, such as projected gross profit from the sales of the systems. This type of benefit can be easily understood and may usually be objectively predicted as a dollar value; however, there are also benefits that are not easily quantified or visualized—we call this type *intangible benefits*, such as increased environmental friendliness, increased customer loyalty, better quality, better service to the community, better employee job satisfaction, better moral values and ethics, and so on. These benefits are difficult to measure objectively. However, it is believed that the intangible benefits of the system can significantly impact the economic feasibility of the system, thus cannot be ignored.

Besides the three feasibility criteria of system analysis, there is another source of feasibility criteria that need to be taken into account in system development: these are legislation, regulations, standards, and codes, which can also be called legal feasibility. There are many regulations and standards developed at the federal, state, or even international level to regulate system engineering development. There are two major types of standards that are related to systems engineering: safety standards and performance standards. These standards include federal and military standards such as MIL-STD-1472D, Human Engineering Design Criteria for Military Systems, OSHA (Occupational Safety and Health Administration) standards for safety; and International Standards, such as ISO (International Organization for Standardization) 9000 (Quality Management Standard). Although some standards are concerned with safety issues, they cannot be ignored in the design and the consequences of not complying with the standards can have a significant impact to system effectiveness, even to the extent of involvement in a product liability lawsuit. For example, ISO has the following requirement for quality management systems:

> The organization shall establish, document, implement and maintain a quality management system and continually improve its effectiveness in accordance with the requirements of this International Standard. (ISO 2008)

When applying the appropriate standard, we have to keep in mind that standards have many limitations that need to be taken into account for the design. As Chapanis (1996) pointed out, standards provide general and minimum requirements for the system, as they are prepared by external agencies; to achieve consensus, they have to be high level and general, as clearly seen from the above ISO example. To implement this requirement, further refinement is definitely needed. Many standards are not precise enough to be used directly; for example, wording such as "simple to use and communicate" means different things for different systems, so system designers have to tailor the relevant standards to the particular system specification, to make the relevant standards applicable to the system to be designed.

2.3.1.3 System Planning

After the proposed system has been shown to be feasible, the corresponding systems engineering program is developed through system planning. This is achieved by preparation of a system engineering management plan (SEMP). According to INCOSE (2012), the SEMP is

> the top-level plan for managing the systems engineering effort. The SEMP defines how the project will be organized, structured, and conducted, and how the total engineering process will be controlled to provide a product that satisfies customer requirements. A SEMP should be prepared early in the project, submitted to the customer (or to management for in-house projects), and used in technical management for the study and development periods of the project, or the equivalent in commercial practice. The format of the SEMP can be tailored to fit project, customer, or company standards.

A typical SEMP document includes the following outline:

- Title, cover page, table of contents, scope.
- Applicable document.
  - This section includes all the documents that initiate the system, including the RFP, relevant standards, codes, and government/nongovernment documentation related to this project.
- Systems engineering process.
  - This section conveys how the organization performs systems engineering efforts, and should include the organization's systems engineering policies and procedures, which contain the organizational responsibilities and authority for systems engineering activities and control of subcontracted activities, define the tasks in the systems engineering master schedule (SEMS) and the milestones of the systems engineering detail schedule (SEDS). These tasks and milestones encompass the following activities.
    - System engineering process planning.
    - Requirement analysis.
    - Functional analysis.
    - Synthesis.
    - Systems analysis and control, including trade studies, cost analysis, risk analysis, configuration analysis, interface and data management, TPMs, review and evaluation efforts.
- Transitional critical technologies.
  - Describes the key technologies (current and emerging) that the system will be using and their associated risks for systems development. Provides criteria for selecting technologies and transitioning these technologies.

- Integration of the systems engineering efforts.
  - Describes how the various systems design efforts will be integrated and how interdisciplinary teaming will be implemented to involve appropriate disciplines and expertise in a coordinated systems engineering effort. Tasks include team organization, technology verifications, process proofing, manufacturing of engineering test articles, development testing and evaluation, implementation of software designs for system end items, sustaining development/problem solution support, and other implementation tasks.
- Additional systems engineering activities.
  - Includes other areas not specified in previous sections but essential for customer understanding of the proposed systems engineering effort. These activities include long lead items, engineering tools, design to cost, value engineering, system integration plans, compatibility with supporting activities, and other plans and controls.
- Systems engineering scheduling.
- Systems engineering process metrics.
  - Including the cost and performance metrics and process control measures.
- Role of reviews and audits.
- Notes and appendix.

Chapter 10 will review SEMP in greater detail. For a more detailed elaboration, template, and example, readers can refer to INCOSE (2012), which can be downloaded free of charge by INCOSE members from www.incose.org.

2.3.1.4 *Requirement Analysis*

A system starts with a need, as we have stated before; system needs come from different sources, with a majority of needs coming directly from customers. Although system design is need driven, these high-level needs often come in different formats; and, moreover, are too general and vague to be implemented directly. System needs must be further refined and analyzed, transformed from an informal format to well-defined system design requirements. A requirement is a formal statement about the intended system that specifies what the system should do, identifying a capability, characteristic, or quality factor of a system to have values for system users. Generally speaking, a system requirement should have the following format (syntax):

A system (or subsystem or components, depending on the levels of requirements) *shall* + verb + object (noun) + (target performance metric) + (requirement-related context or environment conditions). Note, here, that the word "shall" is being used in the sentence and carries a very specific meaning: it indicates the sentence is a requirement statement—that this requirement is mandatory and must be followed. There are occasions that "will," "must," "should," or even "may" are used. "Will" and "must" statements signify the mandatory condition that surround the requirements; "should" and "may" statements are less restrictive—usually they are used for nonmandatory or optional requirements. A "should" statement is a little stronger than a "may" statement; sometimes the reason that "should" is used is because it is considered by the design team as important but not yet determined to be mandatory or difficult to verify. A "may" statement is considered as optional, as recommended by research or previous findings, not a contractually binding requirement. Here are some examples of the different requirement formats:

> Shall statement: The new display shall present the required time of arrival (RTA) status upon request by the pilot.
>
> Should statement: The new display should be two-dimensional in an exocentric format.
>
> May statement: The new display may make it possible for the pilot to select the color of their choice.

System requirements analysis is the most critical design activity in the conceptual design stage; it translates the various formats of the system needs from different sources into well-structured system operational requirements, eliminating ambiguity, resolving conflicts, and documenting requirements so that they can be followed, traced, and verified. Although they vary with different types of systems at different complexity levels, system operational requirements should at least address the following issues (Blanchard and Fabrycky 2006):

- Defining system mission: Identify the primary and secondary missions of the system; namely, what is the overall system objective? What are the main functions that system is intended to provide? Usually a set of operational scenarios (narrative or graphical) are defined to help elaborate the mission profile.

- Defining system stakeholder: Identify the major user classes for the system, together with their main characteristics, including demographic information and skill/training levels for each class.

- Defining system performance and physical parameters: These are the operational characteristics of the system's intended function; for example, how well the system performs under certain operational scenarios. These parameters include system deployment and distribution characteristics, describing how the system is deployed geographically.

- Defining system life cycle and utilization requirements: What is the intended life span for the system? How is the system utilized (i.e., hours per day, days per week, months per year)? This is closely related to system performance parameters, such as system reliability and maintainability measures, as they are both functions with time and utilization frequencies.

- Defining the system effectiveness factors: The system's main technical and economic effectiveness factors are defined at the conceptual design stage; these include life cycle cost, system availability, mean time between maintenance (MTBM), logistic support, failure rate (λ), system operators' and maintainers' skill levels, and so on. Generally speaking, given the system is operational, how efficient and effective is it?

- Defining the environmental factors: These are the environmental conditions for the system's operations and maintenance, including physical environmental conditions (i.e., location, temperature, humidity, etc.), as well as the possible environmental impact from the system. This is essential for system sustainable development, because, during the system life cycle, the system constantly interacts with its ambient environment, especially the natural environment that humans rely on. Environmental friendliness is one of the most important factors for the system's long-term success; this is true even for the system's retirement and recycling of materials. This factor needs to be addressed in the very early stages of system design.

The results of the requirement analysis lead to the preparation of system requirement documents. They define, refine, and record a set of complete, accurate, nonambiguous system requirements, usually in an operational hierarchical structure, based on the different levels of system addressing, managed by database-like model-based systems engineering (MBSE) software (such as DOORS and CORE), and made easily accessible to all designers at appropriate levels. Chapter 3 will explain the procedures for refining high-level requirements into lower-level requirements in greater detail. The purpose of refining requirements into different levels is to avoid ambiguity, provide a clear structure of traceability, and manage the requirements easily.

2.3.1.5 Functional Analysis at the Systems Level

Once the requirements have been captured and organized, the question to be addressed is *what* functions the system needs to have to fulfill these requirements. This question is answered by conducting functional analysis at the system level. Functional analysis is an important activity for system design besides requirement analysis, as it is the first step in describing what the system should do. A system function, according to INCOSE, is a unique action or activity that must be performed by the system to achieve a desired outcome. For a product, it is the desired system behavior. A function can be accomplished by "one or more system elements comprised of equipment (hardware), software, firmware, facilities, personnel, and procedural data (INCOSE 2012)." Many people tend to confuse functions with tasks;

in the context of systems engineering, they are different yet interrelated with each other. A function is an action performed by systems that may or may not involve system users. For example, for an ATM system for banks, "deposit," "withdraw cash from checking account," "check account balance" are instances of system functions; tasks, on the other hand, are activities that system users carry out to *support* certain functions to be performed. We highlight *support* here to illustrate the relationship between functions and tasks; that is, functions come first, then tasks. Tasks are supporting or triggering functions; without functions to perform, there will be no task to act on. For the same example of an ATM, "insert debit card," "key in PIN," and "select amount" are all user tasks; these tasks are necessary for the function "withdraw cash from checking account" to be accomplished. Keeping in mind this precedence of functions over tasks, we can easily understand the relationship of these two analyses, and conduct the right analysis at the right time.

A function is an action of the system; it emphasizes the action by using an appropriate *verb* in the short phrase, such as the "*withdraw* cash from checking account" function for an ATM system or the "*transport* passengers to desired floor" function for an elevator system. Knowing this characteristic of functions helps us to identify necessary functions from the narrative of system operational scenarios; that is, highlighting the verbs is a good start when building the functional model. The method for identifying systems functions, their hierarchical structure, and operational sequences is called functional analysis, which is explained in greater detail in Chapter 4 of this book. A typical functional analysis results is illustrated in the following functional list and the associated Figures 2.2 and 2.3.

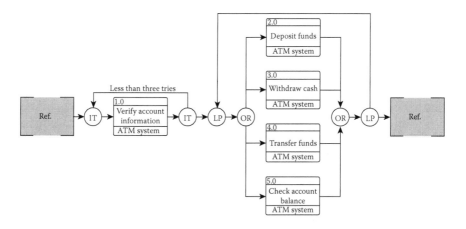

FIGURE 2.2
First-level functional model enhanced FFBD using CORE 9.0. (With permission from Vitech Corporation.)

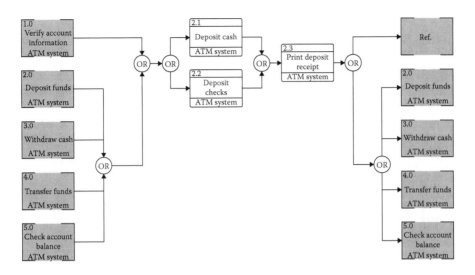

FIGURE 2.3
Second-level enhanced FFBD for deposit function using CORE 9.0. (With permission from Vitech Corporation.)

ATM functions list

0 Perform ATM functions

1.0 Verify account information

2.0 Deposit funds

    2.1 Deposit cash

    2.2 Deposit checks

    2.3 Print deposit receipt

3.0 Withdraw cash

    3.1 Withdraw from checking account

    3.2 Withdraw from saving account

    3.3 Print withdrawal receipt

4.0 Transfer funds

5.0 Check account balance

    5.1 Check checking account

    5.2 Check saving account

    5.3 Print account information

The output of functional analysis is the functional model or functional baseline. A functional baseline defines *what* functions the system should perform and the operational structures of these functions, not *how* the functions are accomplished. It is imperative to keep in mind not to rush into hardware and software solutions at an early stage of the design; in other words, not to over-specify the design in the early stages. Systems engineering is a philosophy of a requirement-driven, top-down process. Design takes a step at a time, and gradually evolves to the final form of the system. In this way, we can ensure that most parts of the design space are explored and the big picture of the design is always considered to achieve an optimal system design solution. Rushing into the detailed system components without properly following the process could cause immature decisions to be made early, causing unnecessary changes in the later design stages.

One essential immediate output from requirement analysis and functional analysis is the identification of the major system parameters and their quantitative target value. Generally speaking, the system performance Y is a function of two types of variables or parameters: $Y = f(X_i, X_d)$. X_i represents design-independent parameters and X_d design-dependent parameters.

- Design-independent parameters (DIPs) are those attributes or variables external to the system design, but influencing the system performance. The design cannot alter the DIP values, but has to take them into consideration. DIPs usually define the environmental conditions for and constraints of systems operations, such as the labor rate, resources costs, demand rate, service rate, customer preferences, interest rate, regional electricity voltage/frequency, and so on. These external factors directly influence the design decisions and system effectiveness. They are independent to the systems design decisions.

- Design-dependent parameters (DDPs) are the attributes/variables that are inherent in the system design; these are the variables that designers can decide on and alter to achieve optimal system performance. Examples of DDPs include system mean time between failure (MTBF), system life cycle cost, output power capacity, acceleration, velocity, weight, dimensions, color, size, and so on. These variables have unique values for the system being designed; in other words, they are dependent on the system itself.

- Technical performance measure (TPMs): these refer to the quantitative values for DDPs. For example, system MTBF of 3000 h, weight less than 40 lb, and so on. These values define the system effectiveness objectives, often included in the system requirements. Quantitative values make it possible for these requirements to be verified.

For requirement analysis and functional analysis, not only are the requirements described in narrative format, but also the objectives pertaining to them need to be stated in a quantitative format, as readers will see later in Chapter 3. TPMs facilitate the efforts of system tests and evaluation. For every requirement and function, a quantitative TPM makes it easier for verification of that requirement. A requirement of "system shall start within six seconds when turned on" is easier to verify than a requirement with vague objectives, such as "system shall be easy to use."

2.3.1.6 System Specification Type A

The results from the design activities described above, including the need identification, system feasibility analysis, requirement analysis, and system function analysis, are integrated and compiled into one major document for the conceptual design stage, *system specification*, or Type A specification or A specs. It specifies the system functional models from the highest level, so is called the system *functional baseline*.

A typical Type A specification includes the following information:

- System general description, including system mission, scope, and applicable documents.
- System requirements analysis results, including the system-level requirements and their refined sublevel requirements. Requirements may be categorized into different classes, depending on the nature of the system; usually, they include system operational requirements, maintenance requirements, system effectiveness factors (such as reliability, maintainability, human factors and usability, system cost-benefit profile, safety and security, etc.); system design and construction considerations (hardware and software technology, data management, CAD/CAM).
- Functional structure of the system, including major functions definitions and hierarchical structure at system level; functional model represented by functional flow block diagram; major maintenance functions and their relationship with the system operational functions.
- Other design considerations and constraints, including system logistics and support concepts; test and evaluation plan; major system milestones for system life cycle.

As the first system configuration baseline, the system specification (Type A) defines the technical, performance, operational, and support characteristics for the system as an entity at the highest level. It includes the allocation of requirements to system functions, and it defines the functions and their relationships (i.e., system decomposition structure and operational flow structure). The information derived from the feasibility analysis, operational requirements, maintenance concept, and functional analysis is included in the Type A specification. It describes design requirements in terms of the *whats* (i.e., the functions that the system is designed to perform and the associated metrics, TPMs).

2.3.1.7 Conceptual Design Review and Evaluation

As stated at the beginning of Section 2.3, the basic methodology involved in the system design process can be summarized in three words: analysis, synthesis, and evaluation. In each of the system design stages, before proceeding to the next stage, intermediate design results are reviewed, tested, and evaluated, to verify that the design is following the right track. These reviews, tests, and evaluations serve as a control point, or decision point, to make sure the design is correct in terms of following the system's operational needs. Systems engineering tries to do things correctly at the beginning; no matter how well the design is laid out and controlled, it is inevitable that any design will have unexpected changes and deviations from the original path, due to a number of reasons, such as changes to requirements, new technology, miscommunication, or lack of in-depth understanding of the design, to name a few. Testing and evaluation is critical in system design, as it serves as a checkpoint to identify the mismatch between the design and the needs; thus, problems may be fixed before it is too late to do so. Throughout the system life cycle and design processes, testing and evaluation are conducted continuously and iteratively, at different levels of detail. The testing and evaluation includes an informal daily check of the work accomplished and a formal test conducted at certain milestones of the system design.

At the end of the conceptual design stage, formal reviews and evaluations are conducted, with representatives from stakeholders (i.e., system users and customers) and designers involved, to determine

- Whether or not the system needs are well understood by all parties
- Whether or not the operational requirements are sufficiently developed for the design
- Whether or not the functional models are complete and reflect the concepts of the system operational requirements

Also called the system requirement review, the conceptual design review uses a variety of techniques and methods, depending on the nature of the system, to gather feedback from stakeholders. For example, in software design, focus groups, questionnaires, and surveys are often utilized by the conceptual design review, while a product design review might use models such as simulations, mock-ups, and trade studies analysis as the basis for evaluation. At the conceptual design stage, since the system only exists conceptually, the review is usually more high level, abstract, analytical, and qualitative in nature. Its purpose is to obtain approval for the system functional model and make sure that there is a correct functional model that reflects exactly what users need of the system, to start the implementation and allocation of the functions to components (the *hows*) in the next stage.

2.3.2 Preliminary Design

The preliminary design extends the results from the conceptual design stage to more detailed levels; namely, the subsystem level and component level. The purpose of the preliminary design is to demonstrate that the design concepts are indeed valid in the sense that they can be implemented through systems components, including hardware and software. Design concepts are further explored, with information about component implementation, translating the *what* aspects of the system (systems functions) to *how* the system requirements are actually fulfilled (function allocations to components).

Generally speaking, the major activities in preliminary design include the following:

- Performing the functional analysis at the subsystem and component levels; taking the functional baseline developed from the conceptual stage, the system-level functions are further decomposed into the subsystem-level functions as well as the operational structure of these systems.
- Developing the specifications for the subsystem and components; this includes components' TPMs, product specification, and processes/materials related to the procuring or producing of these components; preparing preliminary configurations for system physical models, including the interaction between components, and between humans and the system.
- Documenting the design results using engineering design tools and software, including the design data, drawings, and prototype.
- Performing trade-off analysis for the system configuration; optimizing system performance by applying appropriate decision-making models.
- Conducting the system design review to verify that the preliminary design results conform to system requirements.

At the preliminary design stage, *what* functions the system should provide is further investigated. Based on the system concepts developed, systems functions are decomposed to subsystem level; to configure the components to achieve these functions, the question of *how* the system functions are accomplished is first addressed. In other words, the system functional model is translated into system operational models and physical models. These models are instances for the application of model-based systems engineering (MBSE) philosophy.

MBSE is a formalized application of models to design systems, defining system requirements, system elements, verification/validation, and relevant design activities, to support structures and communications in systems engineering efforts throughout the system life cycle. MBSE uses structured modeling languages and semantics to define systems elements and their relationships, such as System Modeling Language (SysML). SysML is inspired by Unified Modeling Language (UML), a commonly used object-oriented design language for software engineering.

SysML was developed in 2003, as a dialect (or profile) of the popular UML. It uses part of the UML construct, removing those constructs that are not needed in systems, and also provides an extension of UML modules for unique systems engineering applications.

SysML including the following diagrams:

1.0 Behavior diagram

    1.1 Activity diagram

    1.2 Sequence diagram

    1.3 State machine diagram

    1.4 Use case diagram

2.0 Requirement diagram

3.0 Structure diagram

    3.1 Block definition diagram

    3.2 Internal block diagram

    3.3 Package diagram

Among these diagrams, the requirement diagram is unique to SysML; the other diagrams are either the same as in UML or modified from UML.

An in-depth review of SysML is beyond the scope of this book; please refer to Friedenthal et al.'s (2009) book for a more comprehensive review of the subject.

Using MBSE, systems can be described via different forms of models, depending on the perspective from which a system is studied. Generally speaking, there are three models that are most commonly used for the system description: functional models, operational models, and physical models.

A functional model of the system describes the system functions architecture. It illustrates the hierarchical relationships among systems functions. Figure 2.4 shows an example of system hierarchical structure.

A system operational model describes the systems in terms of the system operations. Similarly to the system functional flow block diagram, it illustrates the system operations from a temporal perspective, describing the systems operations in terms of timeline relationships. Figure 2.5 shows an example of an operational model of an ATM system.

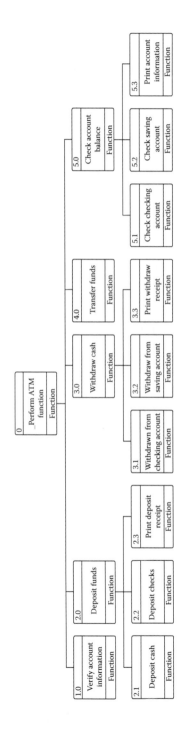

FIGURE 2.4
System hierarchical structure for ATM system using CORE 9.0. (With permission from Vitech Corporation.)

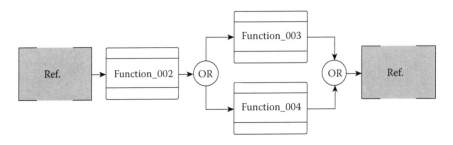

FIGURE 2.5
Illustration of functional operational model. (With permission from Vitech Corporation.)

A physical model describes the physical components of the system, including system hardware, software, and the geographical location information for physical model distribution, as well as environmental information for the components (Figure 2.6).

These three models are instances of MBSE design outputs. They describe and define the systems from different perspectives, and we can say that none of the models is 100% complete; they address the system from one particular dimension of the system characteristics. To get a complete picture of the system, we need to integrate all these models together. It is just like an engineering drawing of a 3D object using a partial 2D picture; one can use the side view, top view, and front view together to get a complete picture of the object.

One also needs to keep in mind that certain models depend on others to be developed first. For example, out of all the models, the functional model is developed first before an operational or physical model is developed, as operations depend on functions and physical components are allocated to functions.

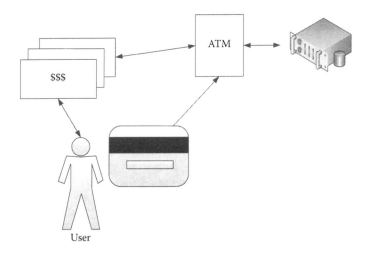

FIGURE 2.6
Illustration of physical model of ATM system.

In the conceptual design stage, the focus is on the system functional analysis; the operational model is developed based on the system-level functional analysis, and it is at the preliminary design stage that the system physical model starts to be addressed; namely, *how* will the functions be accomplished, by what components? The preliminary design allocates functions to components by conducting further functional analysis at the subsystem and component level.

2.3.2.1 Functional Analysis and Function Allocation

The subsystem functional analysis starts with the functional baseline developed from the conceptual design stage. For each of the main system functions, further functional decomposition is conducted and a lower-level functional flow block diagram (FFBD) is developed. The process is very similar to the FFBD in the conceptual design stage, except that it occurs at the lower level. The detailed FFBD method is illustrated with examples in Chapter 4. During this decomposition process, for each of the system functions not only the operational functional sequence is identified, but also the maintenance functions if the operational function fails. In the FFBD, the operational sequence is for the normal conditions of the function and the maintenance functions are for the error conditions of the main function. Considering the failure mode of the function is essential, measures to deal with the system failure (both corrective maintenance to fix a failure if occurs and preventive measures to postpone the failure occurrence) need to be considered in the early stages, as the implementation of the maintenance functions—such as components selection, interface design, and diagnosis system—can have a huge impact on system success, and later changes and modifications should be avoided, just as in any other function.

During subsystem functional analysis, system requirements and related TPMs are also assigned to the sublevel. For example, a reliability of 99% at system level might require a certain level of component performance to achieve that goal. The system itself is not a single object; its performance objectives, defined in TPMs, are accomplished by the components that constitute the system. Allocation of high-level requirements and TPMs to the lower-level components is a necessity to procure the right components. Unfortunately, there are no standards or template to follow to perform these allocations; this is because:

- Different types of requirements necessitate different types of TPMs; some are easy to measure, such as the cost of the components, size, dimensions, and quantitative engineering parameters. These TPMs usually have direct ways to be calculated; some other TPMs are not that easy to measure, especially those involving qualitative parameters, such as system reliability, supportability, and most of the human factor parameters. Most of these type of TPMs rely on empirical data for verification—sometimes even subjective data.

- Analytical models play a very important role when assigning TPMs for quantitative measurements, such as the faulty tree analysis (FTA) to derive the cause–effect relationship of the failure occurrence probability among components at different levels, or the transition functions to study the system control dynamics. However, the majority of the system parameters have no well-defined analytical model to find the optimal solution. In many cases, since the decomposition maps from one (high-level system function) to many (subsystem-level functions), it is nearly impossible to find a unique solution for the assignment. For example, there may be multiple ways of configuring the components that can all meet the system reliability requirements. To find the most feasible solution, one inevitably has to use the most naïve method, trial-and-error, and multiple criteria have to be combined to narrow down the solution space. In the process of decomposition and allocation, often the bottom-up approach is combined with the top-down process; as the top-down process cannot lead to a unique solution, we do need to search for what is available and determine what our choices are. In the later chapters (Section II of this book), we will discuss the different models and methods that can be applied to allocate high-level requirements to various lower levels. Readers have to keep in mind that a good designer is always versatile and flexible, and must master a wide range of models and techniques; experience with similar systems will help a great deal.

In the preliminary design stage, the functional analysis is carried out at a more in-depth level to help find the correct components. For each of the functions, the detailed information for the function is studied, including the inputs to the function (system requirements for the function, material and information needed for the function), the output of the function (product, information, another format of materials, waste and residuals), its controls and constraints (technical, political, environmental, economic) and its mechanism (human factors, computer resources, facility and utility support, maintenance and support). This analysis is further elaborated on in Chapter 4 in greater detail.

The purpose of functional analysis at the subsystem level is to eventually allocate the system functions to components, so that the system functional model can be translated to the physical model. Theoretically, functions can be decomposed into deeper lower levels, all the way to the nuts-and-bolts level; system designers should know when the functional analysis has arrived at the bottom level to stop further decomposition. Usually, the functional analysis stops at the basic configuration level, that is, at the level at which the configuration items can be obtained, either from a supplier or as COTS items. The internal structure of that item is beyond the scope of system design, as long as the item provides the appropriate outputs that support the system functions. Novice system engineers are often confused about the level of depth needed for preliminary design functional analysis. The rule of thumb is to ask the following question: Is there a need to configure the internal structure of the function? If the answer is yes, then another level of decomposition is probably needed; if the answer is no, the function can be achieved by obtaining an external item—then one has arrived at the bottom level of functional analysis. Of course, during this process, since trade-off studies are necessary to obtain the best approach to fulfill a functional requirement, the process of functional analysis usually involves many rounds of iteration and evaluation to point out the most feasible selection of the system components, including hardware, software, people, facilities data, and the combination thereof.

With functional analysis achieved at the lowest level, the lower-level elements are defined based on the functional requirements and TPMs. Subsequently, similar functions are grouped or partitioned into logical subdivisions or packages; this leads to the identification of the major system assemblies, units, and modules. For example, for all the power consumption units, the power input function items can be grouped together, by having one power supply providing power to all the units. This step is necessary, as we will see later; the system package might put constraints on the numbers and sizes of the components inside the system, so having a lower number of components and providing more common functions per component is always a desirable design approach. Usually, this grouping is also dependent on the available standard COTS items; again, the bottom-up search combined with top-down analysis is necessary for successful allocation.

The results of functional analysis at the subsystem level lead to the development of a second system design baseline, the allocation baseline, also called specification Type B; it is a system development specification that includes the technical requirements for the system components/items (such as items of equipment, assembly, data packages, computer software programs, facilities, support items, and so on). These technical requirements are derived from the original high-level requirements, through the process of analysis, allocation, relevant trade studies, and system modeling. Each of the technical requirements for the items includes the performance, effectiveness, and support characteristics that are required from the system-level requirements and necessary for obtaining the actual items for the system.

Another important aspect of the allocation baseline is that interface or interaction requirements are also included. Based on the system functional analysis and allocation to components, each of the functions' inputs and outputs are also addressed. Among them is the necessary human input to activate the functions. It is necessary to conduct an operational task analysis (or operational sequence analysis) as the first step to develop requirements for user–system interfaces. As an inherent part of the system requirements, human factors and ergonomics aspects of the system design are considered at the earliest stage once the human user's role has been identified. Allocation to humans also involves trade studies, as decisions about the allocation can be complex for large systems. Questions to ask include: What is the role of humans within the system functions? What are the advantages and disadvantages of human control versus automated control of particular functions? What is the interaction style between users and the system? What kind of data is involved? What kinds of skills are necessary to operate the system?

The results of the user–system interaction analysis lead to a detailed description of the user–system interface requirements and related components, including input devices, output devices, and the forms of dialogs the interface will contain. We have to keep in mind that, at this stage, the mechanism of implementing the interface is still to be determined. The detailed format of the interaction should evolve together with the other system components, as they are not totally independent of each other.

As part of the interaction design, the working conditions for the user–system interaction should be addressed and described. Some environmental factors such as vibration, heat, and noise may have a serious impact on human performance, thus need to be considered into the design. Human factors engineers and professionals are primarily responsible for working out the design requirements for the user–system interaction and its related environmental conditions.

Based on the design requirements from the user–system interaction and environmental conditions, a statement of personnel requirements should be prepared at the system design level. This statement should include basic information about the system staffing model, including the number of users needed, basic level of skills needed, and training program requirements that are necessary for these skills. These requirements are stated at the system design level and tailored by the details of the design, as the personnel requirements are DDPs, evolving with more details as the system design progresses. It is important to follow the top-down process for these types of requirements as well, because they will change and be modified as system development proceeds in the later stages.

The allocation baseline (Type B specification) breaks the system functions into lower-level functions; thus, the functions can be implemented by system components through an open-architecture configuration structure, providing a foundation through which the subsequent design tasks can be accomplished. Through this baseline, it is ensured that all the subsequent results will follow the original system requirements, providing traceability for design activities. As mentioned earlier, this is the basic design philosophy for systems engineering, in such a way that unnecessary changes can be avoided to a great extent.

2.3.2.2 Design Tools and Analytical Models

A successful systems engineering design is highly dependent on a structured design process, organized documentation, and properly applied models/tools. As the design progresses into the preliminary design phase, and later, the detailed design stage, the level of detail and amount of information increase dramatically; the semantics and interrelationships among different design elements become more complex. Systems designers need to track the relationships, generate different models for evaluation, and manage design changes in a well-controlled and efficient way. This makes utilization of the proper design aids and tools a necessity. After many years of development, CAD hardware and software are now quite mature; the utilization of computer technology in systems engineering has gone far beyond the scope of two-dimensional drawing and graphics design. Nowadays, computer-related technology has become an inherent part of the design; it is hard to imagine a complex systems engineering design project being completed without any computer technology support. The computer technology applied in systems design generally serves two categories of purposes:

> Project management and documentation: In this category, the computer serves as a project management tool, enabling designers to develop a database for the design data, document the design elements, establish and provide traceability among different design elements, generate reports and different models, and aid trade-off studies. Commonly used tools include Rational DOORS by IBM and CORE by Vitech Corporation. These tools provide a well-controlled environment for systems design, to efficiently manage and communicate systems design activities. In Chapter 3, we will introduce the application of CORE in system design at a more detailed level.

> Engineering design aid and prototyping: This category is very common and seen in almost all kinds of systems design. Tools in this category primarily include prototype tools such as AutoCAD, LabView, CATIA, computer-aided manufacturing (CAM) tools, and modeling/simulation software such as ARENA, ProModel, and FlexSim. The application of these tools will be described throughout Section II of this book.

In the preliminary design phase, since the level of complexity greatly increases as the design evolves from functions to subfunctions and components, CAD tools provide great advantages, just as Blanchard and Fabrycky (2006) pointed out. CAD offers advantages including:

1. Enabling designers to address different alternatives in a relatively short period of time, thus making the evaluation and modification of the design more convenient and efficient.
2. Enabling designers to simulate, visualize, and manipulate the design configuration by providing a three-dimensional prototype at low cost and in a relatively short period of time.
3. Enhancing the changing process of the design, in terms both of time and accuracy of data presentation. A well-defined change procedure

can be managed and tracked easily with CAD tools, in conjunction with other tools such as electronic mail and computer office systems.

4. Improving the quality of the design data, in terms of both the volume of the data and its precision/accuracy. With a high quality of data, system analysis and modeling become easier, as does the capability of building large, complex system models.

5. Facilitating communication among design teams and the training of personnel assigned to the project. With a common database, the designers can easily synchronize the design effectively, ensure all design activities are tracked so that everyone is "on the same page" and, moreover, provide a platform for brainstorming, discussion, and training of new personnel, as a design project will usually last years and there is inevitable turnover of designers.

2.3.2.3 Design Reviews

Just as in the conceptual design stage, there are also two types of evaluation at the preliminary design phase. The ongoing informal review and evaluation is conducted on a daily basis, where the designers check their own work, providing technical data and information for the design teams; formal reviews are conducted at the end of the preliminary design stage, to verify that the design results follow the design requirements. Depending on the nature of the system being designed, the type of review and evaluation varies. But, generally speaking, the formal reviews at the preliminary design phase usually include the following:

2.3.2.3.1 System Design Review

This is a formal review usually conducted after a preliminary allocation baseline is completed. The system design review is intended to evaluate how well the system configuration, derived from the allocation baseline, will meet system requirements. This review focuses on the overall system configuration, not the individual components and items. The system design review covers a variety of topics, including system and subsystem functions (including related maintenance functions), function allocations to components, TPMs on subsystems and components, design data and reports (including trade studies results, drawings, part lists, and prototypes), user–system interaction/interface, facilities, environmental conditions, and personnel requirements. A system design review is usually conducted with the involvement of system users to provide feedback on any possible updates and modifications pertaining to the system requirements and configurations.

2.3.2.3.2 Preliminary Design Review

Also called the hardware/software design review, this is usually conducted after a hardware and software components specification is completed, before implementing these components into the design. In the preliminary design, design data related to system hardware and software are reviewed and a variety of test beds are used for the review, usually including the drawings, part lists, trade-off study reports, simulations, mock-ups, and prototypes. Software and hardware performance and the interface between components and users are assessed. The approval of the preliminary design review leads to a design-to systems specification; that is, system components are ready to be procured and system items to be developed, which is the goal for the system detailed design phase.

2.3.3 Detailed Design

With the functional baseline (Type A specification) developed at the conceptual design stage and the allocation baseline (Type B specification) developed at the preliminary design stage, the system configuration is derived in terms of hardware and software components. It is now time to proceed to integrate all the components into a final form of the system; this is the main goal for system detailed design. In the detailed design phase, system engineers will (1) develop design specifications for all the lower-level components and items; (2) develop, procure, and integrate the system components into the final system configuration; (3) conduct a critical system review, identify any possible problems with the system configuration with regard to systems requirements, and control/incorporate changes to the system configuration.

2.3.3.1 Detailed Design Requirements and Specifications

The detailed design starts with the two baselines developed from the previous design stages, the functional baseline (Type A specification) from the conceptual design stage and the allocation baseline (Type B specification). These two baselines are normally at system level, and describe the overall system structure/configuration. At the detailed design stage, all the system components must be finalized, including hardware, software, users, assemblies/packages, system requirements, and specifications, which all need to be further evolved to the lowest level. On the one hand, the analysis process is similar to that of the conceptual design and preliminary design stages; system designers need to iteratively perform the allocation and decomposition analysis, to derive the requirements and related TPMs for the lowest level of components. This is, again, a top-down process; however, since all the components need to be specified and procured or developed in the detailed design stage—and depending on the complexity of the system, the majority of the components most likely need to be procured from external suppliers or using COTS items—it is imperative to understand what is available "out there" before a selection decision can be reached. At the detailed design stage, the top-down process is usually combined with a bottom-up process, to get a complete picture of what is needed and what is available, to obtain the most efficient design solutions.

The analysis of the detailed design will lead to a comprehensive description of the system configuration in terms of system components and operations. With these descriptions, it should be easy to build or install the system with minimum confusion. The detailed design configuration usually consist of the following design baselines:

1. Product baseline (Type C specification): This, according to Blanchard and Fabrycky (2006), describes the technical requirements for any items below the system level, either in inventory internally or that can be procured commercially off the shelf. Type C specifications cover the configuration of the system at the system elements and components level; it comprises the approved system configuration documentation, including the component/part lists, component technical specifications (TPMs), engineering drawings, system prototype models, and integrated design data, which also includes the changes made and the trade-off studies/models for the decisions made for the components.

2. Process baseline (Type D specification): This, according to Blanchard and Fabrycky (2006), describes the technical requirements for the manufacturing or service process performed on any system elements or components to achieve the system functions. It includes the manufacturing processes necessary for system components, such as welding, molding, cutting, bending, and so on, or the service/logistics processes such as material handling, transportation, packaging, and so forth; and any related information processing procedures, including any management information systems and database infrastructure for the design.

3. Material baseline (Type E specification): This describes the technical requirements pertaining to the materials of the system elements/components, including the raw materials, supporting materials (such as paints, glues, and compounds), and any commercially available materials (such as cables and PVC pipes, etc.) that are necessary for the construction of the system components.

These baselines are built on one another, and are derived from the system technical requirements and system analysis, taking into consideration economic factors from the global supply chain, and gradually lead to the realization of the system.

2.3.3.2 CAD Tools and Prototypes in Detailed Design

At the detailed design stage, all the design components have been specified and integrated together. The level of detail and information involved is vast and overwhelming. It is imperative to use computer-aided tools and system project management software to control the design activities and manage all the data involved. The system has evolved to the final form of its physical model; this model needs to be built, verified against the system requirements, and traced against other forms of the system models, including functional and operational models. CAD/CAM tools provide great benefits for detailed design integration and verification; these benefits are similar to those mentioned in Section 2.3.2 (preliminary design phase), providing a well-structured and controlled database and project for implementing the design more effectively. These tools, sometimes combined with system simulation software, enable the design team to create a robust design more efficiently, through quick iterations with different configuration alternatives.

In the detailed design, prototype-based simulation is of importance for validating the design. A prototype is a simulated representation of the system that enables designers and users to visualize, conceptualize, touch and feel, and interact with it to validate the design effectively and efficiently. Prototypes come with different kinds of forms and levels of detail. They range from concept cards, cardboard, hand-sketched flow charts, and storyboards, to near complete complex versions of the system interface or hardware mock-ups. Prototypes are used at all stages of the system design, with different levels of information and different aspects of the system for different purposes. In software interface design, *fidelity* is used to measure the level of detail included in the prototypes. Fidelity of the prototype refers to the degree of closeness of the prototypes to the system counterparts they represent; the closer to the real system, the higher the fidelity. Fidelity is also used in the aviation training and simulation community to measure the quality of the training devices; it has a similar meaning here as with prototypes, as the training devices are often replicas of the real system; for example, the personal training devices (PTD) to train novice pilots. In the early design stages, such as in the conceptual design stages, since the design is focused on the high-level system operational concepts, the prototypes developed at these stages are often of low fidelity. In other words, they do not look like or feel like the final system, but are in a more abstract format, to capture the most important logical relationships of the system functional structure, as they are cheaper and easier to build, and thus provide advantages for exploring different conceptual design alternatives. In the later design stages, especially when approaching the end of the preliminary design stage and detailed design stage, with all the components configured, the prototypes are closer to the final forms of the systems. At this stage, the focus is on investigating the more detailed low-level aspects of the design, such as the physical dimensions, look and feel, and interaction with users.

In detailed design, high-fidelity prototypes are to be developed to serve as the test bed for validating the system components configuration. Prototypes enable designers from different disciplines to come together to communicate in a very cost-effective and time-efficient way. Some of the values and benefits for the detailed design, according to Blanchard and Fabrycky (2006), include:

1. Provide the designers with opportunities to experiment with different detailed design configuration ideas, including (not limited to) facility layout, interface style, packaging schemes, controls and displays, cables and wires, and so on.
2. Provide different engineers with a test bed to accomplish a more comprehensive review of the system configuration, including system functions and operations, reliability and maintainability, human factors and ergonomics, and support and logistics. Reviews are more open and straightforward, with the possibility of interactions between different problem domains.
3. Provide design engineers with a test bed for conducting user task analysis, including the task sequence, time constraints, and skills requirements, which in turn specify the training requirements for system operators and maintainers.
4. Provide a vehicle for designers to demonstrate their ideas and design approaches during the design review.
5. Provide a good tool for training purposes.
6. Provide a good reference for tools and facility design.
7. Serve as a tool in the later stages for the verification of a modification kit design prior to the finalization of data preparation.

Prototypes play a vital role in almost all kinds of design projects; it is hard to imagine a design being completed without any kind of prototype being developed during the design process. In software engineering, there are designs that evolve primarily with the prototypes, as this is straightforward for the users and easy to demonstrate. For example, the *rapid prototyping* approach utilized in software development is an iterative and evolutionary design process that primarily uses prototypes as the main tools to convey the design ideas and get users involved in the early stages of the design.

2.3.3.3 Component Selection

The main activities in the detailed design stage are to further meet the top-down system requirements with bottom-up system component selections; sometimes this is called the *middle-out* approach. Based on the system functional structure and its allocation baseline, components are sought and the system is built. The designer's job is to select the best way to obtain the components (hardware and software) and integrate them into the system as a whole. With regard to the sources of the components, the selection approach has to be based on system specification and driven by requirements, but generally speaking, the following sequence is usually followed:

1. Use a standard COTS item. COTS items are usually easy to obtain from a number of suppliers. The benefits of using a standardized part is because most suppliers specialize in manufacturing those parts, which usually conform with established government and industrial regulations and specifications, such as FAA or ISO9000/ISO9001. These parts are often produced in large volumes at a relatively low unit price. Using standard parts can significantly reduce the cost and increase the efficiency of system development, and even system maintenance/support. The objective of selecting the right components from COTS is to derive detailed requirements for these components through system design analysis, to pick the appropriate supplier for the parts.

2. Modify an existing COTS item to meet the system requirements. If a COTS item cannot meet the configuration requirements completely, a close form of the item obtained from COTS can be modified. These modifications may include adding a mounting device, adding an adapter cable, or providing a software module. The rule for the modification is that it should be kept to a minimum and be simple and inexpensive, to ensure that no new problems or new requirements are introduced to the system design.

3. Design and develop a unique component for the system. If there is no standard component available and it is not possible to modify a standard part to meet our needs, designing a special unique part is our only option. The manufacturing process and materials used for the part, defined in the Type C, D, and E specifications, need to be based on the systems requirements, and it is desirable to use standard tooling/equipment and assembly parts for ease and economy of the installation, operation, and maintenance.

Component selection decisions should be systems requirements driven. The decision-making process usually involves vigorous modeling, simulation, and prototype testing, using systems TPMs as the decision-making criteria. For example, the impact of the selection of components on system functionality performance, economic merits, systems reliability, maintainability, supportability, usability, and even ease of system retirement are all part of the considerations. Just as mentioned above, the selection process is a combined approach, both from the top down and the bottom up; it is systems requirements driven and, meanwhile, involves familiarizing oneself with the available technology and suppliers. Both are essential for the detailed design of the system components.

Another important aspect in detailed design is the implementation of the team effort. Systems engineering requires that all the parties are involved at the early stages of the design, to facilitate communication and stay "on the same page." It is at the detailed design stage that the integration of all the different disciplines reaches the highest level of complexity. With a tremendous amount of data and information, and many personnel involved (even in different parts of the world), the coordination and management of the design team members are very critical for the efficiency of the design activities. With the proper design aids and computer-aided systems management tools, it is now easier to synchronize all the design activities and data within a central database structure. Another important aspect of teamwork is to deal with the design changes. It is inevitable that some changes will occur at the different stages of the design process, due to changes of customer requirements, changes in technology/resources, and so forth. As mentioned above, systems engineering design has traceability incorporated within the design structure; in other words, every design outcome and decision made is not random, but rational, primarily based on the evolvement of the system requirements through the design process. At the later stages of the design, such as the detailed design stage, a small change in the system specification could cause a significant impact on the whole system. As systems engineers, we always try to identify the causes for changes as early as possible, since the further downstream the changes occur, the more costly they will become. Regarding the required changes, a responsible design engineer or staff member cannot just simply implement the change; a proper procedure has to be followed to minimize the impact on the whole system that has been developed. When incorporating changes, the control process or protocol has to ensure proper traceability from the originating baseline to another, through a formally prepared engineering change proposal (ECP), reviewed by the control change board (CCB). Each change proposal is carefully reviewed in terms of its importance, priority, and impact on the system, TPMs, and life cycle costs. Once approved, the responsible teams then prepare the necessary document, models, tools, and equipment, incorporating the changes into the system configuration. All the data, models, and reports are documented, as well as the communication logs and memos. This procedure is illustrated by Figure 2.7.

2.3.3.4 *Detailed Design Review*

At the end of the detailed design stage, with all the components selected, it is time to review the final configuration before going into production. The detailed design review is a critical checking point, since it is the final review for the system design phase; sometimes it is also called the critical design review (CDR). The purpose of the critical design review is to formally review all the system components, hardware, and software with the real users, for compliance with all the system specifications. The majority of the design data is expected to be fixed, and the system is evaluated in terms of its functionality, producibility, usability (human factors), reliability, and

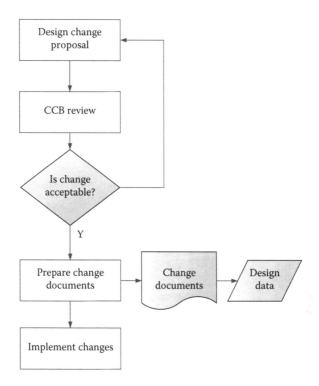

FIGURE 2.7
Change protocol procedure.

maintainability. The critical design review usually involves all the stakehold-ers and the complete set of data, including all the baselines, design drawings, applicable analysis reports and trade studies results, detailed plans on the production, operation, and distribution of the system, and system retirement plans. Human factors professionals need to assess the interaction between the system with the users, in terms of its control and display, dimensions, look and feel, safety features, environment factors, space layouts, training requirements, staffing models, and other human-related factors, under the real operating conditions.

The iterative review, once successful, leads to the approval of the final sys-tem configuration, and the system is released to the next phase, full-scale production and distribution.

2.3.4 System Installation and Deployment, Operation, and Maintenance

After the critical design review has approved the final system configuration, the majority of the system design activities have been completed, and it is time for the system to be installed, delivered, or produced. Depending on the type of system, the system may need to be installed at the customer site—for example, building a house—or the system is handed over to a manufacturing facility, the assembly process/mass production starts, and the finished products—such as a new Boeing 787 aircraft or a new iPad—are delivered to customers. At the system installation and deployment phase, system specifications need to be strictly followed and no major changes or modifications are expected. The focus of system installation is on the efficiency and effectiveness of manpower and cost. Supply chain management (SCM) plays a more significant role in the installation process, by defining the optimum procedure and locations for the system components and resources. A similar focus is carried on to the system operations and maintenance phase, which occurs after systems installation and deployment. During the system operations and maintenance stage, systems designers and engineers are primarily concerned about the follow-up evaluation and feedback from the users, to find out how well the system performs at the user's site and what difficulty users have with the systems. Incidents and accidents need to be documented and investigated, to find their cause and make improvements in the next version of the system. Necessary maintenance and support activities are carried out; for example, recalling faulty components, providing updates/patches, and supporting the system maintenance and warranty activities. These activities are carried out on a continuous basis, possibly until the end of the system life cycle, to provide data for system improvements, materials for training, and, more importantly, for the next version of the system to be more competitive. Techniques that can be used in the system operation and maintenance phase are similar to those in the system design phase, especially in the early conceptual design stage for gathering user needs, such as observation of system operations, surveys and questionnaires, user interviews and focus groups studies, crucial incidents and accidents studies, and related test/evaluation methods. These models will be discussed in more detail in Section II of this book.

2.4 System Engineering Design Process Models

In Section 2.3, we discussed the major steps and milestones of the systems engineering design processes, and provided a basic description and understanding of the activities that are involved in system design. To implement the system design process, it is also important to understand how these different processes and activities are related to one another. In other words, we need to understand the relationships between these activities, how the system evolves from one model to another and the interrelationships between different deliverables. This relationship is captured by systems engineering models. Engineers use these models to provide a management structure to control the efforts that are involved in the system engineering design process.

It is a framework to describe the structure of design activities and facilitate communication among team members.

Over the years, many models have been developed for different types of systems and various levels of complexity. It is hard to say which one is better than the others; we do not intend to emphasize any particular ones, but rather provide an overview for the most commonly used models. It is important to keep in mind that all the models are from the same top-down systems design philosophy, yet each has its unique concentration on slightly different aspects. The models described here include the *waterfall* model, the *spiral* model, and the *vee* model.

2.4.1 Waterfall Model

The waterfall model was originally a model for software project development. It has been adopted and widely used for general systems design models due to its simplicity. The key idea of the waterfall model, as its name implies, is that system design systematically progresses from one phase to another, in a downward fashion. The process follows an iterative manner as well; that is, at any phase, the design can go back to its previous phase to adjust the design results based on the feedback received from the review. Figure 2.8 illustrates the structure of a waterfall model.

The first formal description of the waterfall model was presented by Royce (1970). After several years of use, waterfall models received much recognition as well as criticism. As one of the oldest and most popular models used in systems engineering, the waterfall model offers many advantages for managing system design efforts. These advantages include:

1. The model is linear in nature, which makes it simple to understand.
2. A limited amount of resources is needed to implement the model as the model clearly defines efforts and little interaction across stages is needed.
3. The model provides a clear structure for the system design life cycle, making it easy to document; it also provides a clear structure for system intermediate checkpoints and evaluations.

There are also some criticisms of the waterfall model. The main concern is also one of its benefits mentioned above: its simplicity and linear nature. Many believe that it is impossible for any complex design project to follow the waterfall model strictly; it is difficult to finish one phase completely before moving on to the next phase. For large, complex systems, it is not uncommon that users do not know exactly what they want at the beginning of system design. After seeing a number of prototypes, users usually change their requirements or add something new to them. System designers have little control over such changes. If these changes occur later in the waterfall

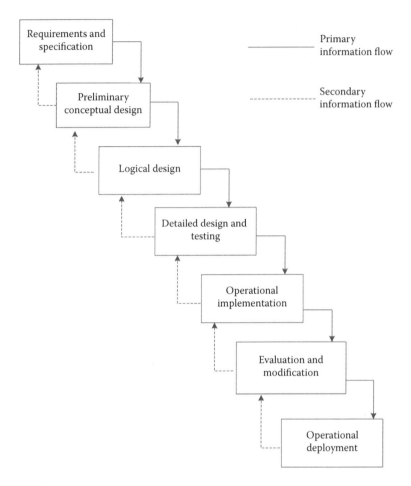

FIGURE 2.8
Waterfall model. (From Sage, A. P. and Rouse, W. B. *Handbook of Systems Engineering and Management*, p.27, Figure 16, 1999. Copyright Wiley-VCH Verlag GmbH & Co. KGaA. Reproduced with permission.)

model, the cost incurred could be tremendous and significant delays could result. It is apparent that the waterfall model is not flexible enough to accommodate constant changes, due to its rigid structure and lack of interaction between different phases. This limitation leads researchers to look for more responsive and iterative variations of the waterfall model; the spiral model is one of these variations.

2.4.2 Spiral Model

The spiral model was initially developed for the software development process; it is based on the concept of fast prototyping in the process, combining both the top-down design process from the waterfall model and the bottom-up process from prototype development and evaluation, to quickly address the constant changes and modifications identified from prototype evaluation in the requirements-driven design process. Figure 2.9 illustrates a basic structure for the spiral model.

The spiral model was first proposed by Barry Boehm in 1986. It was found to be beneficial for large, complex system design, especially for information technology (IT)-related projects. The key idea of the spiral model is continuous development through prototyping to manage the risks. Unlike the waterfall model, the system is not completely defined at the beginning, and each of the phases is not completed prior to progressing to the next one; instead, only the most important requirements are addressed, and the detailed requirements are gradually put in place through iterative prototype feedback with user interaction.

The advantage of the spiral model is *flexibility*. It allows the changes to the design to be implemented at any stage of the design in a relatively quick manner. As the design does not change when completing one phase and moving to the next, but rather is kept open ended during the process, this means that changes may be implemented without impacting the design too significantly. The second benefit of the spiral model is the level of user

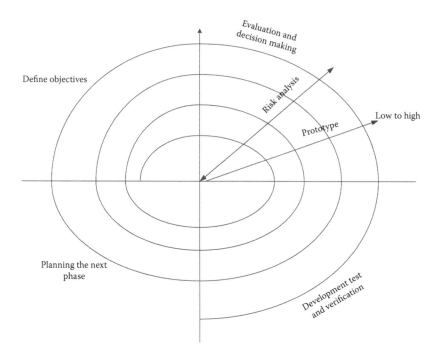

FIGURE 2.9
Spiral model for software system development. (Redrawn from Boehm, B.A. Spiral model of software development and enhancement, *IEEE Computer*, 21, 5, 61–72, © 1988. IEEE.)

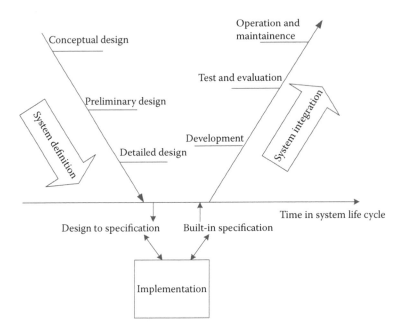

FIGURE 2.10
Vee model. (Redrawn from Forsberg, K., Mooz, H., Cotterman, H. *Visualizing Project Management.* 3rd edn, 2005. VCH Verlag GmbH & Co. KGaA.)

involvement. Users are involved in the design process by providing feedback on the design prototypes on a continuous basis, thus retaining more control over the direction of the project. With constant interaction, users' knowledge and understanding of the system requirements will grow as the design progresses, making interaction and communication more effective in the later stages.

The disadvantages of the spiral model are primarily related to its implementation. The model works best for large, complex systems with possibilities to construct prototypes, such as large-scale software engineering projects. It may not be appropriate for small systems that involve little variation to the requirements, since the requirements are not defined rigidly at the beginning of the spiral model, making such systems development less structural. The model also needs extensive skills in risk assessment and management, requires more in-depth modeling and analysis of risks and uncertain factors. Improper assessment of the risks can lead to a significant increase in system costs.

2.4.3 Vee Model

The vee or V-model is another extension of the waterfall model, but instead of strictly following a linear pattern, the design processes are bent upward after the detailed design phase to form a V-shaped process. The vee model establishes a relationship between the phases of system design and its associated phase of verification and validation, which is achieved by testing and evaluation. Figure 2.10 illustrates the vee process model.

The vee model was first developed simultaneously but independently in Germany and the United States. It is now found widely in commercial and defense systems design. On the left-hand side of the V, the system is defined, leading to the system structure and configuration, just as in the early phases of the waterfall model; on the right-hand side of the V, system integration and verification of subsystems and components occur, resulting in operations being verified and validated at the complete system level.

The vee model integrates users and designers by linking the system definition phase and system components verification and validation phases; it focuses on the systems life cycle, just like the waterfall model, but provides a mechanism to quickly address the responsibility between the design and verification; thus, the changes can be implemented more efficiently. Criticisms of the vee model primarily concern the application aspect: it is argued that the model does not address the supporting structure of the design, but focuses on the project rather than the organization, and the concepts of maintainability and supportability of systems are hardly covered. Separate efforts are needed to define these concepts.

It is clear that all the systems design models have their advantages and limitations. When choosing a model in practice, we should keep these advantages and disadvantages in mind. Sometimes it is not uncommon that combinations of different models are used or necessary adaptions are required for the specific needs of the system design. There is no one model that is superior to the others, and designers have to be flexible, making the model serve the design, not vice versa.

2.5 Summary

In this chapter, we have looked at the system life cycle and systems design processes. Systems engineering is a top-down process; it is driven by systems needs and requirements, gradually bringing the system into being via well-defined design phases. It is the only proven method that will efficiently and effectively design a complex system. In this chapter, we introduced the systems life cycle; described the major forms of systems for different system life cycle stages; described the elements that are involved in the system design process, including system need, requirements, functions, components, prototypes, models, and their relationships with regard to different phases of the life cycle; described the system engineering process, defining the main objectives for each of the phases in the design processes, including the main design activities, major milestone baselines, product, and design evaluation approaches applied in each of the phases; and finally, we reviewed the most common models for the systems design process, including the waterfall model, spiral model, and vee model, and their unique features for different types of systems.

PROBLEMS

1. What is a life cycle? Describe the general life cycle stages for a system.
2. Describe the waterfall model, spiral model, and vee model, and their unique characteristics.
3. What is a system requirement? What is a function? Give an example of a system function.
4. Describe the system user classes.
5. List the major system design processes and describe the purposes and major output for each of these processes.
6. Describe the system functional model, operational model, and physical model. What is the relationship between these models?
7. Define model-based systems engineering (MBSE).

Section II

Systems Methods, Models, and Analytical Techniques

3

Systems Requirement Analysis

In the previous chapters, we have learned about the basic systems concepts, system life cycle, and major processes that are involved in systems engineering design. No matter what kind of system is being designed, from a small appliance to a complex system like a Boeing 787, they all share the same characteristics; that is, every design is driven by systems requirements. Requirements driven is perhaps the most significant characteristic for systems engineering. In the early years of man-made system history, the importance of the system requirements was often ignored; designers used to start their configuration process with little knowledge of what users really wanted, or started with partial understanding of the system needs, hoping to incorporate more needs during the design process. With many failures, costs, and delays involved in the later changes of system needs, lessons were learned the hard way. We now have realized that without having a complete picture of what users need and formally translating that into systems requirements, it is nearly impossible to design a complex system within its time and budget limits. The requirements are vital to the success of the system design. Although systems requirements were briefly introduced in Chapter 2, we feel that it is necessary to revisit the concept of system requirements here in greater detail for readers to grasp the idea of performing requirements analysis (RA). This chapter is perhaps one of the most important chapters in the whole book, as it lays out the foundations of systems design, so that the system is built effectively and efficiently. System requirements analysis (SA) is perhaps the determining point for systems success; as the old proverbs tell us, "the pen follows where the mind reaches," or "when conceiving ideas first, writing is easier." In Chinese culture, it was said that "grinding the ax first will facilitate cutting the firewood." These proverbs all imply the same meaning; that is, spending effort in early planning should not be neglected, just like RA in system planning.

In this chapter, we will

1. Define system requirements in greater detail; describe the nature and type of requirements and their characteristics.
2. Introduce the format of system requirements. What is the common structure for system requirements and major syntax components that should be included in a system requirement statement?

3. Define the characteristics of a good requirement. What should be included and what should be avoided in writing a system requirement?

4. Describe the sources for obtaining the original requirements; the fundamental techniques used in requirements capture and collection process. Basic techniques include user interview, questionnaire and survey, observation, study of similar systems, documentation and historical reports, critical incident/accident analysis, and standards. A comparative analysis is provided for these techniques so that we know when to use which method in the requirements collection process.

5. Introduce the basic concepts of RA; describe the major analysis and trade-off models that are used in RA, including affinity diagram, quality function deployment (QFD) model, and house of quality (HOQ).

6. Introduce the concepts of requirements management; illustrate the process of systems requirement capture, organization, and documentation. Learn how to use the software tool CORE to manage systems requirements.

On completion of this chapter, we expect readers to have become familiarized with the concepts and models that are related to systems requirements, know how to write a good requirement, know how to organize requirements, and perform necessary analysis of the different types of requirements using appropriate models.

3.1 What Is a System Requirement?

In Chapter 2, we defined the system requirements briefly: A system requirement is a necessary attribute or function that a system must have or provide. It is a formal statement to identify a capability, characteristic, or quality that a system must have to provide value and utilities to system users.

From Chapter 2, we learned that system requirements are important for systems design. The requirements are the driving forces for the design activities; requirements define the need to develop the systems, describe the system functionality, and also regulate system engineering efforts for the development of the system. For many years, the role of the systems requirements was not given enough attention; many designers, especially the engineers who were doing the "real" design work, such as hardware developers and software programmers, believed that the actual design work was the real systems engineering process, while the requirements-related work was primarily documentation, and could be accomplished within the design work. Not enough attention was given to the requirements-related efforts, including spending the time to capture the complete set of requirements, understanding and organizing the requirements, and resolving and trading off conflicting requirements. For simple systems with small numbers of requirements involved, the role of RA might not be as critical as in complex system design. But, for a complex system design that involves hundreds or thousands of requirements, we cannot expect these requirements to come to us naturally as we carry out the design work. There are often misconceptions about "stated" requirements and "real" requirements. Stated requirements are provided by systems customers; for example, via commonly used requests for proposals (RFP) and statements of work (SOW). These requirements are usually stated in the customer's language and at a very high and abstract level. The real systems requirements, which are stated in precise designer language, define the systems functionality and utility. These are the requirements that drive the system design activities. There are often huge differences between user-statement requirements and real system requirements; the real system requirements are based on the original user requirements, but user-statement requirements cannot design the system themselves; usually, they are incomplete, vaguely stated, in users' naïve language, and, sometimes, controversial in nature. These users' requirements need to be organized, filtered, translated, expanded, and formatted into a complete, technically sound, and accurate set of system requirements to be implemented. This translation process, often referred to as the requirements analysis (RA), usually occurring in the early design stages, is most critical for understanding the needs of the customer and forming a foundation for proper communication between designers and customers.

It is well understood that a little planning goes a long way and systems requirements analysis is the foundation of system planning. The quality of systems requirements and their analysis are of utmost importance for overall system success; their impact reaches all the way to the end of the system life cycle.

Now we know what a requirement is and why RA is important, and from the previous chapter, we have learned that a system requirement should have the following syntax/format:

> System (or subsystem) XYZ+SHALL+verb (action)+desired outcome/performance level+operational/environmental conditions

This is the general format/syntax for a properly written requirement. Although the format might vary for different type of requirements, there is one important characteristic that all system requirements should have; that is, it is a *SHALL* statement. A SHALL statement is different from *should*, *would*, and *may* statements; it means that it is mandatory and contractually binding for this need; as a requirement, the designed system must comply with it. The other words, such as *may* or *would* statements, usually convey options to the designers; the strength of these statements is not as strong as the SHALL statement.

3.1.1 Types of Requirements

Systems requirements can be categorized into one of the following four major categories:

1. *Functional requirements*: A system functional requirement usually defines a desired function that the system should provide or a function that the system must carry out. Sometimes, a functional requirement is also referred as a WHAT requirement; for example, the automated teller machine (ATM) shall enable customers to deposit cash or checks and withdraw up to x dollars at a time from their checking accounts on a 24 h a day and 7 days a week basis.

2. *Performance requirements*: Unlike functional requirements, performance requirements specifies HOW WELL the system or systems function shall be performed; for example, the automobile system shall start the engine within 6 s after the driver turns the ignition key under normal weather conditions; the system shall have an operational availability (A_o) of 99.99%.

3. *Constraint requirements*: A constraint requirement provides limitations on the design or construction of the system. Constraint requirements are usually part of the results of system feasibility analysis and strategic planning, regulating the design space and pointing in a certain direction for the design. For example, "The automobile system shall use both grid electricity and *B20* biodiesel as the energy source."

4. *Verification requirements*: Verification requirements are those that define what and how is intended to verify or test the functional requirements or performance requirements for acceptance of the system/components. Verification requirements specify the verification events, methods, and procedures to prove the satisfaction of the system requirements. Usually, systems functional requirements and performance requirements contain the verification criteria within the requirements themselves. If there is a special need for the verification, or a unique verification method/technology to be applied, the verification requirements need to be defined. Slightly different from the functional requirements and performance requirements, a verification requirement contains four basic elements: a verification objective, a verification method (inspection, analysis, test, or demonstration), the verification environment, and success criteria. For example, system operational availability shall be calculated using the definition given by MIL-STD-721, conducted by a

government-certified contractor; the demonstration shall show that the system, under the defined operational environmental conditions, shall have an operational availability of at least 99%.

There are also some other types of categorization; for example, depending on the sources, requirements can be classified as originating requirements (those that come directly from customers), derived requirements (requirements obtained from RA to further refine the originating requirements), and design decision requirements (not explicitly from customers, but from the design team analysis results, necessary for fulfilling customer requirements). Based on the group of people involved, requirements may be classified as end-user requirements, management/business requirements, or maintenance/support requirements.

All these different types of requirements overlap with each other; they describe the system needs from a different perspective, just as the different system models describe the system from different angles. In this text, we will be touching on different types of requirements in different sections for their intended purposes—such as functional requirements for functional analysis and user requirements when talking about human factors—in later chapters.

3.2 Characteristics of Systems Requirements

Requirements are the key to project success, as the quality of the requirements determines the quality of the understanding of the systems need. Writing requirements may look straightforward, but they are a real challenge. Writing good requirements requires engineering knowledge of the system, skills in dealing with humans, and, most of all, the ability to think critically and creatively. As a systems engineer, one should know what a good requirement should look like. Although different types of systems vary a great deal in complexity and in nature, the good news is that no matter what type of system you are working on, the criteria for good requirements are essentially the same. Understanding these criteria not only helps us write good requirements, but also validates any other requirements that other people develop. Generally speaking, a good requirement should have the following characteristics:

1. *A good requirement is correct.* A requirement being correct means that it is technically feasible, legal, achievable (within the organization's capability), and complies with the organization's long-term strategic planning goals.
2. *A good requirement is complete in and of itself.* In other words, a good requirement statement should contain the complete information that pertains to the requirement itself.

3. *A good requirement is clear and precise.* The requirement should not contain any ambiguous meanings that cause confusion for the people who read them. Imprecision arises from a number of reasons in a requirement, including

   a. The use of undefined terms or unfamiliar acronyms
   b. The use of imprecise quantifiers and qualifiers
   c. The use of indefinite sources and incorrect grammar

4. *A good requirement is consistent with other requirements.* A requirement statement should use the unified language standard, syntax, and glossary across all the requirements for the system, to avoid conflict between requirements. By conflict we mean the writing style and format, not mainly the content. Sometimes, requirements in different categories have contradicting goals; for example, customers want to cut their costs but also desire more functionality and reliability in the system. These types of contradiction can usually be resolved by conducting a trade study to achieve a break-even point or a balance between the requirements.

5. *A good requirement is verifiable.* For any requirement we develop, we should always ask this question: Is there any way we can determine that the system meets this requirement or not? If we cannot provide a good answer to this question, then the requirement is considered not to be verifiable. Generally speaking, a requirement can be verified in one of three ways: (1) by inspection for the presence of a feature or function, such as verifying the existence of a functional requirement; (2) by testing for correct functional behavior; for example, verifying the performance requirements for a certain function; and (3) by testing for correct nonfunctional attributes, such as verifying the usability of the system interface. For the requirements to be verifiable, the conditions and states for functional requirements must be clearly specified. For nonfunctional behavior, precise ranges of acceptable quantities must be specified.

6. *A good requirement is traceable.* Any requirement should be upward and downward traceable; we should know where it comes from and where it goes from here. As we mentioned in the system engineering design concepts, systems design is requirement driven, and every requirement has a source, either originating from customers or derived by the design activities based on the customer needs. For this reason, every requirement has a rationale associated with it; that is, we know why there is such a requirement. We can also say that a good requirement is necessary; there are no redundant requirements, and every requirement can be traced back to its source and is addressed by at least one function, feature, or attribute. If this requirement is not being addressed by any functions, the validity and necessity of this requirement need to be questioned.

With the above characteristics, we can easily know the requirement's source, understand what users are involved, and develop a clear idea of what we must do to fulfill this requirement and how to verify it afterward. When writing requirements, we should really spend time and effort to make sure all requirements follow these good criteria. The following list provides a general set of guidelines for developing good requirements (Telelogic 2003):

1. Avoid using conjunctions (and/but/or) to make multiple requirements. Make sure one requirement focuses on one subject at a time. A requirement containing multiple elements might cause confusion and difficulties for verification.
2. Avoid let-out phases, such as "except," "unless," "otherwise," and so on.
3. Use simple sentences. For every requirement that addresses one single aspect of the system, a simple sentence is usually sufficient to express the requirement. Using a simple sentence makes it easier to be understood, thus minimizing any ambiguity.
4. Use limited vocabulary. Ideally, the vocabulary size should be fewer than 500 words. A limited vocabulary is easier for communication and translation into technical language. This is more critical for systems that involve international audiences.
5. Include information about the type of users in the requirements, if possible.
6. A focus on the results is to be provided for the users in the requirement statement.
7. The desired results/outcomes should be stated in verifiable criteria.
8. Avoid using fuzzy, vague, and general words to minimize confusion and misunderstanding. Examples of such words are "flexible," "fault tolerant," "high fidelity," "adaptable," "rapid" or "fast," "adequate," "user friendly," "support," "maximize," and "minimize." ("The system design shall be flexible and user friendly" is an example.) Fuzzy words also cause difficulty for verification; for example, if a web page is to be user friendly, what is meant by user friendly? Rather than inserting "user friendly" into the requirement, it is better to rewrite the requirement in a more verifiable way, such as, "the web page shall be completely loaded in less than 7 s for a T1 internet connection speed" and "the web page shall enable the user to find the information by exploring no more than 5 levels of categories."

Here are some examples of some good written requirements:

A single pilot shall be able to control the aircraft's angle of climb with one hand.

The pilot shall be able to view the airspeed in daytime and night-time conditions.

Requirements are the foundations of system design; developing a complete set of good requirements provides a good starting point for project success. Making a good start can save us a lot of time and effort. Writing requirements is a process of translation from the gathered user needs into a well-structured technical document. During this translation process, it is inevitable that individual experiences will impact the understanding and quality of the requirement statements; as a matter of fact, in the field of requirement engineering, there is no absolute right or wrong for any requirements. By using the characteristics of good requirements and following the guidelines to write them, we can ensure the requirements developed may be easily understood and thus facilitate the design process, translating the requirements to system functions and eventually the final system. In the next section, we will discuss sources of requirements and how requirements are captured and organized before they are analyzed.

3.3 Requirements Capture Technique

Some key points need to be kept in mind for an effective process of requirement capture and collection. These key points are essentially nontrivial for large, complex system design, as there are hundreds and even thousands of requirements involved in the design. These key points include

1. Identify the stakeholders for the system.
2. Identify the requirement sources with users' involvement.
3. Apply a variety of data collection techniques.
4. Use a computer-based software tool to document and manage the requirements.

Good requirements come from good sources. To determine the requirement sources, we first need to identify the stakeholders of the system. As discussed in Chapter 2, there are two major types of system stakeholders: First, there are system users, as mentioned in Chapter 2—mainly systems users/operators and maintainers—these are the direct user groups; besides those who directly interact with the system, there are also people who indirectly influence the system users, such as administrators, supporting personnel, and those who design the systems. Every user group plays a role in overall system requirement development, more or less; they each address a different aspect of the system. For example, end users might primarily concentrate on system functionality and usability; the system designers, on the other hand, might consider the long-term sustainability and environmental friendliness of the system, while the management primarily focus on the profitability and life cycle cost (LCC) of the project. These are all important factors for design project success. To develop a complete picture of the system requirements, we need to reach out to all these stakeholders.

There are many sources from which to gather the requirements. The original requirements come directly from the customers. Government customers, such as the Federal Aviation Administration (FAA) and Department of Defense (DoD), have a formal procedure to initiate system needs for systems developers to start the requirements collection process. An RFP is commonly used for government system acquisition and development.

> *An RFP is a formal document that an organization issues to invite the potential vendors/developers/contractors to bid for the project.*

A typical RFP should include the following information:

- Introduction.
  - Organization background: Describe the background of the organization or company who is issuing the RFP.
  - Statement of purpose: Define the mission for the project; solicit proposals to meet these mission requirements and describe detailed proposal requirements; outline the general process for evaluating proposals and selecting criteria.
  - Scope of the system and terms/conditions: Define the scope of the system to be developed; specify any terms and conditions that are related to the life cycle and payment for the system.
  - Other related issues: Communication, coverage and participation, ownership agreement, disclosure and extension, any legal disclaimers, and so on.
  - Communication information: Point of contact (POC) information, phone/fax number, email address, and so on.
- Technical requirements.
  - This section defines the background of the system to be acquired and the technical objectives that are to be achieved by the system. Information on project deliverables are specified and technical measures are provided to measure system success. If the system is part of a larger system, the preferred technical approach and desired interface with other systems are also specified; these will identify the limitations to the technological approach.
  - The format of the proposal to bid on the RFP should be explained in detail, including major sections, page limits, and page format

if desired (such as font type, font size, line space, page margins, etc.). Many organizations publish their own standard for proposals, such as the DoD and the National Aeronautics and Space Administration (NASA), which have their own proposal guidelines, with which every proposal to is required to comply. Usually, a link to these guidelines is provided in the RFP for the proposers to follow.

- Business requirements: This section specifies the major milestones of the deliverables, the expected budget of the project, and the deadlines for the submission of the proposal.

- Evaluation criteria: Describe the major criteria that will be used to evaluate the proposal and the number of awards expected for this project. Typical evaluation criteria include technical capability of the proposer's organization, technical soundness of the proposed solution/design, economic feasibility of the proposed budget, overall project management and control, and personnel qualification. Sometimes a points system is used to compare the competing proposals.

When an organization is trying to bid for the project, the RFP is a good source for the overall requirements, but, generally speaking, the requirements in the RFP are usually not sufficient to completely understand the requirements of the proposed system; additional information and sources are usually necessary to collect more requirements that have been omitted from the RFP, such as the relevant standards and codes for similar systems, research findings, published design guidelines, and, of course, surveys of the customers for whom the systems are being designed.

A successful bid leads to the award of the project, and as a common requirement before any work starts and funding is released, a SOW is developed by the organization awarded the contract to submit to the customer. The SOW, once approved by the customer, is a contractually binding document that specifies the relationships and responsibilities of both the contractors and the customers, defining the major milestones and deliverables for the life cycle of system development. A SOW should be developed based on the requirements and structures in the corresponding RFP; usually it includes the following sections:

1. Background of the project
2. Scope of the project
3. Background of the organization
4. Period of performance
5. Technical approach
6. Major milestones and deliverables

7. Evaluation standard and criteria
8. Design team, business personnel, and their qualifications
9. Payment method and schedules

Requirements extracted from RFPs are usually very high level; they cover the minimum requirements for the system to be acquired or designed. As mentioned above, to obtain the complete set of requirements, we cannot solely rely on the RFP; research is needed to extend the scope of the requirements covered in the RPF to a wider range of sources. These sources, including standards/codes/regulations, published guidelines and data, and even customer surveys, are necessary for us to obtain hidden and implicit requirements, and to acquire a comprehensive understanding of what is really needed for the desired systems. These sources are particularly important for projects that are not RPF initiated. In the following section, we will discuss these various sources that could be used to collect potential requirements, as well as some of the techniques that are commonly used in capturing requirements.

1. User interviews: This is one of the most popularly used techniques and sources for collecting requirements concerning product design and quality improvement. Interviews involve asking users a number of questions; each question addresses a certain aspect of the system to be designed. The interview can be conducted face-to-face or in a remote fashion, such as using the telephone or instant messenger (IM). The interview can be broadly classified as structured, unstructured, or semistructured, depending on how well the questions were predefined and organized. The structured interview looks for answers for specific questions of interest, but lacks flexibility to explore other issues that are beyond the scope of the predefined questions. On the other hand, the unstructured interview enables users to express their opinions freely without being limited to a set of predetermined questions. This can often lead to some issues that have not been thought of by the designers, but the problem with unstructured interviews is that they are time consuming and often inefficient in helping users to think, as some users are better at giving opinion only when specific questions are asked. To help users to provide answers and inputs, sometimes a scenario can be used to aid in the interview session. As interviews are typically conducted one-on-one, focus groups and workshops may be used as an alternative to interviews to achieve group dynamics and interaction effects on the users. The use of groups can gain many benefits that individual interviews do not have; for example, through discussion and brainstorming, the focus group can help different users to gain a consensus view that certain items are required, solving any disagreement or conflict

between different opinions. When controlled and structured well, focus groups and workshops can be very effective in collecting a set of validated user requirements.

2. Survey and questionnaire: These are very popularly used in a wide range of applications and fields, such as social studies, behavioral science, and market analysis. Surveys and questionnaires use a series of questions to seek inputs from a large pool of users. A questionnaire may use closed or open questions. Closed questions have limited answers for users to choose, such as multiple-choice questions or a selection from a Likert scale, while open questions do not limit users' answers; users can enter their own words in the answers, like an essay question in an examination. Unlike interviews, surveys and questionnaires can be done off-line; they can be distributed to the users and completed without being monitored or administered. Compared to interviews, surveys and questionnaires can reach a wider range of people within a short period of time, without committing too many resources; usually mail or email is sufficient. However, since they are conducted in an asynchronized way without the questioner being present, there is no guarantee that the user will respond to the survey. The response rate is the most critical issue for requirement collection via surveys and questionnaires. A typical response rate for a survey and questionnaire study is around the 20%–40% range. The design of the questionnaire is the most critical factor impacting the response rate; many techniques have been developed to improve the clarity, flexibility, and friendliness of the questionnaire design. Another concern for the questionnaire design is its validity; in other words, we have to make sure the users actually take time to read the questions, and think about them, before writing down their answers. Some techniques, such as asking the same questions in different places, have been used to validate users' answers. Nevertheless, the questionnaire has been one of the most popular methods used in gathering users' opinions, despite the disadvantages of low response rates. Questionnaires are often used in combination with other techniques; for example, during an interview session, a short questionnaire is often used at the end of the interview to obtain users' feedback about anything that might have been omitted from the interview session.

3. Observation: This involves spending time with users, observing how they do certain functions and how they perform tasks. When asked about their expectations of and preferences regarding systems, users are sometimes not good at telling the complete story about the whole system; the mental model of users is only related to what they do on a daily basis, and based on their experiences, the input given by users will only cover a subset of the whole spectrum of the system

requirements. Observation can fill in the gaps in the details of user input, and provide a better understanding of the user's needs more objectively and completely. Sometimes called naturalistic observation, the key to the success of observation is in the users' natural working environment, trying not to disturb them when making the observations. To obtain greater insight into how the users feel, the observers can also take part in users' tasks, participating in users' functions, to get a first-hand experience of system functions. This is called an internal or inside observation. This is in contrast to a typical outside observation, where the observers adopt a passive role just to observe, and no active participation is involved. Since the observer is certainly limited in his/her cognitive resources, multiple observers are sometimes used to avoid missing features and subjectivity of personal understanding of the tasks. Technologies such as data logging and video/audio recording are used to aid data collection during observation, if permitted by users.

4. Study of documents and reports: Historical documents and reports are good sources for eliciting requirements. Information on how systems were used and misused through historical sources can provide great insights for items that need to be included or improved in the next version of the system. These documents include, but are not limited to, system manuals, government and industrial standards and regulations, research articles and published guidelines, marketing and sales reports, and critical incident and accident reports. Standards, codes, and regulations should be one of the primary sources to be consulted for requirements, as many of the requirement sources, including the RFP and user inputs, do not necessarily include a complete reference to the standards. It is the designers' responsibility to consult these published standards to comply with them, to make the design more efficient, and most importantly, to avoid any product liability lawsuit in the future. For example, in designing cars, the Federal Motor Vehicle Safety Standards and Regulations by the Department of Transportation should be consulted to include any important features in the design of the vehicle. There are many standards, and weeding through them can be time consuming, but the impact of ignoring any important standard can be significant. In systems engineering, the typically used standards include Occupational Safety and Health Administration (OSHA) standards, military standards and handbooks (MIL-STD/MIL-HDBK, DoD standards), NASA standards (NASA-STD), American National Standard Institute (ANSI) standards, Institute of Electrical and Electronics Engineers (IEEE) standards, and International Organization for Standardization (ISO) standards. Research findings in the published literature can also serve as an important source for

eliciting requirements; these research studies will provide guide-lines for researchers to look at potential variables that are involved in the design and help to design better methods to capture system requirements. Another source for potential requirements are histori-cal reports on critical incident and accident events. Most companies and organizations have an established way of reporting any inci-dents, accidents, and defects when using the system. These events inform us of potential ways that users could make mistakes or errors when interacting with the systems. These mistakes, errors, and acci-dents, which occurred in the past, reveal the situations in which the system causes difficulties. Investigating these situations will reveal to designers what the causes and sources for these difficulties are and, as an important set of the system requirements, investigations of these incidents and accidents, and related factors such as environ-ments and user interactions, will lead to a more reliable system and a better system maintenance design.

5. Study of similar systems: At the current time, it is very difficult to think of any system that does not have any sorts of predecessors. On the other hand, there are always competitors with similar systems as well. Looking at predecessor systems or the current similar sys-tems of others will certainly help designers to gain understanding of what should be included in their own design. To study similar systems, a combination of techniques can be used to collect data, including observation, interviews with users who have experience with different systems, questionnaires, user task analysis, and criti-cal incident/accident studies, to look at data and information related to systems functions, operability, maintainability, user types/skills, and usability issues. Besides the main functional factors, some other useful analysis products should also be derived from the analysis of similar systems, including environmental factors for different sys-tems and human factor issues that could be significant in making the systems successful.

A good review of some of the methods can be found in chapter 7 of Rogers et al. (2011). Readers can refer to their book for a side-by-side comparison of the different data collection methods and their advantages and disad-vantages. The list above is by no means an exhaustive set of sources for obtaining systems requirements; they are just the most commonly used in systems engineering design. We have to keep in mind that, as every system is different, so are the requirements sources. A good system designer will not only think critically, but creatively as well. Some requirements will not be salient, and it takes great effort, resources, skills, and time to develop a good set of original requirements. This is the most important phase in the system design, as we have expressed many times in previous chapters; a good planning phase is the key to find the system stakeholders and the most appropriate requirement capture methods. With the complete set of require-ments captured, the next step to conduct an in-depth RA to implement the requirements in the design.

3.4 Requirements Analysis (RA)

Requirements, especially those originating from users, come in different formats and are of different qualities; some are in users' natural language, others are in a variety of media, video/audio, graphs/drawings, notes, and many more. For the design to implement these requirements, we need to elicit and record them in a strictly defined structure and syntax, as defined in the system requirements definition section. Moreover, many require-ments are ad hoc, and do not follow a well-defined arrangement automati-cally; it is the designers' responsibility to document, manage, and control these requirements according to engineering design standards, so that they may be further translated into systems functions. This is not to men-tion that no matter how complete the original requirements are, they still will be far from defining every factor of the system. Many of the unspo-ken requirements will have been taken for granted by the users; sometimes they are thought as common sense or just so straightforward that they are ignored in the collection of users' views. Designers need to derive those hid-den requirements or sometimes even make decisions or recommendations for those factors of which users have little experience. RA is an important design activity, following the original requirements elicitation, to analyze the captured user requirements, for the requirements to be implemented in the design. Generally speaking, RA usually involves the following major activities:

1. Rewrite the user original requirements into the correct syntax for-mat, the "shall" statement; make sure all requirements are written in a way that is clear, complete, verifiable, and unambiguous.

2. Organize the requirements into different categories and hierarchies; map the requirements to systems architecture; derive the require-ments that are not included in the original requirements using appropriate design models.

3. Perform trade-off studies to prioritize the requirements, ranking them so that the most critical requirements are given more attention in the design.

4. Document and record requirements using requirements manage-ment tools, establish semantics for relationships, and provide ratio-nality and traceability for each requirement.

The purpose of RA is to prepare the requirements for the next step of the system design, which is functional analysis. The results of RA will facilitate the translation from system requirements to system functional structure; they serve as the bridge between users' language and designers' technical language. The most commonly used methods used in RA include affinity diagrams, scenarios, use cases, and QFD.

3.4.1 Affinity Diagram

An *affinity diagram* is a tool to organize a large set of captured requirements into a hierarchical structure based on their relationships. For thousands of years, people have been grouping different things into a limited set of categories, but it was not until the 1960s that the term was first used and a process was developed by the Japanese anthropologist Jiro Kawakita. The affinity diagram is also called the KJ diagram for that reason.

The affinity diagram is a good way to consolidate different ideas into a well-organized structure. It is well suited for RA with large amounts of collected data; especially after the interviews, focus groups, and brainstorming sessions, there are usually many difficulties processing the data collected. Besides the large volumes of data, the data might contain situations that are unknown or unexplored by the design teams; it might present circumstances that are confusing or disorganized. This happens more often when meeting participants with diverse experiences or incomplete knowledge of the area of analysis.

With the affinity-diagram-based analysis method, the design team can address the issues mentioned above by:

1. Sifting through large amounts of data in a relatively short period of time.
2. Encouraging shared decision-making, communication, and criticism to facilitate new patterns of thinking. With group efforts, design teams can break through traditional ways of thinking, enabling the team to develop a creative list of new requirements ideas in a very effective way.

The procedure of conducting an affinity diagram analysis is quite straightforward; with all the data collected, all you need are cards or sticky notes, a pen, and a large surface (a board, wall, or table) on which to place the cards.

- Step 1: Generate and display ideas. Record each requirement with a marker pen on a separate card. Randomly spread all the cards on the work surface so all cards are visible to everyone. The entire design team gathers around the cards and moves on to Step 2.
- Step 2: Sort and group ideas. It is very important that every individual member works on his/her own in this step; no collaboration/

conversation/discussion is allowed. Every member takes time to look for the requirements that are related to each other in some way. Put all related cards in the same group and repeat until all cards have been considered. Note that a single-card group is allowed if this particular card does not fit into any other groups; if cards belong to multiple groups, make duplicates for those cards so that a copy can be placed in every group. Any member of the team may move a card to a different group if he/she disagrees with where it is currently placed. After all the members have done the grouping, move on to Step 3.

- Step 3: Agree on and create groups. It is now time for the team members to discuss their grouping results and the structure of the grouping diagram. By solving any potential conflicts and disagreements between the different members, changes can be made to the final grouping results. Repeat the process until a consensus has been achieved. When all the requirements are grouped, select a heading for each group and place it at the top of the group.
- Step 4: Subgrouping. Repeat Steps 1–3 on a large group to develop subgroups, or combine groups into "supergroups" if desired.
- Step 5: Document and record the final group structure.

Figure 3.1 illustrates a sample affinity diagram.

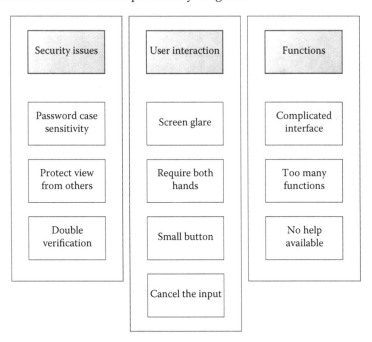

FIGURE 3.1
Affinity diagram.

3.4.2 Scenarios and Use Cases

One of the objectives of RA is to understand what users really need and derive a complete picture from the lists of requirements, describing how the system will be used when it is built from the users' perspective. It is desirable to tell a complete story, especially for the verification of the system requirements with users, so that they may gain an impression how the system will fulfill their needs, without needing to understand the technical background. There are different ways to describe the systems; among those, *scenarios* and *use cases* are the most commonly used tools. Especially for software-related systems, use cases are almost a standard design method (as in Unified Modeling Language (UML) for software engineering). Scenarios and use cases are not mutually exclusive; they are often used in combinations to capture different perspectives of the system's behavior, and are also applied in different stages throughout the system life cycle.

A scenario is an informal description about how the system will be used by users. By understanding and compiling the requirements collected, system designers develop scenarios to tell a story about user task activities and systems functions, to explore and discuss any potential issues with the intended use of the system described. A scenario is usually written in plain narrative language, providing a natural way for system users and stakeholders to understand how the system will fulfill their needs. The focus of the scenario is user tasks and system responses, together with information necessary to describe the tasks, including environmental conditions, context, and any technological support information.

The scenario provides a natural way for explaining to the stakeholders what will be designed and how the systems will be used, facilitating communication between the users and designers, providing a good starting point for users to understand, and exploring the intended functions, context, and constraints, identifying any missed requirements and misunderstandings between the users and designers.

The procedure to develop a scenario is quite straightforward; it requires the designers to understand the requirements and the existing behaviors involved for such a system. Based on their understanding, designers describe a story of the system usage from the users' standpoint, either in a first-person or a third-person perspective. For example, the following scenario is developed to illustrate the start-up of a hybrid vehicle:

> Roger *gets into* his vehicle, he first *puts in* his car key and *starts* the vehicle, and the seat *configures* itself to Roger's preset position. Roger then *enters* his trip information from home to his workplace into the dashboard menu; the onboard computer *accepts* this input, and *provides* information about the GPS route, traffic and weather information, and the estimated time of arrival. Roger *presses* "accept" regarding this recommended route and *starts backing* his car out of the garage, and then *starts driving*.

Note that, in the above example, all the verbs have been italicized; as we mentioned earlier, the scenario provides a starting point for requirements to be analyzed, and, eventually, system requirements need to be translated into functions and functional structure. There is a big gap between the requirements and functions, as they are stated in different formats; requirements are general, stated in natural language, while functions are more technical and very system dependent. But when a scenario is portrayed, all the potential system requirements are embedded within it; one can start by looking at all the verbs to derive system functions, as every system function involves some kind of action, which usually starts with a verb.

Another method used commonly in SA is use case analysis. Unlike the scenario, which focuses more on user tasks, use cases concentrate on the interaction between users and systems. Use cases were originally developed in the software engineering industry, in the object-oriented-design community. In UML-based software design, the use case diagram is used to model the dynamic part of the system, to describe the system behavior by defining the relationships between the users (actor) and system/subsystem elements.

As indicated by its name, a use case in system engineering is a sequence or list of action steps, allowing users and systems to interact. A use case is associated with actors; actors are the users of the system, through proper interaction, and they achieve their goals through use case actions. One use case is usually defined for one system function, this function could be a normal use of the system, or a maintenance function for handling system failure.

A use case analysis includes three elements: identification of the actors, a list of the system actions, and the interaction relationship between the actors and system actions; a use case diagram is developed to illustrate these three elements.

There is no standard procedure to develop a use case; it varies with different systems. A commonly used method includes the following steps:

1. Identify the actors and their goals.
2. Define the scenarios to achieve the goals. The scenario is described as a numbered list of steps that involves both actors and system (subsystem or system components) to achieve the goals. Each step is written as a simple straightforward statement of the interaction between the actors and the system.
3. Make notes of extension of the use case: extension is the condition that results in different courses of actions or alternatives from the normal course of actions.

For example, a possible use case for a check account balance for an ATM system looks like this:

1. The use inserts his/her bank card.
2. The system prompts for the user's account personal identification number (PIN).

3. The user enters the PIN.

4. The system verifies the PIN.

5. The system displays a menu of choices.

6. The user chooses the "check balance" option.

7. The system displays the account type menu of choices (checking or saving).

8. The user chooses an account type from the menu.

9. The system retrieves the account information.

10. The system displays the account information.

11. The user quits from the account information.

12. The system returns to Step 5.

An extension for this use case:

    4.1 If PIN is not valid

        4.1.1 The system displays error message

        4.1.2 The system returns to Step 3

A diagram for the "check balance" use case is illustrated in Figure 3.2.

The most common use of the use case diagram is to model the context of the system, the active roles of users, and the translation of the user requirements into system functions. One can easily identify the use cases from the scenarios, and with use cases, the system functional model can be derived, as each use case involves one or more system functions to achieve the particular objective. Besides system functional model development, use case analysis also provides a starting point for deriving requirements for system interface design, as use case specifies the boundary between the users (actors) and the system itself. For more information, please refer to Booch et al. (2005). for a more in-depth introduction to UML design methodology, including use case analysis.

3.4.3 Quality Function Deployment (QFD)

One important aspect of SA is to translate users' requirements into quantitative technical performance measures (TPMs). Since all requirements are not equally important for the system, it is critical to prioritize the requirements, ranking them in order, so that the most important features are given more attention. An excellent tool to aid in the establishment and prioritization of the TPMs that relate to requirements is *quality function deployment* (QFD). QFD originated in Japan and was first introduced at the Kobe shipyards of Mitsubishi Heavy Industries in 1972 (Shillito, 1994). The need for QFD was driven by two related objectives (Revelle et al. 1998). These objectives started with the users (or customers) of a product and ended with its producers. The two objectives of QFD are

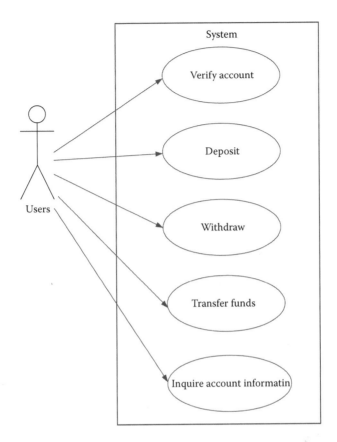

FIGURE 3.2
Example of use case diagram.

1. To convert the users' needs (or customers' demands) for product benefits into substitute quality characteristics at the design stage
2. To deploy the substitute quality characteristics identified at the design stage to the production activities, thereby establishing the necessary control points and checkpoints prior to production start-up

QFD is an interdisciplinary team process to plan and design new or improved products or services in a way that (Shillito, 1994)

- Focuses on customer requirements
- Uses a competitive environment and marketing potential to prioritize design goals
- Uses and strengthens interfunctional teamwork
- Provides flexible, easy-to-assimilate documentation

- Translates general customer requirements into measurable goals, so that the right products and services are introduced to market faster and correctly the first time

There are three basic structural techniques used to analyze and structure qualitative data in QFD. These tools are used to build a matrix of customer information and product features/measures. The voices of customers, are gathered by observations and interviews, which are then organized to construct a tree diagram.

Then, the same procedure is used to generate product features and measures, also called the voice of the company (VOC). These measures are also arranged into a tree diagram. The two trees are then arrayed at right angles to each other, so that a matrix diagram can be formed in the middle. This matrix provides a structure to systematically evaluate the relationship between the items in both dimensions. The relationship between rows and columns can then be coded by symbols. The intersections of the two trees form a matrix that serves as the basis for constructing the first QFD matrix, termed the house of quality (HOQ).

The HOQ is a structured communications device. Obviously, it is design oriented and serves as a valuable resource for designers. Systems engineers may use it as a way to summarize and convert requirements into design specifications. The HOQ, through customer needs and competitive analysis, helps to identify the critical technical components that require change. The critical issues will then be driven through other matrices to identify the most important aspects, manufacturing operations, and quality control measures to produce a product that fulfills both customer and producer needs within a shorter development cycle time.

Figure 3.3 shows the structure of this matrix.

The next vital step is to complete the relationship matrix of user voices versus design features. The final analysis stage relies heavily on the use of the relationship symbols at the intersections of WHAT and HOW. Specifically, we are looking for direct relationships in which the design features satisfy the user voices. There are four types of symbols used in coding these relationships:

⊙ Very Strong (8)

● Moderate (6)

○ Weak (4)

△ Very weak (2)

Based on the relationship identified, the last step of HOQ analysis is calculation of the weightings for the design features. A commonly used traditional deterministic model uses normalized technical importance (NTI) ratings to identify critical characteristics (Cole, 1990).

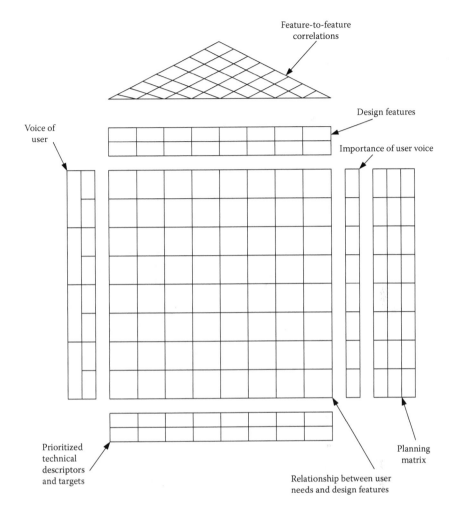

FIGURE 3.3
House of quality (HOQ). (From Liu, D. Web design using a quality function deployment methodology, 2002. PhD dissertation, University of Nebraska, Lincoln.)

$$\text{NTI}_i = \frac{\text{Individual Rating}}{\text{Maximum Individual Rating}} \quad (3.1)$$

$$\text{Individual Rating}_i = \sum_{j}^{n} (\text{Relative Importance})_{ij} \times (\text{User Rating})_j \quad (3.2)$$

$$\text{Maximum Individual Rating} = \underset{i}{\text{Max Individual Rating}}_i \quad (3.3)$$

where user rating j denotes the importance of the jth user needs perceived by the user and relative importance ij represents the relative importance of the ith design feature with respect to the jth user need in the HOQ matrix (strong $=8$, moderate $=6$, weak $=4$, very weak $=2$).

As shown in the above three equations, the individual rating reflects the overall importance of a design feature after taking into account its relative importance and the user self-stated-importance rate. Critical features can be singled out based on the individual rating obtained. The NTI rating uses the ratios of the individual ratings over the maximum individual ratings to facilitate the identification process.

There are many ways to rate the importance of the requirements; for example, one can simply have users rank all the requirements in order. For a small number of requirements without many levels of structure involved, the simple ranking method might work well; however, if the volume of the requirements becomes large and involves many hierarchical structures (such as user requirements, usability requirements, functional requirements, environmental requirements, etc.), it is difficult for users to rank them at the same time across different categories. It is simply beyond human cognitive capability to do so; we need to rely on some kind of mathematical model to aid the decision-making process. In Chapter 6, some models will be introduced, including the analytical hierarchical process (AHP) model.

Figure 3.4 illustrates how HOQ is used in RA.

3.5 Requirements Management

3.5.1 What Is Requirements Management?

Requirements management is one of the most important design activities within project management. A well-implemented requirement management plan can significantly increase the efficiency of the project for almost all the design activities, from functional analysis to design verification, and even facilitate communication among stakeholders. Late changes in design due to poor requirements management can cause significant cost and time delays. If we think requirements are the foundation of the system, then requirements management is the laying out of the foundation into a structure that we need to build our house on (systems functions).

What is requirements management? Generally speaking, requirements management is a process of documenting, categorizing, analyzing, tracing, and prioritizing the system requirements into a structured database, enabling effective communication between requirements and other elements of the system design and controlling changes to requirements. Requirements management is a continuous process throughout the design process; its primary

Row #	Weight Chart	Relative Weight	Customer Importance	Maximum Relationship	Customer Requirements (Explicit and Implicit)	1 Filtration Size (micron)	2 UV Dosage (mW/cm^2)	3 Pumping Distance (ft)	4 Flow Rate (gpm)	5 Weight (lbs)	6 Max Size (inches)	7 Max Drop Distance (feet)	8 Nominal Power for Operation (W)	9 Runtime on batteries (hours)	10 Canister Filter Life (gallons)	11 Max Press. During Operation (psi)	12 Maintenance Cost ($)	13 Setup Time (minutes)	14 Filter Replacement Time (minutes)	15 Cost ($)	16 Max Influent Salinity (ppm)	17 Aux Power Input (V)	18 Water Resistant (feet)
1	▬	11%	18	9	Must produce clean water for 750–1000 people	●	●		●				○	●	●			●	○	○	●	●	▽
2	▬	10%	17	9	Low Power Shutoff		○	▽		▽			●	○					▽				
3	▬	9%	16	9	Portable Backpack within airline luggage size/weight requirement					●	●			▽						●			
4	▬	9%	15	9	Durability					▽		●			▽	▽	○			●			●
5	▬	8%	14	9	Must be able to operate sustainably from solar energy		○	○	●	▽	▽		●	●		▽	▽	▽		●			
6	▌	8%	13	9	Non-technical Operation				●									●	●	▽			
7	▌	7%	12	9	UV Filter Change Indication		●			○	▽			▽					○	▽			▽
8	▌	6%	11	9	Canister Filter Change Indication	●				●	▽			▽	●	●	●		○	▽			▽
9	▌	6%	10	9	High Pressure Emergency Shutoff	▽			▽	▽			▽				●			▽			▽
10	▏	5%	9	9	Easy to maintain										○	○	○			●	○		
11	▏	5%	8	9	Parts Accessibility									●			●			●			
12	▏	4%	7	9	Battery Level Indication		○			▽			●	●						▽		●	▽
13	▏	4%	6	9	Possible Connection to External 12V Battery					●	▽		●	●		○		▽		▽		●	▽
14	▏	3%	5	9	Possible AC input to charge batteries					▽				●				▽		▽		○	▽
15	▏	2%	4	1	Possible Auxiliary DC Output to charge devices					▽										▽			▽
16	▏	2%	3	3	Water Resistant & Able to Float				○			▽								○			●
17		1%	2	9	Ergonomic					●	●							●	●	○			▽
20		1%	1	9	Aesthetically Pleasing						○									●			
				Target		0.5 micron	>30,000	50 ft. radially, 5 ft. deep	>2 gpm	<50–60 lbs	L x W x H<115"	<5 feet	>80 W	12 hours	3000 gallons	55 psi	$500/month	<30 minutes	<15 min.	<$5000	0–500 ppm	12V Source	Up to 3.3 ft. @ 1.1 atm
				Competitor 1 XYZ					1.2	26					11,000	60							
				Competitor 2 ABC		0.5	14,000		1	52	51				10,500	65		1				12	
				Max Relationship		9	9	9	9	9	9	9	9	9	9	9	9	9	9	9	9	9	9
				Technical Importance Rating		158.48	224.56	102.92	284.8	159.06	104.68	80.702	282.46	302.34	219.3	153.8	308.19	125.15	198.25	453.22	94.737	78.363	137.43
				Relative Weight		5%	7%	3%	9%	5%	3%	2%	9%	9%	7%	5%	9%	4%	6%	14%	3%	2%	4%
				Weight Chart																			

FIGURE 3.4
Example of house of quality.

concern is to effectively control information integration related to systems requirements and to maintain the integrity of the requirements for the system life cycle.

The purpose of requirements management is to assure a well-documented requirements body, recording the necessary information pertaining to each requirement, including its sources, origins, types, rationale, and TPM, and establishing the integral hierarchical structure among the requirements, to trace, verify, and meet the needs and expectations of its users, customers, and internal or external stakeholders. As part of the project planning efforts, requirements management begins with the analysis and elicitation of the objectives, the constraints of the systems, and the missions of the organization. After requirements are collected and initially analyzed, requirements management then provides planning for requirements so that they can be translated and verified in the design, standardizes the format, and records the attributes for each requirement, integrating requirements with the organization and documenting the relationship with other information delivered against requirements. For example, a lower-level requirement refines the higher-level requirements, an external file documents the original sources of the requirement, and all requirements are the basis for particular system functions.

3.5.2 Why Requirements Management?

1. Systems engineering is requirement driven; requirements are needed throughout the whole system design process and even the life cycle. System requirements need to be retrievable at any point of the system design. For a large, complex system, it is not uncommon that the volume of the requirements becomes very large; it is unsurprising to have tens of thousands of requirements collected in multiple levels and categories. It is hard to record and manage these large numbers of requirements simply using paper documents. With large volumes of requirements and complex relationships between requirements—and between requirements and other system elements—it is not possible to depend on simple methods such as paper documents or the human memory to reliably manage the requirements; database-based systems management tools are needed to manage the tremendous amount of information related to requirements, so designers can be released to focus on the design activities. A well-selected systems requirements management tool can significantly increase the efficiency of the system design project, thus saving time and money for the design.

2. From the nature and sources of the requirements, it is not uncommon for a large, complex project for requirements to undergo a constant modification process. Changes may occur at any stage of the system design; as we mentioned in Chapter 2, changes can be frequently initiated for a number of reasons—for example, trying to incorporate a new technology, the customers have changed their needs, a new design error is identified, a new environment is required, there are new laws/ regulations, and so on. For most of the time, making the changes themselves does not involve too much difficulty; it is the impact of the changes that causes a lot of problems. As we know that system design is driven by requirements, the interrelationships between the design elements are very complicated; making a small change in one can have a significant domino effect: it could cause a series chain reaction and have a great impact across the system's hierarchical structure. The process of implementing the inevitable changes must be formalized and strictly controlled. Most of the contemporary requirements management tools are computer software based in a database structure; thus, they can easily determine the related system elements and their relationships, so that a change proposal will impact within

very little time, significantly reducing the time and effort required to prepare the change implementation, thus reducing the cost involved. Requirement management has been proven to be the most effective way for implementing changes related to the design project.

3. Requirement management provides requirement traceability. Traceability is one of the key characteristics of MBSE. In the context of systems engineering, traceability refers to

> [the] ability to describe and follow the life of a requirement, in both forward and backward directions (i.e., from its origins, through its development and specification, to its subsequent deployment and use, and through all periods of ongoing refinement and iteration in any of these phases). (Gotel and Finkelstein 1994)

Requirement traceability is concerned with the bidirectional relationship between systems requirements, and the evolving relationships between requirements and other system elements, such as system functions and components. Traceability is one of the most important factors to ensure the rationality of the system design; among hundreds or thousands of design activities and efforts, traceability makes sure all these activities are conducted in a well-planned and rational manner. Traceability is one of the most important reasons that requirement management is necessary. As mentioned in the requirement collection section (Section 3.3), system requirements come from various sources and in various formats; during the system life cycle, system requirements will undergo various evolutionary stages, and, eventually, all the requirements will be fulfilled through some kinds of system functions, with some components. During the process of translating requirements into the final system physical models, we want to make sure that (1) all the requirements have been addressed at least once by some system elements, functions, or components; and (2) any design decisions or components have a reason or origin for that decision. System design usually takes months, even years; during the long transition period, we want to document every single change and evolution for each of the requirements, to ensure that design follows the right track. Maintaining such a traceability relationship also facilitates the verification process in the later stages. We want to make sure there are no so-called orphan requirements that are not supported by any design components. As the name indicates, traceability traces each requirement throughout the life cycle of the system. As one of the most important functions in the requirements management tool, each requirement is documented and its relationship within the body of requirements and between other system components will be established and recorded. For example, within the requirement body, high-level original requirements may be refined by detailed lower-level requirements that are most likely derived from design teams or decision-making models (sometimes this is called *horizontal traceability*, since they occur within the requirement body itself); every requirement is a basis for at least one system function, and documented by some external sources (or *vertical traceability*). In the next section, we will give an example of how to use CORE by Vitech Corporation to develop this relationship. One has to keep in mind that these traceability relationships are usually bidirectional. For example, high-level requirements are refined by lower-level requirements; at the same time, lower-level requirements refine the high-level requirements. Which term to use is dependent on the perspective in which the traceability relationship is looked at. Traceability is documented using the relationship links within the requirement management tool, and these are attached to each requirement for the lifetime of the system. Traceability can be presented by means of a traceability matrix, a hierarchical flow chart, reports, and tables, based on the needs of the analysis.

3.5.3 Requirements Management Using CORE

3.5.3.1 Requirements Management Software Tools

There are many requirements management software commercially available; some of the most popular ones include Rational DOORS and Rational RequisitPro by IBM (previously DOORS by Telelogics), and CORE by Vitech Corporation. For a complete list of the requirements management tools available, please refer to the INCOSE requirement management tool survey, sponsored by the INCOSE Tools Database Working Group (TDWG). The list is published and constantly updated on the INCOSE website (http://www.incose.org/productspubs/products/rmsurvey.aspx). All these tools are based on the fundamental features of requirements management, yet offer different focuses on specific systems or aspects. For example, some are optimized for software systems, and some are configured for easier compliance with government contracts and projects, such as the Department of Defense Acquisition Framework (DoDAF). Regardless of the differences between these tools, they are similar in terms of basic capabilities and features for requirements management. Learning to use one tool will make it easier to transfer to another tool at a later time.

Here, we use CORE (version 9) to describe the basic steps of how the requirements are documented and managed. CORE is developed by Vitech Corporation, serving the systems engineering community since 1992. CORE is more than a requirements management tool; it is a fully integrated model-based approach for collaborative system design, designed by systems engineers. According to Vitech, CORE provides a solution to synchronize requirements, analysis, and architecture, to maintain consistency, reduce risks, and to deliver both technical and management insights to the system design. CORE provides a comprehensive method of tracing requirements throughout the design stages and easily enables designers to build different systems models (such as functional or physical models) based on the designers' need for a better understanding of the dynamics of the system. CORE provides basic simulation features for the system functions, and is capable of producing on-demand documentation, graphs, and charts automatically in various output formats. CORE provides a university version of the software for instructors and students to download and use at no charge. CORE is compatible with the Windows operating system; it applies a typical Windows style of interface with menus and toolbars. It is relatively easy for anyone who has Microsoft Windows experience to learn to use the software.

As mentioned in Chapter 2, CORE uses a system definition language (SDL) to specify the system and its elements. SDL is a formal structured language using standard English to define the system structure. It is closely linked to and highly resembles the systems concepts, as described in Chapter 2. SDL is built on an element-relationship-attribute (ERA) schema, augmented by graphical structures with semantic meanings.

- *Element*: An element in CORE corresponds to the design objects/entities. For example, requirements, functions and components are all elements. An element is defined by using an English-language noun.
- *Relationship*: A relationship defines a link between two elements; it is a binary, two-way relation between two design elements. For example, requirement X is a basis for function Y, and by the same token function Y is based on requirement X. CORE establishes and maintains this two-way communication link automatically once one direction of the relationship is specified by the user.
- *Attribute*: An attribute is associated with an element; it defines the element further by specifying its critical properties. Attributes are like adjectives modifying nouns (elements); they are defined together with the element. For example, a requirement's name, number, and type are all attributes of the requirement element.

The ERA structure comes naturally for system requirements management and analysis, as most of the management tools are based on a database structure. This has made many other features simpler, such as searching for specific elements by attributes, or producing relationship charts.

In the next section, we will use a set of sample projects to illustrate the application of CORE in requirements management.

3.5.3.2 Requirements Management Example Using CORE

Figure 3.5 illustrates a sample requirement element in CORE.

In CORE, requirements can be manually inputted or, if there is an external document, such as a Microsoft Word document, CORE allows the user

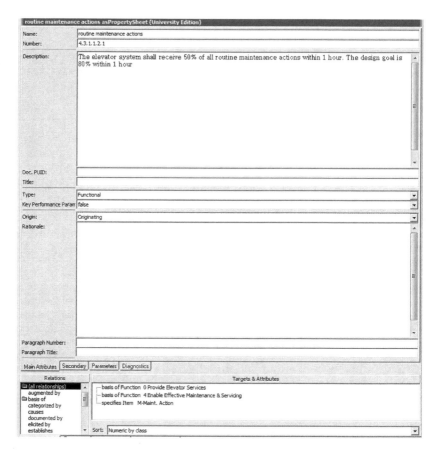

FIGURE 3.5
Sample requirement element dialog window in CORE 9.0. (With permission from Vitech Corporation.)

to extract those requirements directly from the document, so that the user does not need to type them in. This greatly reduces the workload, as many organizations have already collected and documented the original requirements; so, an importing tool in the software is important, not only for the purposes of efficiency, but also, it minimizes the chances of making errors during the transition process. Importing from the original document is a must-have feature for almost all requirements management software. Here, we use CORE to illustrate the steps of extracting the requirements using Document Parser; it is assumed that the features of other software would be similar.

Document Parser is user friendly and efficient for importing requirements directly from a document; it will recognize the statements that contain "shall," "will," or "must" from the text and automatically organize them into different requirement elements, which makes it easy for designers to review and edit them. It has also the ability to link to the "Document" elements where the original document is linked, thus facilitating the establishment of traceability among elements. To use "Document Parser", click on "Tools" from the main menu and then choose "Document Parser". Load the document file that you want to import; the file will be loaded into the left-hand pane in the Document Parser, as seen in Figure 3.6.

Using Document Parser, the text from the loaded document can be extracted directly into any class of elements. The example in Figure 3.6 illustrates parsing the requirement class from the document; select the "Requirement" class from the class drop-down menu and click on the Document Parser icon from the toolbar to parse the document into requirements. All the requirements recognized are marked and numbered, and other statements are marked as "debris" to separate them from the requirements. These debris statements may contain text that is applicable for refining the requirements. For more details, readers can refer to the CORE user manual (guided tour) and resource library at www.vitechcorp.com.

Figure 3.7 illustrates a possible relationship map for a requirement.

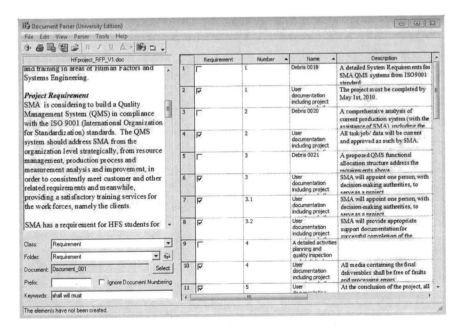

FIGURE 3.6
CORE Document Parser window.

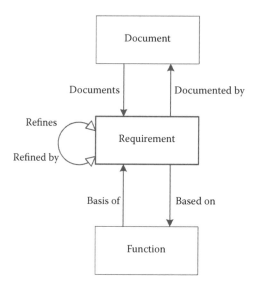

FIGURE 3.7
Requirements relationship map to functions and documents.

3.6 Summary

In this chapter, we examined systems requirements in greater detail. As mentioned many times in Chapter 2, requirements are the driving forces of systems design; spending time and effort to develop a good complete set of system requirements can save designers time, cost, and effort in the long run. In this chapter, the syntax of the requirement was defined and the nature of system requirements and their categorization were given in greater detail. Commonly used categories for requirements include functional requirements, performance requirements, constraint requirements, and verification requirements. The characteristics of a good requirement were given to show how to write a good requirement. Generally speaking, a good requirement shall be correct, complete, clear and precise, consistent with other requirements, verifiable, and traceable. Some general guidelines may be utilized for writing requirements; this was followed by several examples of well-written requirements.

In the second section of this chapter, we reviewed the techniques for capturing requirements, including the sources of requirements, the RFP structure, user interviews, surveys and questionnaires, observations, study of documents, standards, critical incident/accident reports, and the study of similar systems. Each technique was reviewed in detail, and their advantages and disadvantages were presented, for us to use the right technique for the right set of requirements.

A large portion of the chapter was devoted to the methods and models for RA. Major activities in RA were reviewed, and three models that are commonly used in RA were presented, including affinity diagram, scenarios, and use cases, and finally QFD. For each of the techniques, examples were given to illustrate the application of the technique in RA.

In the last section of the chapter, requirement management was briefly introduced. Requirements management is essential for the success of the system design, as it efficiently organizes the large volume of requirements, enables an effective and controlled process for requirement changes, and, most importantly, provides traceability within the system life cycle, which makes design efforts more effective and easy to understand. Examples using CORE were given at the end to show how a software tool can be utilized to implement management of the requirements.

PROBLEMS

1. What is a requirement? What are the characteristics of a good requirement? What are the sources of a requirement?

2. What is an RFP? What is an SOW? What are the purposes of the RFP and SOW?

3. What is requirements analysis (RA)? What are the major methods involved in requirements analysis?

4. Describe the QFD approach. What is the major benefit of using the QFD approach? In your opinion, what are the limitations of the QFD method?

4

Functional Analysis and System Design Models

In Chapter 3, we learned everything about requirements, including what system requirements are, how to write good requirements, where to obtain them, and how to manage them. Requirements define the system, but they cannot design the system. Requirements tell us what is desired for the system, but will not tell us how to make it happen. In other words, requirements are design independent; that is to say, the details to achieve those requirements are yet to be determined and there may be more than one way to fulfill those needs. Systems engineering uses requirements to drive the design in the right direction, but the use of a well-structured design process and a series of methods, models, and activities, leading to the final form of the system, bring it into being. System requirements need to be translated into technical specifications. This process of translation requires designers to have the necessary knowledge of the nature of the system and use appropriate models at the right time. When designing systems, especially complex ones, there are hundreds of thousands of factors and variables involved. It is nearly impossible to study the relationships of these factors to the design without using some kind of model. As the representation of the system, models are essential for designers to concentrate on the most critical factors of the design, simplifying the situation by ignoring the nonrelevant factors, and thus enabling them to provide a solution for the problem to be addressed. This chapter is intended to review the most commonly used models for system design once the requirements are developed; these models, although we call them systems engineering design models, are by no means solely developed for systems engineering. There are no such things as systems engineering models; systems engineering uses any models that are deemed useful for design purposes. As systems engineering is a relatively new discipline, it is an applied field; most of the models utilized in the field of systems engineering are borrowed from other fields, such as social studies, psychology, and operations research/simulation, to name a few. In this chapter, we will attempt to give a comprehensive review of some of the models that may be used in systems design; more specifically,

1. Define models, review the benefits of using models and categorize models.

2. Describe the functional modeling approach; define functions, how to derive functions, and introduce the functional flow block diagram (FFBD).

3. Introduce function allocation model; introduce the processes that are involved in function allocation modeling.

4. Review task analysis, cognitive task analysis, timeline analysis, link analysis.

It is worth mentioning here that the list of models reviewed in this chapter is far from complete; these are some typical models that most of the designs will utilize. There are some other models that we will cover in later chapters, as we do not want to make parts of the book redundant. There will be later chapters that are dedicated to specific models; for example, there will be a chapter solely on system simulation modeling. By exposing readers to these models, we are hoping that they will get a general idea of what kinds of models are available, what the inputs and outputs of these models are, and how these models can be applied to systems design.

4.1 What Is a Model?

In systems engineering, we study systems, trying to design a system based on needs, observing and measuring systems performance, and controlling system behavior. When studying a system, many believe that the best method is to use the system itself. It is beneficial if the nature of the system allows us to investigate it; it is direct and straightforward, but, unfortunately, studying the system itself is not possible most of the time. There are several reasons for this impossibility; firstly, the system may not yet exist. For a system design, especially a new one, most of the time we only have concepts; there is no real system available for us to manipulate. Secondly, sometimes it is not feasible to play with the system itself. Studying the system involves manipulation of and experimentation with the system variables; for example, changing the system layout, eliminating some factors or adding new components. These changes require inevitable interruptions to the current system and can be costly if the new design does not work the way we want it to. Moreover, sometimes it is even dangerous to experiment with the real system itself; for example, it may involve exposure to extreme environments, poisons, radiation, and so on. Under these circumstances, interacting with the real system is out of the question.

An alternative to studying the real system is to use a model of the system. A model is a representation of the real-world system; it is a duplication of the real system in a simplified and abstract form. A model may sound intriguing to most people, but, in fact, we use models and modeling techniques in our everyday life. For example, if we want to describe something, or tell a story, we need to use symbols, language, pictures, or numbers to let others develop a mental picture of the things we are describing or understand the story we are trying to tell. All the artifacts you are using, language, symbols, and pictures, are instances of models. Models are being used at every moment of our lives. Using models to describe the nature and interpret the causal relationships between different factors and variables has become a standard approach in every scientific and engineering discipline. With more technical data and high complexity involved in systems, models have become the inherent and critical part of designing, evaluating, documenting, and managing systems. It is hard to see any system being designed without using any type of models.

Using models in systems engineering can bring us many benefits. First, models allow us to study system concepts without directly interacting with the systems themselves. Some systems do not exist, except in their conceptual stages; some systems have a larger scope, beyond the capability of design teams, such as the social system and environmental systems, while other systems involve dangers that prohibit direct human interaction. Because models are duplicates of systems, they concentrate only on the most critical factors; ignoring the irrelevant factors enables us to focus on the most important aspects of system behavior. By doing this, we can make complex system relationships simpler and easier to be investigated. Second, it is much quicker to build a system model than to build the system itself, not to mention a lot cheaper too. Models range from simple sketches to scaled-down system mock-ups, and are a lot easier to develop, especially with modern technology and advanced computer hardware and software. During the process of system design, along with the evolvement of the system, many of the system parameters are not at the optimum operational level; constant testing, evaluation, and system modification are necessary, and the effect of system modification will not be identified until it is implemented. Interrupting the operating system without confidence in the model can cause lots of problems. Once the model is built and proven valid, that is, it truly represents the correct system behavior relationships, this enables designers to experiment and manipulate it with minimum effort, and various design ideas and modifications can be applied and tested without disrupting the real system, while maintaining a certain level of accuracy of prediction of the effects of the design changes on the system. In other words, even if our experiments fail or we mess things up here and there, we can start over again without costing much to the real system. Building systems models is one of the most important activities involved in systems engineering; it allows engineers to analyze the nature of the problems, leading to their solutions, thus achieving the technical goals of the design and bringing the greatest economic benefits.

Characteristics of Systems Models

The following is a list of the fundamental characteristics specifically of systems models:

1. Systems models represent the system and its behavior in a valid, abstract, and effective way.
2. Systems models use the minimum number of factors (variables) to describe the nature and characteristics of the system; these factors are considered most critical for the nature of the system and other factors can be ignored without affecting the validity of the model. In this way, we can simplify the problem so that it is solvable.
3. Systems models describe the relationships between the factors, and system behaviors, effectively and accurately.

4.2 Model Categorization

4.2.1 Classification Based on Model Format

From the format of the model representation, models can be classified as physical, analogue, schematic, or mathematical models.

4.2.1.1 Physical Models

Physical models are the geometric duplicate of the system, usually in a scaled-down format; they concentrate on the physical, dimensional aspects of the system, including the geometric shape, orientation, color, and size; for example, a three-dimensional mock-up for a product prototype, a ground simulator for an aircraft cockpit system, or a layout plan for a plant facility. Physical models are used primarily for demonstration purposes and sometimes for experimenting with the system. For example, a building mock-up is utilized as a template for the layout of the departments, personnel, and machinery for a better flow. The most important aspects in a physical model are its physical dimensions and the spatial relationships between the components, such as the size and orientation of these components and the physical interaction between them.

4.2.1.2 Analogue Models

Analogue models, as the name implies, describe the relations between the system components. Unlike physical models, where there has to be a proportional duplication of the physical dimensions, in analogue models, although these are still physical in nature, the geometric dimensions are of little importance, but rather the interrelationships among the components are emphasized. For example, electric circuits are utilized to represent a mechanical system or a biological system. In analogue systems, it is not surprising to see that dots are being used as symbols to represent some system components.

4.2.1.3 Schematic Models

A schematic model uses a chart or diagram to illustrate the dynamics or structure of the system; unlike physical or analogue models, schematic models may not look like the real system physically. By proper coding of the system elements, using the appropriate symbols and constructs, a schematic system is intended to illustrate the dynamics of current and future situations or hierarchical static structures in that system. A typical example of the schematic model is an instructional diagram for a basketball play, for a coach to illustrate the ideas of offense or defense. A hierarchical chart of an organization is another common example of the schematic model.

4.2.1.4 Mathematical Models

Mathematical models describe systems using mathematical concepts and language. Based on observed system behavior and data, mathematical models formulate the system into a set of assumptions, variables, formulas, equations, and algorithms to illustrate the relationships between system variables. For example, designers often use linear programming models to optimize system resources allocation; or, a set of differential equations are used to describe the dynamics of system control. Mathematical models are very powerful tools to study the underlying fundamental laws and theories behind system behavior. They reveal the basic cause–effect relationships of systems variables, so that systems can be measured, controlled, predicted, and optimized. They have been commonly used in systems engineering. Of all the mathematical concepts being used in systems, operations research is one of the most popular mathematical modeling tools. It is an applied mathematical field that aids designers to make better decisions, usually concerning complex problems. As a matter of fact, the majority of the models covered in this book belong to the discipline of operations research. Operations research originated in military efforts before World War II for the optimal deployment of forces and allocations of resources; its modeling techniques have since been extended to address different problems in a variety of industries.

In developing a mathematical model, a typical procedure involves the following steps; first, the objective of the problem is identified by understanding the background to the problem and collecting the necessary information about it; second, based on the objective of the problem, a set of assumptions or hypotheses are developed to simplify the situation and prioritize the relevant factors. Assumptions make it possible for a system to be modeled mathematically and quantitatively. This is a critical step, as the validity of the assumptions will impact the validity of the model directly; a valid set of assumptions not only makes a mathematical model possible, but also captures the most critical and essential aspects of system behavior so that the model can reliably predict it. Third, based on the objectives and assumptions made, the most appropriate mathematical tools are chosen to develop the models. We are entering a wide field of applied mathematics; many tools are available for us to use to solve different problems, including algebra, probability and statistics, geometry, graph and network theory, queuing theory, game theory, and mathematical programming, to name a few. Mathematicians have prepared a rich field for us to explore. A good designer should be educated in advanced mathematical tools, so that the most appropriate tool will be selected for the right problem. Fourth, it is important to note that solving the model takes great effort. It is difficult to obtain an analytically definite solution for many complex models; for the purpose of system design, an approximate answer or near-optimum solution is often sufficient. Many techniques, such as graphical approach, numeric analysis, and simulation are very useful for reaching an empirical answer. Computer technology plays a vital role in solving models, making the solution process much more efficient. Last but not least, the results are implemented back into the system. Implementing the results requires another set of critical thinking skills, as many assumptions are made within the models; sometimes, we will find that the solutions are too ideal for real system implementation. For a feasible application, additional analysis is needed for practical purposes, such as sensitivity analysis and error analysis.

4.2.2 Classification Based on the Nature of the Variables

Based on the nature of the variables in the model, systems models can be classified as deterministic or stochastic models.

4.2.2.1 Deterministic Models

A deterministic model describes the relationships between deterministic variables; in other words, there is no randomness involved in the state of the models. Not only are the variables not random, the relationships are also fixed. With the same conditions, a deterministic model is always expected to produce the same results. Deterministic models can be complex in nature, and although there are definite answers that explain model behavior, such behavior is sometimes hard to obtain. For example, in control systems, some deterministic models are represented as differential equations, and it is difficult to express the state of the system explicitly at a particular point in time. Numeric analysis is one of the most popular approaches to approximate the answers in such a system. We can find many examples of deterministic models in our daily life; for example, most of the models in statics belong to this category, such as Newton's three laws of motion. One particular type of deterministic model that needs our special attention is the chaotic model. This model is considered determinist because, theoretically, if the initial conditions are completely known for a dynamic system, its future state can be predicted exactly according to deterministic relations governed by a set of differential equations. But, in reality, it is impossible that the initial conditions of a system are known to the degree of precision required; thus, it is impossible to predict the future trajectory (behavior) of a chaotic system. That is why a chaotic system displays traces of random-like behavior.

4.2.2.2 Stochastic Models

In contrast to deterministic variables, there are random variables, which take a possible value from a sample space without certainty. Models that address random variables are called stochastic models, sometimes also referred to as stochastic processes. Probability and statistics are the most commonly used mathematical tools for developing and solving stochastic models. Randomness and uncertainty is everywhere in our daily life, as also seen in system design. Risks caused by environmental uncertainty can have a significant impact on system design efficiency and effectiveness. Understanding, analyzing, and controlling the level of randomness in the system design is very critical for system success. This implies that stochastic processes are widely applied throughout the system life cycle. Stochastic models take random input and produce random output. They use large samples to overcome individual indeterminacy, giving a likelihood of an outcome rather than a definite answer for given input values. In systems engineering, discrete stochastic models are very useful for analyzing system behavior and assessing design risks; these models include time series models, as seen in forecasting models, queuing models for a production process, and human factors performance data and modeling. We will be discussing each of these models in later chapters.

4.2.3 Other Types of System Model Classification

There are many other classifications of systems models, depending on the different perspectives of looking at them. For example, based on the scope of the models, there are macromodels and micromodels. *Macromodels* address issues in a large population and across a wide range of areas; for example, macroeconomics models to investigate financial situations in a region or a country. *Micromodels*, on the other hand, are usually focused on a small scope or problems in a single area; for example, models for a factory or for a production line are models at a microlevel, as these are very narrow and focused on a small area.

Based on the functions of models, systems models can be further classified as *forecast models, decision models, inventory models, queuing models, economic planning models, production planning models, sales models,* and so forth. These models are usually very specific, and we will also introduce these models in later chapters.

Model categories help us to understand the different formats, functions, and scope of the models, which, in turn, helps designers to choose the right models to solve problems. These model categories are not mutually exclusive; they overlap a great deal. For example, a mathematical decision-making model can be stochastic in nature and micro in scope. No matter what kind of model is being developed, there are some common features or things that need our attention across all different models:

1. Models need to have a satisfactory level of validity. Validity refers to the level of closeness of the model to the real-world objects. The closer the model, the higher validity it has. Model validity reflects the accuracy and precision of the models; it depends on different factors, including the prior assumptions of the model and the variable selection/logic involved in the model. A valid model is believed to contain the most important and critical variables for the problem; thus, it is safe to use the model to replace the real-world objects for analysis and prediction of system behavior.

2. Simplicity is preferred over complexity. When constructing models, given that the validity requirements are satisfied, simple models are desired for system analysis. Simple and clear models are easier to understand and quicker to build. If a simple model is sufficient to address the problem, there is no need to seek more complex models. In developing models, we should always start with simple methods, avoiding the use of intriguing mathematical concepts; remember that this is an applied field, so generalization is of little use. We should avoid developing models with no possibility of solving them. As a general rule, more variables are always believed to be more accurate than fewer variables, but we have to be careful about variable selection, and it is not always the case that the more the better; a balance between the complexity and numbers of variables should be sought to maintain a certain level of precision yet ensure a simple and clear model format.

In the following sections, a number of models that are used in system design and analysis are reviewed in sufficient detail. These models are most commonly found in almost all kinds of contemporary complex system design and systems engineering activities. We hope that, from the review, readers can acquire a comprehensive understanding of how these models can be applied in various stages of system design, what input and output factors are involved, and what special attention should be given when applying these models.

4.3 System Design Models

As mentioned before, systems engineering design is a process that translates intuitive user needs into a technically and economically sound system. This process involves many creative and critical thinking skills; we cannot develop a complex system based solely on recommendations, guidelines, and standards. Rigorous models are necessary for a precise transformation from the user's needs to technical specifications. During the systems design process, almost all the different types of models can be found to have an application at certain stages. As a matter of fact, proper applications of modeling are essential for the success of the system design.

Before we get into the details of models, there are some common characteristics about system design models that need to be elaborated for readers to better understand them and their application within system design.

1. System design models and methods address applied research questions. Basic research addresses fundamental research questions; these questions are basic in nature and often lead to general ground theories. These theories may be generalized to a wide range of fields. For example, the study of human skill performance is basic research; the theory behind such performance can be used to explain many kinds of human skill acquisition, such as learning musical instruments or learning how to fly an aircraft. Applied research, on the other hand, addresses specific research questions; for example, how does the location and size of the control button affect the user performance for a specific product interface? Applied research does not care as much about the generalization of the findings from the model; its focus is only on the specific product it is studying. The scopes of the applied research results are narrow and usually cannot be applied to other types of product for this reason. Systems engineering models only serve one specific system; their validity remains within the scope of the system. Although the finding may also make sense for other systems, there is no external validation guarantee for other types of systems unless the results are validated in those systems. Since basic research is conducted in ideal, well-controlled laboratory conditions, its findings have general meaning and wide applicability. Systems engineering models are usually conducted in actual design fields, with high internal but low external validity.

2. Systems engineering models focus on prediction more than description. Scientific research is concerned more about describing the relationships between factors in nature. This is even more so for basic scientific research, as its purpose is to discover truths from evaluating nature around us; in other words, its purpose is the

description of natural phenomena to try to give a scientific explanation for them. Systems engineering models, on the other hand,
focus more on the prediction of behavior of systems and their
components. For system design, most of the time when models
are applied, the system is just a concept; the physical system does
not yet exist. The purpose of modeling is to predict the future performance of the system, given certain inputs from the design and
current system conceptual configurations. According to Chapanis
(1996), application of models in systems development is much more
difficult than that in basic research; since, in basic research, as the
purpose of modeling is to discover and describe, even if a mistake
is made, others can always carry out follow-up procedures based on
the previous findings, with better data and analysis, to overcome
the deficiencies in the previous studies; thus, the mistakes will be
eventually corrected, because science is "self correcting." In systems
development, however, it is often not possible that many chances are
given to create a design. Once a severe mistake is made, fixing it will
significantly impact the system design efficiencies, causing delays
and unexpected costs, sometimes even causing unrecoverable
system failure. The accuracy of prediction is of utmost importance
in systems development. When predicting system performance,
applied systems models also rely on the findings from scientific theories; in other words, we cannot totally separate the prediction from
the description. As mentioned in Chapter 2, engineering originates
from science; designing a man-made system requires resources from
Mother Nature—a good engineering model is always based on the
solid foundation of scientific discoveries. Knowledge of both is a
must for a good system designer. We should keep this mind; proper
integration between basic scientific models and engineering design
models is the key to the success of system design.

3. System engineering models are not unique. Systems engineering is
a relatively new field; all the models used in system design are borrowed and modified from other fields. This is due to the multidisciplinary nature of systems engineering. None of the models applied
in systems engineering are unique; they were developed in other
scientific or engineering fields, and have been adapted for systems
engineering design purposes. For example, task analysis has long
been used in psychology and social sciences to understand human
behaviors; now it is used widely in systems engineering, especially
to understand the interaction requirements between users and systems. In using the models, system designers should keep an open
mind and be flexible. Since practical and specific answers are the
concern, we should make effective use of all models appropriately in
the design process.

4. Systems models are used iteratively throughout the design process; during the process, it is not uncommon that one particular model is applied multiple times in different stages. For example, functional modeling is first applied after the requirements are completed to design the system-level functional structure. This is the typical usage of functional modeling at the conceptual design stage. In subsequent stages, functional models are also used to design the subsystem-level functional structures, with similar procedures but different starting points and levels of detail. This analysis is conducted iteratively until the final system functional structure is completed. Sometimes, results from the previous models are revisited or reexamined for verification purposes. For this reason, it is difficult to say that a particular model is only used in certain stages; it is very possible that a model will be used at every design stage, depending on the type of system being designed. By the same token, it is also possible that some models are not used at all for one system but used extensively in another system design. Similarly, designers need to keep an open mind when selecting appropriate models for the right system at the right time, by tailoring specific models to the system design needs.

5. Systems models in design follow a sequence. Applications of certain models in systems design depend on the completion of other models, as some models take the results from others as input. This implies that certain models need to follow a particular series sequence. For example, functional modeling has to wait until requirement analysis is complete and task analysis is usually conducted after functional modeling is complete, as tasks are designed to serve certain functions. Understanding the prerequisite requirements for each model helps us to correctly construct the model at the right time with the correct input, thus increasing the efficiency of the system design.

In the following sections, commonly used systems design models are reviewed; please note that this is not an exhaustive list of models for system design, as some models are covered in other chapters. The intention of this review is to illustrate how different models are applied in the systems engineering context, for readers to get some basic understanding about the benefits and limitations of using models in systems design.

4.3.1 Functional Models

Functional modeling and analysis is one of the most important analyses in systems design; its intention is to develop a functional structure for the system, based on the systems requirements. Regardless of what kind of system is being designed, ultimately it will provide functions to meet user needs. In other words, user needs are fulfilled through system functions. Functional modeling and analysis is the very next step after systems requirements analysis, to identify what system functions shall be provided, what structure shall these functions should follow, and how these functions shall interact together in an optimal manner to achieve users' needs efficiently. Functional models provide a picture of *what* functions that system should perform, not *how* these functions are implemented. In model-based systems engineering (MBSE) design, we want to evolve the system from one model to another, following a strictly defined methodology and a tightly controlled process, to minimize unnecessary changes and rework. This top-down approach will ensure a natural transition between models; going beyond the model's scope and over-specifying the details before its maturity often cause rework issues. It is a common mistake that, in system functional models, physical models are blended within the function, leading to partial understanding or even an incomplete picture of the functional structure, narrowing the potential problem-solving space for future models.

4.3.1.1 What Is a Function?

Functional models address system functions. What is a system function? As defined in Chapter 2, a system function is a meaningful action or activity that the system will perform to achieve certain objectives or obtain the desired outputs to support the overall system mission or goals. The common syntax for defining a function is as follows:

Verb + nouns + (at, from, to ...) context information

A system function has the following characteristics:

1. A function is an activity or an action that the system performs, driven by the system requirements and regulated by technology and feasibility; a function is a desired system behavior. Including a verb is usually a typical way to identify a function; for example, "deposit check" or "withdraw cash from checking account" are two functions for an ATM system, while "user friendly" is definitely not a function, since there is no active verb in the phrase. It may define a property of a function, but not a function itself.
2. A function can be performed at different levels; in other words, functions are hierarchical within the system architecture. A function may be accomplished by one or more system elements, including hardware, software, facilities, personnel, procedural data, or any combinations of these above elements. A higher-level function is accomplished by a series of lower-level functions; that is to say, higher-level functions are *decomposed* by lower-level functions.

3. Functions are unique; there are no redundant functions unless redundancy is one of the functions to be achieved. For example, in enhancing system reliabilities, redundant functions are often provided, to be activated when the original function fails. Other than for this reason, functions should be unique, as space and resources are limited in system design.

4. Functions interact with each other. From the same level of functional structure, each function takes inputs from the outputs of other functions, while providing outputs to yet other functions. All the functions work together cohesively to support the overall system's highest-level function: the system mission.

System functions are derived based on the requirements; to develop an accurate functional model, designers need to understand the systems requirements, and, with help from the requirements analysis, to develop a functional behavior model that will effectively accomplish the system requirements. It is a translation process that turns users' voices into a well-defined system functional structure. The development of system functions and their architecture relies heavily on designers' knowledge, skills, and experiences with similar systems. Usually, a good starting point for functional modeling is scenarios and use cases, as these describe all the possible uses of the system, and all the functions are embedded within one or more scenarios if the scenarios are complete. An intuitive way for deriving the functions, especially for a new system, is to highlight all the verbs in the scenarios, and formalize and organize these verb phrases into a functional format. There is no standard method for functional development, as every system is different and everyone has different preferences for how to approach the system. As a rule of thumb, one should always start with the highest level of functions, the major function modules for the system mission, from the very top level (Level 0); the lower-level functions are specified for each of the major function modules. Through this decomposition process, perhaps with much iteration within each level, the complete functional model can be developed.

4.3.1.1.1 Input

Functional modeling uses information from system requirements analysis, including scenarios, use cases, analysis of similar systems, and feasibility analysis.

As mentioned earlier, taking this input information, designers start to identify the highest level of functions and decompose the higher-level functions into lower-level functions through an iterative method. Functions are formally defined, including the desired technical performance measures (TPMs) for the functions, such as power, velocity, torque, response time, reliability, usability, and so on; function structures are developed and traceability between functions and requirements, and between functions, is recorded.

124   *Systems Engineering: Design Principles and Models*

4.3.1.1.2 Results

The output of the system functional modeling is development of a complete list of (1) systems functions and their hierarchical structure and (2) the interrelationships between system functions at different hierarchical levels, represented by the *functional flow block diagram,* or FFBD. The FFBD describes the order and sequence of the system functions, their relationships, and the input/output structures necessary to achieve the system requirements.

4.3.1.2 Functional Flow Block Diagram (FFBD)

A functional flow block diagram (FFBD) is a multilevel graphical model of the systems functional operational structure to illustrate the sequential relationships of functions. An FFBD shall possess the following characteristics:

1. *Functional block*: In the FFBD, functions are presented in a single rectangular box, usually enclosed by a solid line, which implies a finalized function; if a function is tentative or questionable at the time the FFBD is constructed, a dotted-line enclosed box should be used. One block only is used to represent one single function, with the function title (verb + noun) placed in the center of the box. When one function is decomposed to a new sequence of lower-level functions, the new sequence should start with the origin of the new sublevel FFBD by using a reference block. A reference-functional block will indicate the number and name of that function. Theoretically, each FFBD should have a starting reference block. In Chapter 2 we presented an example of an FFBD; Figure 4.1 illustrates another example of a functional block diagram.

2. *Function numbering*: An FFBD should have a consistent numbering scheme to represent its hierarchical structure. The highest level of systems functions should be Level 0 (zero); functions in the highest level of the FFBD should be numbered as 1.0, 2.0, 3.0, and so on. Note that the level number (0) is placed after the sequence number only for the highest level. The next level down after Level 0 is the first level of the FFBD; functions at this level should be numbered as 1.1, 1.2,

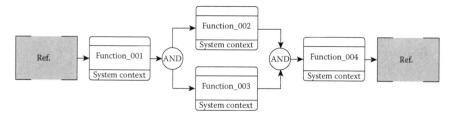

FIGURE 4.1
Example of enhanced FFBD using CORE 9.0. (With permission from Vitech Corporation.)

2.1, 2.2, 3.1, 3.2, 3.3, and so on. It can easily be seen that the numbers at this level only contain one dot, but are different from Level 0 as they have no zero in the numbering. The numbers easily inform us of the hierarchical relationships between functions at different levels; 1.1 and 1.2 are subfunctions of Function 1.0, while 3.1, 3.2, and 3.3 are subfunctions of Function 3.0. Using the same convention, functions at level 1 can be decomposed one step further to Level 2, with two dots in the numbering. For example, Function 1.2 can be further expanded to sublevel functions, 1.2.1, 1.2.2, 1.2.3, and so on if necessary. From this numbering system, it is easy to visualize the hierarchical relationships between different functions; we can immediately identify that 2.2.3 is a sublevel decomposition function of 2.2, and 2.2 is a child function of 2.0. Another benefit of using such a numbering system is to help designers gradually derive the system function structure step by step, in a very organized way, and avoid putting different levels of functions into the same level of the FFBD. Putting different levels of functions into one diagram is called recursive; it is a common mistake often found in FFBD development. Using the numbering system, it is easy to identify this error. Many software packages can detect the recursive error once the numbering system is in place.

3. *Functional connection*: In an FFBD, functions are connected by lines between them. A line connecting between two functions implies that there is a functional flow between these two functions; that is, one function takes the output of another function as its input, either material, power, or information. If there is only a time-lapse between two functions and no other relationships, there should be no lines connecting the functions.

4. *Functional flow directions*: Usually, for any function, a typical flow (go path) always goes in to its left-hand side (input flow) and goes out from its right-hand side (output flow). The reverse flow (from right to left) is used for feedback and checking flows. The top and bottom sides of the function are reserved for no-go flow; that is, the flow when the function fails. The flow exits from the bottom of the function and enters the diagnosis and maintenance functions (no-go paths). In the FFBD, no-go paths are labeled to separate them from go paths. Usually the flows are directional and straightforward; the directions form the structural relationships for the functions, such as series structures, parallel structures, loops (iteration loops, while-do loops, or do-until loops), and AND/OR summing gates. These constructs, together with their functions, define the functional behavior model for the system.

5. *Numbering changes*: An FFBD is an iterative process; it is inevitable that deletion and addition of functions will occur during this process. With the deletion of existing functions, it is simple just to remove the

function from the FFBD; for the addition of a new function, a new number needs to be assigned to the new item on the chart. The rule of thumb is to minimize the impact to the rest of the FFBD; although the new function might be placed in the middle of the FFBD, we do not need to renumber the functions to show its position, but we just choose the first unused number at the same level of the FFBD for the new function. In that way, the rest of the functions are not impacted (Blanchard and Fabrycky, 2006).

6. *Stopping criteria:* The development of FFBD follows a top-down approach, as previously mentioned; it starts at the highest level of the functions and then, from there, decomposes each of the highest-level functions to its next sublevel, and so forth. It is imperative to have a stopping criterion for this decomposition process, because without it, decomposition can literally go on forever. While there is no well-established standard for such a criterion, it usually depends on the nature of the system and level of effort of the design. According to INCOSE (2012), a commonly used rule is to continue the decomposition process until all the functional requirements are addressed and realizable in hardware, software, manual operations, or combinations of these. In some other cases, decomposition efforts will continue until the FFBD goes beyond what is necessary or until the budget for the activity has been exhausted or overspent. In terms of the design tasks, however, pushing the decomposition to greater levels of detail can always help to reduce the risks, as this will lay out more requirements for the suppliers at the lower levels, making the selection of the components a little easier. We will revisit this process later in the functional allocation model.

The FFBD-based functional model is the backbone of the type A specification, which we discussed in Chapter 2. It defines the functional baseline and system behavior model, which serves as the first bridge between user requirements and technical system specifications. The success of the FFBD will lead to the final success of the system design.

We will use an ATM example from Chapter 2 to illustrate the FFBD structure. The following list shows the functions for a simple ATM system.

1.0 Verify account

2.0 Deposit funds

    2.1 Deposit cash

    2.2 Deposit check

3.0 Withdraw funds

    3.1 Withdraw cash from checking account

    3.2 Withdraw cash from saving account

4.0 Check balance

    4.1 Check checking account balance

    4.2 Check saving account balance

5.0   Transfer funds

The traceability of these functions is illustrated in Figure 4.2.

An FFBD for simple systems can be developed simply by using paper and pencil, as the development process is very straightforward. However, for large, complex systems, the FFBD can easily become large and deep in the number of levels, and the interrelationships within the FFBD can become very complex. In this case, a computer-aided tool is desirable to manage such a large FFBD. Many systems engineering management software packages provide a capability for conducting FFBD analysis. As a matter of fact, the FFBD is a standard procedure for all kinds of systems design. In the next section, we will use Vitech's CORE to illustrate how functional modeling is conducted.

In CORE, functional modeling is also called functional architectural modeling. Its purpose is to define system behavior in terms of functions and their structure, and to define the context in which functions are performed and control mechanisms activate and deactivate the system. Some of the basic functional model-related elements include

1. *Use case*: In CORE, use cases are "precursors to the development of system scenarios"; they are extremely useful for gaining insights into the system's intended uses, and thus help to lead us to the integrated system behavior models. As mentioned earlier, use cases and scenarios in the requirement analysis stages tell the stories of how the system will be used once built; they are one of the primary sources for extracting the functions. In CORE, a use case is defined by *use case elements*; as in UML, each use case involves one or more actors, defined by components in CORE; each use case is *elaborated* either by a *function* element, a *programActivity* element, or a *testActivity* element, depending on where the use case affects the system behavior (system design, program management, or verification/validation). A *use case* can be extended to other *use cases*. For example, for the ATM example, a use case could be "withdraw funds from the checking account"; the component would be "users" and would be elaborated by the "withdraw" function.

2. *Functions*: In CORE, functions are actions that the system performs, just like the functions we defined in systems engineering; they are transformations that accept one or more inputs and transform them into outputs, providing inputs to other functions. A *function* is *based on* some *requirements*, it *elaborates* some *use cases*, and it is *performed by* one or more system *components* (hardware, software, or human).

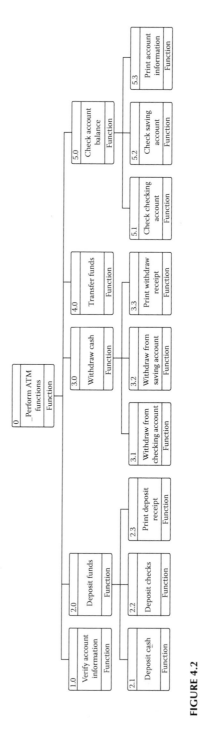

FIGURE 4.2
Hierarchical structure of ATM function model using CORE 9.0. (With permission from Vitech Corporation.)

In the ATM system, "verify account," "deposit funds," "withdraw funds," and "check balance" are all system-level functions.

3. *Item*: In CORE, items are the resources or constraints necessary for functions to be performed, representing the flows for the functions. These resources include the materials, power, information, and user input. Items can be *input to* a *function, output from* a *function* or simply *trigger* a *function* to start. For example, a typical item for the ATM system's "verify account" function is the passcode for the account; without the user input of this information, "verify account" cannot succeed.

4. *DomainSet*: In CORE, DomainSet defines the conditions for the FFBD flow logic, such as defining the number of iterations or replications in a control structure, including the while-do, do-until, and IF-THEN conditions for the function decision logics.

5. *State/Mode*: A state/mode in CORE represents certain states or phases in system operations. Recent development of MBSE requires integration with standard design methodology, such as SysML, which has been developed based on UML. System requirements specify different states/modes into which the system may evolve, which can be exhibited by systems components. States/modes are optional for system functional modeling if such items as phase structures are not necessary for the design.

Figure 4.3 illustrates the interrelationships between the elements in the system functional model, extending Figure 3.7.

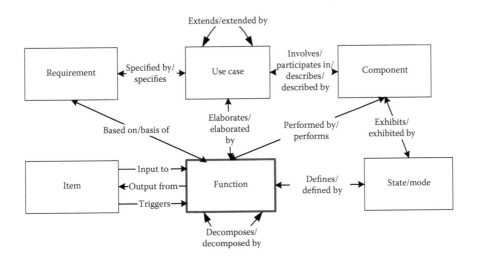

FIGURE 4.3
Functional modeling relationship matrix.

TABLE 4.1

Function ERA Structure

Element	Attribute	Relations	Target Elements
Function	Description	Allocated to/performs	Component
	Doc. PUID	Based on/basis of	Requirement
	Duration	Decomposed by/decomposes	Function (at a lower level)
	Number	Decomposes/decomposed by	Function (at a higher level)
		Defines/defined by	State/mode
		Inputs/input to	Item
		Outputs/output from	Item
		Triggered by/triggers	Item

Defining functions in CORE is very straightforward, just as we described in requirement definition; when defining a function, we should also follow the element-relation-attribute (ERA) format. Table 4.1 illustrates the basic attributes and relations for a function in CORE.

Within CORE, the basic constructs for FFBD include

1. Sequential construct: This is the most commonly used construct; it connects two or more functions together based on their input/output relationships. See Figure 4.4.
2. Iteration: in CORE, iteration is used for a repeated sequence; it is similar to a while-do loop, as the iteration condition is tested at the beginning. For example, in the ATM system, if the user inputs the wrong passcode, he/she has two more attempts to enter the correct code; after three unsuccessful attempts, the system will lock the card and disable any further function. The condition (fewer than three unsuccessful attempts) is defined using DomainSet. See Figure 4.5.
3. Select (OR) construct: A select OR construct defines an exclusive OR path; only one path is enabled at a time. See Figure 4.6.
4. Parallel (AND) construct: An AND construct defines a parallel path, where all the branches must be enabled at the same time. See Figure 4.7.

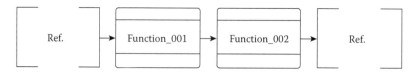

FIGURE 4.4

Sequential construct using CORE 9.0. (With permission from Vitech Corporation.)

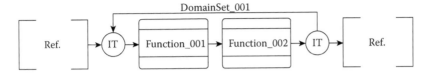

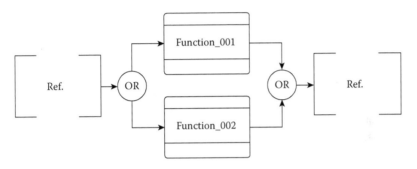

FIGURE 4.5
Iterative FFBD construct using CORE 9.0. (With permission from Vitech Corporation.)

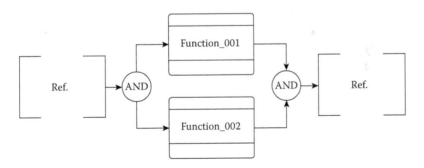

FIGURE 4.6
Selective OR construct using CORE 9.0. (With permission from Vitech Corporation.)

FIGURE 4.7
Parallel AND construct using CORE 9.0. (With permission from Vitech Corporation.)

5. Loop construct: This is similar to a do-until structure in computer program languages; it will repeat until a predefined condition is met. Unlike the iteration construct, the loop construct tests the condition after each iteration. Every loop is associated with a loop exit construct to allow exit from the loop. See Figure 4.8.

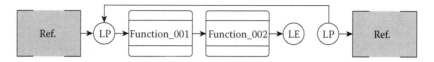

FIGURE 4.8
Loop construct using CORE 9.0. (With permission from Vitech Corporation.)

6. Replicate construct: The replicate construct is the structure short-hand notation for identical processes that operate in parallel. The coordinates between these functions are defined in DomainSet in the coordinate branch. The replicate construct can be replaced by the parallel construct plus the coordinate branches. See Figure 4.9. A typical example of the replicate situation is a manager overseeing multiple checkout lanes in a grocery store.

When developing a FFBD, it is common to have different kinds of construct together within one diagram to elaborate the complex functional structure. For example, the FFBD in Figure 4.1 illustrates the high-level functional model for the ATM system.

The rest of the FFBD can be developed using similar methods until the lowest level of the FFBD has been reached. In developing the FFBD, although it is very straightforward and sometimes even intuitive, this also implies that there are no hard and fast rules to follow about how to conduct FFBD analysis; designers' experiences and subjective judgment play a significant role in this process. There is a high degree of subjectivity in functional modeling, and there is no easy solution for this. Group decision making, more research findings about the system to be designed, and multiple rounds of iteration through verification and feedback are, perhaps, the only way to overcome this subjectivity. As is easily seen from the outcome of the FFBD and functional modeling, the system at this stage is still general in nature; that is, we only know *what* needs to be provided by the system, not *how* to achieve the functions. There is nothing said about the implementation of the system

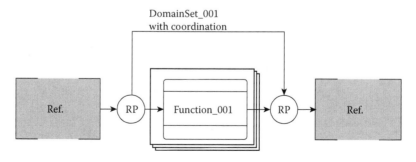

FIGURE 4.9
Replicate construct using CORE 9.0. (With permission from Vitech Corporation.)

functions. Implementations of *how* functions are achieved need a further analysis, that is to say, allocating functions to systems components, which will be reviewed in the next section.

4.3.2 Functional Allocation

In Chapter 2, we briefly introduced function allocation methods and the functional allocation baseline developed from the allocation. Functional modeling produces a detailed structural and operational definition of the system functions, that is, *what* functions need to be performed and *how* these functions are structured to achieve the system mission. To implement these functions, systems elements are needed to carry out these functions; we need to know *who* is doing *what* functions. The typical system elements can be categorized into three basic forms:

1. Hardware/physical elements: These are the tangible/physical components for building the system, whether static or dynamic, such as the facility, the system frame, parts, wires, and so forth. The allocation outputs are the quantitative and qualitative physical configurations of the hardware component.

2. Software elements: These include the computer code and programs that are executed to control the system's physical components. Software elements are responsible for the information flow and data management of system operations. The allocation output is the software configuration for the component, including the input/output specification and performance specifications, also including certain software platforms and interface structures.

3. Human elements: These include the system operators/users and maintainers. These are the people that directly interact with the system and fix it if something goes wrong. The allocation outputs of the system functions to the human elements are the human staffing model (number of people and their scheduling), the operation and maintenance procedure, human system interaction specification and the skills and training requirements for the human elements.

Functional allocation starts with the results from the functional analysis, the function lists, and the FFBD, and just like functional modeling, there is no well-defined template or standard procedure to follow to produce a good allocation; knowledge of the system, familiarity with cutting-edge technology, experience with systems engineering, start-of-the-art design performance capability of hardware and software, and understanding of human capabilities and limitations, plus designers' critical thinking skills and flexibility, are all possible inputs for conducting an allocation analysis. There is no guarantee that the first attempt will lead to a successful allocation baseline; like any other systems engineering analysis, allocation analysis is an iterative process, going through many rounds of iteration, and only from users' feedback, verification and validation from simulations, and analysis and prototype testing may the most feasible allocation baseline be achieved.

Although there is no template for function allocation analysis, there are some general guidelines that can be followed. The transition from an FFBD-based functional model to a function allocation model always starts with a detailed FFBD analysis at the lowest level and function resource analysis for the lowest-level functions.

4.3.2.1 Resource Requirements for Functions

When all the functions are identified, we need to determine how these functions are accomplished. This is achieved by looking at the resources for each of the functions—that is to say: What are the inputs and outputs for the function? What are the controls/constraints for activating the function? And what types of mechanism are involved in the function? Example control/constraints include technical, social, political, and environmental factors; examples of mechanisms include human, materials, computer, facility, and utility support and so on. This process is performed for every function to determine the best way of achieving it. A graphical structure for this process is presented in Figure 4.10.

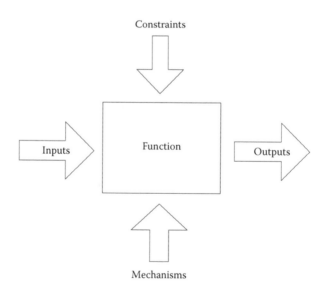

FIGURE 4.10
Identification of resource requirements for function. (Redrawn from Blanchard, B. S. and Fabrycky, W. J. *Systems Engineering and Analysis*, 4th edn. Upper Saddle River, NJ: Prentice Hall, 2006.)

Through this process, every function's detailed resource requirements, as well as the quantitative technical performance measures (TPMs), can be determined. Based on the requirements, designers can seek the most feasible components (hardware, software, human, or a combination of these) to accomplish this function.

4.3.2.2 Allocation of TPMs

In performing the allocation procedure, a mixed process of top-down and bottom-up approaches is usually applied. First, the requirements and TPMs are allocated to the lower-level functions. This is generally a top-down process, and depending on the nature of the system structure, TPMs are allocated accordingly. For some quantitative measures, such as system reliability (usually measured by the failure rate or the mean time between failures [MTBF]), the allocation will be more vigorous, since the relationship between the components structure (e.g., series or parallel) and system reliability is well defined mathematically (the detailed models for reliability and related concepts will be described in Chapter 5). For many other TPMs, the allocation process is not that obvious; there will be a high degree of subjectivity involved. For example, to determine the human factors and usability issues in the lower levels, there are no equations for us to follow; the allocation is largely dependent on the personal experiences and capabilities of the person performing it and his/her understanding of the system and components. There is no shortcut for such an allocation other than iteratively reviewing and improving the design with teamwork and user involvement.

For most of the allocation, it is nearly impossible to achieve an optimal solution; with such a large degree of uncertainty and high level of complexity, it is not easy to formulate the problem into a well-structured optimization problem and provide a solution for it. Most of the time, we are seeking the most feasible solution within our understanding of the system functions, knowing the feasibility of current and emerging technology, and with help from the suppliers' catalogue and global supply chain management resources. It is an iterative process that involves intensive decision making, trade-offs, prototyping, design, synthesis, and evaluation activities. It is believed that with this evolving design cycle, a feasible allocation baseline can be gradually achieved.

The product for the allocation analysis is the identification of the various system elements in terms of hardware components, software components, and human components, together with the data/information and TPMs associated with each element, or type B specification (the allocation baseline). The eventual goal of the function allocation is to know *who/which* is doing *what* function, *how* they are accomplished and by *how much* (the TPMs), providing a basic configuration for the system elements, so that system construction may be carried out in the next step.

Once the lowest level of system components and elements are identified and TPMs are allocated to those elements, the next step is to realize these components by configuring the assemblies for the system. When trying to work out the system components configurations, there are some limitations to be considered; one of them is that of the physical dimensions. There are certain requirements regulating the size and number of components to fit in a limited space. Layout and packaging design are issues that need to be considered in configuring the system structure, as mentioned in Chapter 2. Further development is needed to specify the assembly selection (Type C specification: product baseline), manufacturing process/procedures for these assemblies (Type D specification: process baseline), and materials specifications for the assemblies (Type E specification: material baseline). In developing these baselines, traceability has to be ensured to make sure the baselines conform with the systems requirements and design constraints.

In responding to these elements needs, designers need to conduct trade studies to select the most feasible alternative for realizing these components. Based on the functional allocation results, starting with the lowest level of the element architecture, elements providing similar functionalities are investigated together, and based on the current technology and manufacturing capabilities and the suppliers' catalogue, possible elements with similar functionality are grouped together as the potential assembly of the system. This process, together with the trade studies results, evaluation, and testing, are carried out iteratively, until a feasible assembly plan is obtained (Type C: product baseline). Figure 4.11 illustrates this development process.

With the current globalization and supply chain environment, it is not always economically efficient to manufacture everything ourselves in-house; as a matter of fact, there are multiple suppliers that can provide similar components, and commercial off-the-shelf (COTS) items are the most cost-effective solutions for most of the system components selections. As mentioned in Chapter 2, when selecting a specific component fulfilling the function requirements, a series of trade studies and decision making is necessary to achieve the best feasible selection. Many of the decision-making models are conducted under uncertain situations and involve risks based on the predictions of future system performance. Decision making under risks and uncertainty are essential for systems design, as the design process constantly involves the optimization of limited resources. We will be covering the decision-making models in greater detail in Chapter 6.

As for the selection of the components, there are no standards for us to follow, as each system is different and involves different types of items but, generally speaking, there are some rule-of-thumb guidelines for the designer to follow; these guidelines have been practically proven to be efficient for saving costs and time. When selecting a component, we should follow the following order or sequence (Blanchard and Fabrycky 2006):

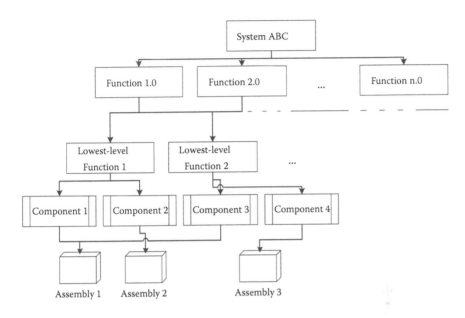

FIGURE 4.11
Grouping components with similar functionalities into logical assemblies.

1. Select a standard component that is commercially available. COTS items usually have multiple suppliers, and those suppliers are usually specialized and mature in terms of their manufacturing capabilities at an economically efficient stage of production. They are also certified and comply with quality standards, such as ISO standards, so we can take advantage of vast volume and high-quality items with the least cost involved. These suppliers often come with a high level of service for maintenance and repair of the components, which is another advantage when seeking more efficient and cost-effective system support.

2. If such a COTS item is not available, the next logical step is to try to modify an existing COTS item to meet our own expectations. Simple modifications such as rewiring, installing new cables, adding new software modules, and so on, enable a function to be accomplished with minimum effort, without investing a huge amount of money to develop a manufacturing facility within the organization.

3. If modifying an existing COTS item is not possible, then the last option is to develop and manufacture a unique component. This is the most expensive alternative and should be avoided if possible. As the last resort, we should not attempt to do everything all by ourselves; with the existence of a globalized supply chain, outsourcing and contracting should be considered for such a development, to take advantage of differences in costs of resources worldwide, including materials and labor costs.

The decision made concerning the design elements is to be documented in the Type C specification (product baseline) and the Type D specification (process baseline) if a manufacturing or developing process is involved.

4.3.2.3 A CORE Example of Detailed Functional Analysis

In CORE, the functional allocation model starts with expansion of the FFBD model to an enhanced version (the eFFBD model), by incorporating the resources and constraints information into the functions. To develop the eFFBD, we start with the FFBD results, adding the necessary resource information by defining the input/output of the functions (items) and assigning the functions to components. Their relationships can be seen from Figure 4.12 (functional relationship charts).

Figure 4.12 illustrates an example for an ATM system eFFBD with items included.

Table 4.2 gives the ERA definition of the items and components elements.

Similarly to the functional model, the functional allocation model in CORE is conducted iteratively and also follows a top-down approach. Based on the allocation of TPMs and functional decomposition, plus the assessment of feasible technology, the components are allocated level by level, starting first with the high-level components and following the path of functional decomposition; the lower levels of components are then derived. This process continues until the lowest level of assembly is achieved; that is to say, COTS items will be obtained for that assembly, which is the stopping rule for the decomposition.

For communication between system components with external elements outside of the system boundary, *interface* and *link* elements are used to define such relationships. An interface element identifies the external components

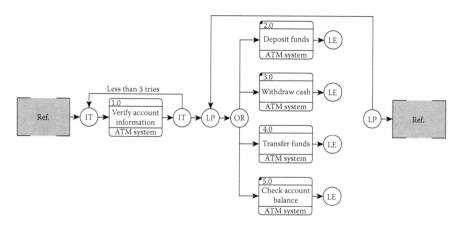

FIGURE 4.12
eFFBD of ATM functional model using CORE 9.0. (With permission from Vitech Corporation.)

TABLE 4.2

Item Element ERA Definition

Element	Attribute	Relations	Target Elements
Item	Accuracy	Decomposed by/	Item
	Description	decomposes	
	Doc. PUID	Input to/inputs	Function
	Number	Output from/outputs	Function
	Precision	Triggers/triggered by	Function
	Priority		
	Range		
	Size		
	Size unit		
	Type		
Item	Abbreviation	Built from/built in	Component (lower level)
	Description	Built in/built from	Component (higher level)
	Doc. PUID	Performs/allocated to	Function
	Purpose		
	Number		
	Type		

TABLE 4.3

Interface and Link ERA Definition

Element	Attribute	Relations	Target Elements
Interface	Description	Joins/jointed to	Components
	Doc. PUID	Comprised of/comprises	Link
	Number	Specified by/specifies	Requirement
Link	Capacity	Decompose/decomposed by	Link
	Delay	Through	
	Delay units	Connects to/connected to	Component
	Description	Specified by/specifies	Requirements
	Doc. PUID	Transfers/transferred by	Item
	Number		
	Protocol		

with which the system communicates, and the details of the interface are captured in the link element definition; that is, what kind of data is involved, what kind of hardware connection, software subroutine, and so forth. The relationships between the interface, links, and components are listed in Table 4.3.

For more information on these elements, readers can refer to the system definition guide published by the Vitech Corporation, which is available for download from www.vitechcorp.com.

4.3.3 Task Analysis Model

In system engineering design, task analysis is commonly used, primarily for specifying human–system interaction requirements and system interface specifications. It is used to analyze the rationale and purpose of what users are doing with the intended system, and to try to work out, based on the functional objective in mind, what users are trying to achieve and how they achieve the functionality by doing which tasks.

Task analysis has been a popular tool for applied behavior science and software engineering since the 1980s; it gained popularity due to its ability to include humans in the loop and its straightforwardness and simplicity. System designers use this method to investigate the human components allocation, especially their skills and the staff model based on the task requirements. In conducting task analysis, it is often found that the concepts of "tasks" and "functions" are confused; some functional models are conducted using a task analysis, and vice versa. In the previous chapters, we have briefly discussed the difference between these two. Here, for readers to better understand the task analysis model, let us spend some time again to distinguish between these two terms.

A function, as we have stated many times previously, is an action for which a system is specifically fitted, used, or designed to accomplish a specific purpose. In other words, it is a system action; it is what the system should do or perform. It is usually a more abstract goal or objective—although it involves a verb—but not an overly detailed activity. A task, on the other hand, provides such detail. According to the Merriam-Webster dictionary, a task is "a usually assigned piece of work often to be finished within a certain time." In the systems engineering context, a task is an activity, usually performed by a system component, including hardware, software, or humans, in a timely manner to accomplish a particular function. A task is performed for a purpose; that purpose is a system function. So, system functions come first, and tasks come second, to serve the functions and enable them to be accomplished. Examples of functions in the ATM system design would be the "deposit" function or the "withdraw" function, and the tasks for these functions would be "insert card," "input passcode using keypad," "select a menu option," and so on.

Now we know the difference between functions and tasks, this, in turn, implies that task analysis usually comes after the completion of the functional model, since tasks are dependent on functions. Just as Chapanis (1996) has pointed out, models in systems engineering follow a sequential order, in the sense that certain models use outputs from other models as inputs. Task analysis models are based on functional model structures; the tasks derived are associated with certain system functions, while functions are the rationale for task activities.

With this difference in mind, let us look at the definition of a task analysis model. Task analysis is a procedural model to identify the required tasks and their hierarchical structure, and produce the ordered list for all the tasks at different hierarchical levels. Based on this, task analysis is also called hierarchical task analysis (HTA).

4.3.3.1 Input Requirements

Starting from the functional model, inputs for task analysis include the function list, architecture, and the FFBD, supplemented with the understanding of the function and technological requirements, research findings from the literature, observations from the users, and expertise and experiences from the subject matter experts (SME).

4.3.3.2 Procedure

Designers take a team approach by integrating the information together for task analysis. Designers usually start from the highest level of the FFBD for each of the functions; based on the input, output, and resources/constraints information, designers use their experience and knowledge of the system and its functions, listing all the required tasks, describing them, putting them in order, and decomposing them into subtasks if desired. This procedure goes on until all the tasks are identified and no further decomposition is needed.

4.3.3.3 Output Product

The major output of an HTA is an ordered list of the tasks that users will perform to achieve the system functions. Additional information is also gathered for the tasks, including

1. Information requirements for each task
2. Time information for each task, including the time available and time required
3. User actions, activities, and operations
4. Environmental conditions required to carry out the task

Here is a sample task analysis for the ATM verification function:

0. Verify user account information
1. Insert bank card in the card reading slot
   1.1 If card read successfully, go to 2
   1.2 If card read fails, take out the card, and repeat 1
2. Enter PIN
   2.1 If accepted, go to 3

2.2 If rejected, repeat 2 until user has tried three times

2.3 Choose "cancel," go to 4

3. Choose account menu

    3.1 Go to "withdraw" function

    3.2 Go to "deposit" function

    3.3 Go to "inquiry" function

    3.4 Go to "transfer fund" function

4. Take card

From the above example, one can easily obtain the communication and data requirements for each task, and there are models available for the prediction of the time-required information. For example, in software engineering and human computer interaction, techniques such as GOMS (goals, operations, methods, and selection rules) and KLM (keystroke level model) are widely used for prediction of time information for a procedural task. These models provide an intuitive way to estimate the time and workload requirements for certain computer-related tasks, but they are subject to severe limitations; one of them is that they are deterministic models that do not account for errors and individual differences in experience and skills. Any unpredictable factors could skew the results, so that in the real-world context, justification and allowance are necessary for the application of these models.

Recently, another variation of task analysis has increasingly been used in systems design, which is called cognitive task analysis (CTA). Cognitive task analysis is an extension of the traditional HTA, with more focus on human cognition. Traditional HTA concentrates primarily on the physical activities of humans; these activities have to be observable for them to be recorded. However, in complex system interaction, many of the activities, especially mental activities, are not easily observed. For tasks that involve many cognitive activities, a slightly different approach than the traditional one is necessary to capture the cognitive aspects. There are five common steps involved in a typical CTA analysis:

1. Collect preliminary knowledge: Designers identify the preliminary knowledge that is related to the system, to become familiar with the knowledge domain. Usually, the research findings from the literature and consultation with SMEs are utilized for the elicitation of such knowledge.

2. Identify knowledge representation: Using the preliminary knowledge collected, designers examine it and allocate the task hierarchical structure, based on the traditional HTA structure. Designers may use a variety of techniques such as concept maps, flow charts, and semantic nets to represent the knowledge for each of the tasks.

3. Apply focused knowledge elicitation methods: Based on the different representations of knowledge, designers may now choose the appropriate techniques for knowledge elicitation. Multiple techniques are expected for a better articulation of the knowledge. Techniques include interviews, focus groups, and naturalistic decision-making modeling; these are the most commonly used methods, and most of the techniques are informal.

4. Analyze and verify the data: On completion of the knowledge elicitation, the results are verified using a variety of evaluation techniques, including heuristic evaluations and walkthroughs by experts, or formal testing involving real users to demonstrate the validity of the knowledge requirements for each of the tasks.

5. Format results for the system design documentation: The results of the CTA are translated into design specifications so that they will be implemented in the design. Cognitive requirements, skill levels, and users' mental models of task-performing strategies are derived to be translated to design specifications, especially the human–system interface aspects, so that systems functions are accomplished in an efficient way.

HTA and CTA are utilized widely, due to the simplicity and intuitiveness of the methods involved. The procedures involved in tasks analysis are very straightforward; with minimum training effort, a person can perform this analysis with little difficulty. However, good benefits also come with great challenges. The nature of the task analysis model requires, first, that the tasks to be investigated are observable, or can at least be partially observed. CTA uses knowledge elicitation of such mental tasks; this knowledge is also based on the research results concerning the cognitive resources for observed mental activities. Researchers and designers use their expertise and understanding of the system functions to develop the task and decompose it into subtasks to support the functions. This, in turn, implies another challenge for task analysis, which is the high level of subjectivity. There are no well-developed templates or standards for task analysis; the quality of the analysis is solely dependent on the capabilities of the person who is conducting it. Moreover, with a different person, it is entirely possible that the task analysis results would be different, and both sets of results may work for the design. To overcome the personal bias and distortion caused by subjectivity, task analysis usually takes a team approach by having a group of people involved to achieve a consistent outcome from the team. More research findings, data quality, and iterative checking/balance also help. These will facilitate the task analysis results to be more compatible with users' true behaviors. For these reasons, task analysis models are most commonly used for design purposes, and are not suitable for evaluation, as there is no standard for right or wrong results from the task analysis.

4.3.4 Timeline Analysis Model

The timeline analysis model follows naturally from task analysis; it provides the supplemental time-required information for the task analysis, so that time-related workloads for various tasks can be identified.

In timeline analysis, graphical charts are often used to provide a visualization tool to lay out the sequence for all the tasks, and based on the research findings of each model, the time information is plotted on the charts to illustrate the temporal relationships of the tasks performed. Timeline analysis is very similar to the Gantt chart model, which is used to present project activity schedules. Figure 4.13 illustrates a simple time line analysis for a withdraw function for an ATM system.

4.3.5 Link Analysis Based on Network and Graph Model

Link analysis is a model derived from network and graph theory; it is concerned with the physical arrangement of the items so that the efficiency of operations between the items is optimized. These items include workstations, instrument panels, machines, offices, or any work areas involved. Network and graph theory is an important area of operations research, and since the 1970s, the application of network and graph theory has bloomed due to the increasing demand for large, complex facilities and layout planning. A graph is represented by a set of vertices and edges (sometimes called nodes and arcs/lines). A network is a graph with numbers associated with each edge or arc; these numbers could be the distance, cost, reliability, importance, frequencies, or any related parameters. Network models can be used to solve a variety of problems, including the decision tree, shortest path, minimum/maximum flow, matching and assignment, traveling algorithm, location

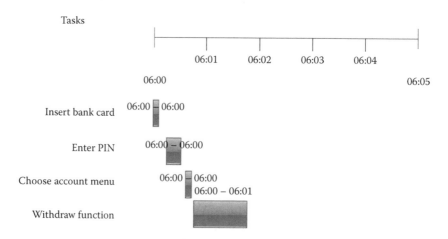

FIGURE 4.13
Timeline chart of a simple ATM system.

selection, and project scheduling, to name a few. In industrial engineering, optimization using networks and graphs is commonly found to be applied to facility layout design and planning. In this chapter, we will only select several fundamental models to get readers started on the familiarization of network optimization models. For more in-depth discussion of graph and network theory, one should refer to a more advanced text on this subject such as Evans and Minieka (1992) or Tompkins et al. (2003).

The application of graph and network theory in systems engineering is also referred to as the link analysis model, as it studies the linkage between different elements within the system. Link analysis enables designers to visualize the spatial relationships of systems elements and quantify the parameters involved for these relationships; thus, the overall effectiveness of the links can be optimized.

4.3.5.1 Input

The primary inputs required for a link analysis are data from the functional analysis, functional allocation, and task analysis, plus knowledge of existing and emerging technology, similar systems, and any data collected during the requirement analysis stages, including the observation, interviews, and surveys from users.

4.3.5.2 Procedure

Regardless of what specific network model is being used, a general approach for link analysis involves the following steps:

1. Identify the objective of the link analysis. What is the purpose of conducting such an analysis? What is the measure of effectiveness and efficiency for this particular problem; that is, are we trying to find the best layout to minimize the total traveling distance among elements? Or, are we trying to put the system elements together in such a way that the total importance index between elements is maximized?

2. List all the elements for the link analysis, including the personnel and items. Based on the relationships between all the elements and the objectives, quantify the links between the elements, including the distances, frequencies, cost, and so forth.

3. Measure the priority/importance score for each of the links if necessary.

4. Develop the from-to matrix for the vertices (for the elements) and the arcs, based on the values assigned.

5. Apply the appropriate algorithms to place the elements in the layout until all the elements have been assigned.

6. Evaluate the final layout by calculating the score for the measures of effectiveness. Implement the layout based on the results.

To illustrate the above procedure, we will give one example based on graph theory to explain how the algorithm works.

Example 4.1

Five stations are being put in a limited space; the frequencies between the stations and their corresponding importance (or weightings) are given in Figure 4.14.

The weighted-frequency matrix is derived by multiplying the frequency of the pair with its corresponding importance score (or weighting). For example, the weighted importance between Stations 1 and 2 is 10, which equals the frequency score between 1 and 2 (2) multiplied by the corresponding importance score (or weighting) for the link between 1 and 2 (5). Find a layout for the five stations that maximizes the total sum of the weighted frequencies for the chosen links.

The procedure for graph-based link analysis is described as in the following steps:

Step 1.   From the weighted-frequency relationship matrix, rank the all the arcs/links in descending order, as shown in Table 4.4. From the list, select the station pair with the largest weighted-frequency score. Ties are broken arbitrarily. For this example, Stations 4 and 5 are chosen to be placed in the layout first as they have the largest score (16) (Figure 4.15).

Step 2.   Now we need to find the third station to enter the layout graph. The remaining stations are compared in terms of the sum of the scores with respect to the selected pair from Step 1. Select the station with the highest score, as

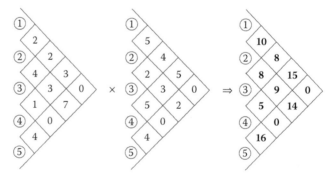

Frequency matrix   Importance matrix   Weighted-frequency matrix

FIGURE 4.14
Relationship matrix chart for the example.

TABLE 4.4

Ranked Weighted-Frequency
Score for All Links

Link	Score
4–5	16[a]
1–4	15
2–5	14
1–2	10
2–4	9
1–3	8
2–3	8
3–4	5
1–5	0
3–5	0

[a] Indicates the chosen station to enter the layout.

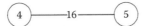

FIGURE 4.15
Step 1 of the procedure.

TABLE 4.5

Step 2 Calculation

Candidate Station	4	5	Sum
1	15	0	15
2	9	14	23[a]
3	5	0	0

[a] Indicates the chosen station to enter the layout.

seen in Table 4.5. For this example, Station 2 is selected as it has the highest sum of scores (23) with respect to Stations 4 and 5 (Figure 4.16).

Step 3. Now we need to find the fourth station to enter the layout graph. The approach is similar to Step 2, except now we have three stations already in the graph. This step involves assessing the sum of the scores of the remaining stations by comparing the sum of the scores when placing the station in one of the faces of the graph. A face of the graph is defined as a bounded region, enclosed by three or more station nodes. For this step, there is only face, 2–4–5. Table 4.6 lists the sum of scores for the two unassigned stations. Based on the comparison, Station 1 is chosen to be placed in the face of the graph (Figure 4.17).

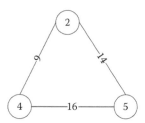

FIGURE 4.16
Step 2 calculation.

TABLE 4.6

Step 3 Calculation

Assigned Stations	2	4	5	Sum
1	10	15	0	25[a]
3	8	5	0	13

[a] Indicates the chosen station to enter the layout.

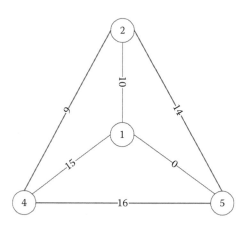

FIGURE 4.17
Step 3 results.

Step 4. Now we have only one unassigned station (3) left. The task now is to determine in which face to place Station 3. There are four faces available in total from the results of step 3: 1–2–4, 1–2–5, 1–4–5, and 2–4–5 (do not forget this face). The sum of the scores for putting Station 3 in the four different faces are compared, and we choose the placing which has the highest total. See Table 4.7 for the comparison (Figure 4.18).

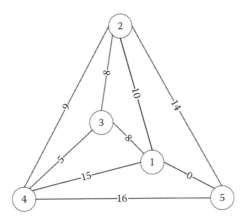

FIGURE 4.18
Step 4 results.

TABLE 4.7

Step 4 Results

Faces	Sum of Scores When Station 3 Is Placed in the Face
1–2–4	$8+8+5=21$[a]
1–2–5	$8+8+0=16$
1–4–5	$8+5+0=13$
2–4–5	$8+5+0=13$

Note: Total weight frequency score is $23+25+21+16=85$
[a] Indicates the chosen station to enter the layout.

The steps above give an algorithm to find the layout with the maximum score. If the object is to find the minimum score for any possible reason, the above steps would still be the same, except to change the rule to select the lowest score instead of the highest score. The procedure is straightforward and very similar to finding the highest score; as an exercise, readers can try to find the layout for a minimum-weighted score for the above example. We have also included some exercise questions at the end of the chapter for further practice.

In some cases, obtaining the final link structure is not sufficient; designers need to translate the link diagram to fit a limited area. Imagining that the stations are actually departments, if distances are not taken into consideration, how would we translate the final results into a rectangular floor plan? There are many procedures that have been developed to aid in this planning, such as Muller's systematic layout planning (SLP) procedure

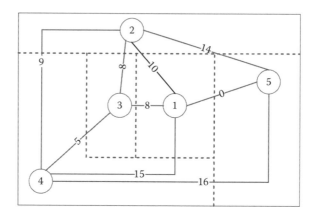

FIGURE 4.19
Results of layout assignment.

(Tompkins et al. 2003). These procedures are quite straightforward and intuitive; they are not difficult to understand. Readers can refer to other texts for a more detailed explanation of those procedures. A same layout planning for the above example is illustrated in Figure 4.19.

4.3.6 Center of Gravity Model for Facility Location Planning

Project managers and system designers often face the decision of selecting a location for the system facility. There are several options when selecting a location; one can expand the existing facility if there is adequate space for it—this option is more desirable because it costs less than other alternatives. A second option is to add a new location. For example, it is economical to build a facility near a source of raw materials to cut down the cost, or to outsource the manufacturing process. This option has more impact than simply adding extra capacity, since it involves more investment and has more influence through interaction with the new environment and society. The decision shall be made based on the system mission and requirements, to bring in more benefits and save costs, with other constraining factors in mind, such as the political, legal, and financial conditions of the selected location.

There are many models for location planning; selection of the model depends on the different objectives and constraints in which the planning is involved. Here we introduce a model called the center of gravity method. It is widely used in operations management to find the location of a facility that minimizes total travel time or costs to various destinations.

The center of gravity method uses a map to show all the destination locations. Using an (x, y) coordinate system, each of the destination locations can be identified on the map, such as the one in Figure 4.20.

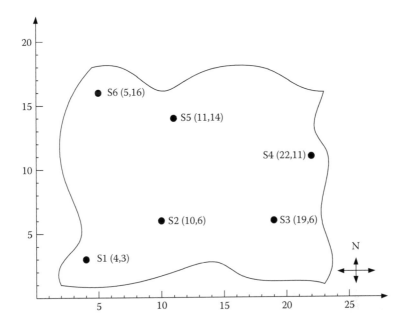

FIGURE 4.20
Coordination system for all the destination sites.

If the quantities to be shipped to every location are equal, then coordinates for the center can be easily determined by the following formula:

$$\bar{x} = \frac{\sum x_i}{n} \tag{4.1}$$

$$\bar{y} = \frac{\sum y_i}{n} \tag{4.2}$$

where:
(x_i, y_i) is the coordinate pair for the ith destination location
n is the total number of locations

So, for the above example represented by Figure 4.20, the center of gravity is located at

$$\bar{x} = \frac{\sum x_i}{n} = \frac{x_1 + x_2 + x_3 + x_4 + x_5 + x_6}{6} = \frac{4 + 10 + 19 + 21 + 11 + 7}{6} = 12$$

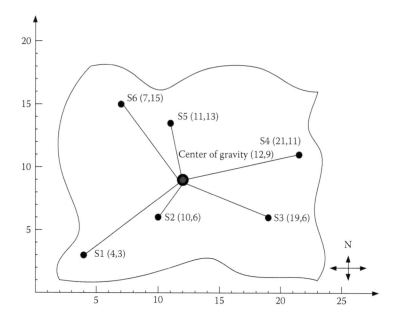

FIGURE 4.21
Center of gravity for shipment of equal quantities.

$$\bar{y} = \frac{\sum y_i}{n} = \frac{y_1 + y_2 + y_3 + y_4 + y_5 + y_6}{6} = \frac{3 + 6 + 6 + 11 + 13 + 15}{6} = 9$$

as shown in Figure 4.21.

If the frequency (or quantity) of the shipment to each location Q_i is not equal, then a modified weighted average is used to determine the center of gravity.

$$\bar{x} = \frac{\sum Q_i x_i}{\sum Q_i} \qquad (4.3)$$

$$\bar{y} = \frac{\sum Q_i y_i}{\sum Q_i} \qquad (4.4)$$

For the above example, suppose the daily shipment frequency and quantities per shipment are as shown in Table 4.8. The weighted-average-based center of gravity is

TABLE 4.8

	Frequency	Quantities per Shipment	Total Q
S1	1	600	600
S2	2	400	800
S3	2	450	900
S4	5	250	1250
S5	3	400	1200
S6	4	100	400

TABLE 4.9

Probability P_{1j}	Shipment Amount Q_{1j}
0.2	250
0.3	400
0.1	600
0.4	800

$$\bar{x} = \frac{\sum Q_i x_i}{\sum Q_i} = \frac{4(600)+10(800)+19(900)+21(1250)+11(1200)+7(400)}{600+800+900+1250+1200+400}$$

$$\cong 13.54$$

$$\bar{y} = \frac{\sum Q_i y_i}{\sum Q_i} = \frac{3(600)+6(800)+6(900)+11(1250)+13(1200)+15(400)}{600+800+900+1250+1200+400} \cong 9.19$$

Please note that the above example assumes a deterministic shipment amount. In the stochastic case, the shipment quantity is a random variable, and follows some distribution function. In this case, the expected value is used for the quantity. For example, for S1, the shipment has the discrete distribution shown in Table 4.9. The expected value for the shipment quantity for S1 is given by the following formula:

$$E(Q_1) = \sum_{j=1}^{4} P_{1j} Q_{1j} = (0.2)(250)+(0.3)(400)+(0.1)(600)+(0.4)(800) = 550$$

The formulas for Equations 4.3 and 4.4 become

$$\bar{x} = \frac{\sum E(Q_i) x_i}{\sum E(Q_i)} \qquad (4.5)$$

$$\bar{y} = \frac{\sum E(Q_i) y_i}{\sum E(Q_i)} \qquad (4.6)$$

A comprehensive review for the probability and statistical models is given at the end of this book in Appendix I.

4.4 Summary

Models are the abstract representation of the real system; they play an important role in systems, as they make complex systems simpler, so that the relationships between critical variables may be investigated. In this chapter, we defined the models in the systems engineering context and identified the major characteristics and benefits of using models in systems design. Models used in systems design can be classified into different categories, based on different perspectives. Based on the format, models may be categorized as physical models, analogue models, schematic models, or mathematical models; based on the variable types, they may be categorized as deterministic or stochastic models; based on the model's scope, there are macromodels and micromodels; based on the model's functions, there are forecast models, decision models, queuing models, and so forth. For each of the categories, descriptions of their significance and examples were given for a better understanding.

In terms of systems engineering models, it is noted that most systems design models are applied research from nature and focus on the prediction of system behavior; they are borrowed from other disciplines and modified for systems design purposes. These systems engineering design models are used iteratively in the system life cycle, with various levels of detail for different iterations.

Systems models and analysis are the main theme for this whole book; in almost every chapter, different types of models are reviewed. In this chapter, some commonly used design models were discussed, while trying not to overlap too much with the remainder of the book. Two major design models, the functional analysis and function allocation models, were reviewed in great detail. System functions were defined, and their characteristics, syntax, and the FFBD graphical method was explained. As a hands-on exercise, we used CORE as a platform to illustrate how to use software to conduct FFBD analysis. An ATM system was used as the example of function analysis. Function allocation and the relationship between different design baselines (Types A, B, C, D, E) were discussed at the end of the FFBD analysis section. Following functional modeling, task analysis and cognitive task analysis (CTA) were introduced to obtain the interaction design specifications between the users and the system. The input and output for these models and general procedures are given for conducting a task analysis.

The last section of this chapter was dedicated to link analysis. The two most commonly used link analysis models were reviewed: graph-based layout planning and center of gravity models to determine facility location. The procedures were elaborated with numeric examples and their application to systems were discussed.

PROBLEMS

1. Define a model. What are characteristics for a model?

2. What are the benefits of models? What are the possible limitations of a model?

3. List all the models categories based on format, variables, scope, and functions. Give one example of each.

4. If you were a systems engineer, explain why you would want to use models in your design and what the unique characteristics of systems engineering models are compared to other disciplines, say psychological models?

5. Define functions and describe the characteristics of systems functions.

6. Develop the functional models (functions and their architectures) for the following elevator system scenario:

   A passenger elevator is located in a four-story building in a College of Arts and Sciences Building. It has two doors on each floor, opening on two sides for passengers to access the elevator. The elevator operates on a 24/7 time basis, unless it fails, or there is an emergency, or it is time for a monthly maintenance check, performed by a certified technician.

7. Download CORE from vitechcorp.com, and use CORE to develop an FFBD for Problem 6.

8. From Problem 7, pick one function, conduct the resources management analysis for this function, and identify the input/output and constraints for this function. Define "item" for this information and develop an eFFBD using CORE.

9. Based on your experiences and observations with elevators, develop an HTA to transport a passenger upstairs.

10. What are the differences between task analysis and cognitive task analysis (CTA)?

11. Describe how traceability is ensured between system design baselines (i.e., Type A to B to C, D, and E)?

12. A manufacturing company produces one part for a major automobile customer in five locations throughout a country. Raw materials will be distributed to these five locations from a centralized warehouse within the country. The weekly quantities to be delivered to these five locations are assumed to be same. The following table shows the coordinates for the five locations. Use the center of gravity method to determine the coordinates for the location of the warehouse. Draw the five locations and the warehouse on graph paper.

Location	x	y
L1	2	3
L2	8	2
L3	7	7
L4	9	5
L5	4	13

13. A grocery supplier is trying to determine the best location for its new distribution center within the area. The coordinates and daily shipping amount (in lbs) from the center to the six stores is listed in the following table. Determine the coordinates for this new center using the weighted-average center of gravity method

Location	x	y	Amount (lbs)
L1	2	2	100
L2	3	5	150
L3	5	4	125
L4	9	6	200
L5	7	8	200
L6	4	10	350

14. A grocery supplier is trying to determine the location for its new distribution center within the area. The amounts shipped to the nearby six stores are random variables. The coordinates and daily shipping amount (in lbs) and probability distribution from the center to the six stores are listed in the following table. Determine the coordinates for this new center using the weighted-average center of gravity method.

Location	x	y	Amount (lbs) and Probabilities
L1	2	2	P(100)=0.5, P(150)=0.5
L2	3	5	P(50)=0.1, P(100)=0.5, P(150)=0.4
L3	5	4	P(100)=0.3, P(125)=0.7
L4	9	6	P(200)=1
L5	7	8	P(80)=0.2, P(200)=0.2,P(300)=0.6
L6	4	10	P(100)=0.3, P(200)=0.3, P(350)=0.4

15. Shown below is a relationship matrix for five departments, conduct a graph-based link analysis to maximize the total adjacent importance score.

Rate	Score
V (very important)	8
I (important)	6
M (moderate)	3
U (unimportant)	1

16. Consider the layout for a five-department design (sizes are ignored). The material flow is given in the following table.

	A	B	C	D	E
A	—	25	0	6	20
B		—	5	15	30
C			—	20	0
D				—	10
E					—

a. Develop the final adjacent graph using the graph-based link analysis procedure to maximize the total adjacent flow. What is the final total flow amount?

b. Fit the layout of the five departments from (a) into a rectangular block.

5

System Technical Performance Measures

In previous chapters, we have described the systems design processes and some basic models that are utilized within the design process, including the functional analysis and functional allocation models. Models play a significant role in systems design, and rely on variables and parameters to produce valid results. Taking the functional analysis model as an example, for each of the functions in the functional flow block diagram (FFBD), there are performance or constraints parameters that regulate the function to serve the overall system mission. These design-dependent parameters (DDPs) are usually expressed in quantitative format, which are also called technical performance measures (TPMs).

Systems engineering is requirement driven, but, as mentioned before, requirements will not design the system, it is the technical specifications derived from the requirements that will lead to the realization of the system. Based on system requirements, TPMs provide detailed quantitative specifications for system configurations, regulate system technical behavior, and are necessary for designers to obtain the system components, construct the system, and, moreover, to test and evaluate system performance. Developing a precise, accurate, and feasible set of TPMs comprehensively is essential for the mission success of the system design.

In this chapter, we will review some of the most popular system design parameters and the TPMs relevant to them, and describe the fundamental models to analyze and integrate these TPMs into the systems design life cycle. More specifically, we will

1. Provide a comprehensive overview of system DDPs and TPMs and describe their significance for system design and evaluation.
2. Review some commonly used TPMs for systems engineering, including reliability, maintainability, producibility, supportability, usability, and system sustainability. The definitions as well as the characteristics of these parameters are given, and the mathematical modeling for these parameters is reviewed in detail.
3. Life cycle consideration and design integration, and implications for the DDPs, are illustrated so that readers will have guidelines for how these DDPs and TPMs are developed and applied within the various stages of the system life cycle.

On completion of this chapter, readers will have some basic understanding of the general pictures of the technical side of system parameters, know the scope and challenge to develop these parameters, and be familiar with common used analytical models and concepts, so that they may use the right model in the future practice of system design.

5.1 Technical Performance Measures (TPMs)

System technical performance measures, or TPMs, are the quantitative values for the DDPs that describe, estimate or predict the system technical behaviors. TPMs define the attributes for the system to make the system unique so that it can be realized. Examples of TPMs include systems functional parameters (such as size, weight, velocity, power, etc.), system reliability (i.e., mean time between failures [MTBF]), system maintainability (i.e., mean time between maintenance [MTBM]), usability (i.e., human error) and system sustainability.

Table 5.1 illustrates an example for a typical TPM metrics for an automobile. TPMs are derived from requirements analysis; recall that in Section 3.4 on requirement analysis, we discussed the method of using quality function deployment (QFD) to derive and prioritize the TPMs. The development of the TPMs from requirements ensures that the system attributes and behaviors comply with the ultimate users' needs. TPMs provide estimated quantitative values that describe the system performance requirements. They measure

TABLE 5.1

Sample Functional TPMs for an Automobile Design

Design Parameters	TPMs
Acceleration: 0–60	15 s
Acceleration: 50–70	12 s
Towing capacity	≥680 kg @ 3.5%, 25 min@ (45 mph)
Cargo capacity	90 m^3
Passenger capacity	≥5
Braking: 60–0	<50 m
Mass	≤2300 kg
Starting time	≤15 s
Ground clearance	≥180 mm
Fuel economy	7.5 L/100 km (32 miles/gal)
Petroleum use	0.65 kWh/km
Emissions	Tier II Bin 5
Range	≥200 mi – 320 km
WTW GHG emissions	219 g/km

the attributes or characteristics inherent within the design, specifically the DDPs. The identification of TPMs evolves from the development of system operational requirements and the maintenance and support concept. During the system design processes, one of the largest contributors to "risks" is the lack of an adequate system specification in precise quantitative forms. Well-defined TPMs will ensure that (a) the requirements reflect the customers' needs, and (b) the measurements (metrics) provide designers with the necessary guidance to develop their benchmark.

Another advantage of using TPMs to balance cost, scheduling, and performance specifications throughout the life cycle is to specify measurements of success. Technical performance measurements can be used to compare actual versus planned technical development and design. They also report the degree to which system requirements are met in terms of performance, cost, schedule, and progress in implementation. Performance metrics are traceable to original requirements.

Nevertheless, the types of parameters and TPMs involved in different systems vary a great deal; development of TPMs primarily relies on the clear understanding the nature of the system and the knowledge and experiences of the developers. It is impossible to review every single type of TPM within one chapter, as there is a tremendous amount of information involved in various types of parameter. Specialized models and methods are required to develop specific parameters; for example, physics for the system's power, acceleration, and velocity. Here, some of the commonly shared parameters are reviewed. These parameters are contained within almost all types of system; they include reliability, maintainability, producibility, supportability, usability, and sustainability. We hope that, by reviewing these basic TPM concepts, readers will gain a comprehensive understanding of the most common parameters that will be involved in the design of most systems, and know how to apply the appropriate methods and models to derive those TPMs accurately. Thus, to that extent, this chapter can be thought of as an extension of Chapter 4, to introduce more systems-design-related models, which are more specific to system DDPs.

5.2 Systems Reliability

5.2.1 Reliability Definition

Generally, system reliability can be defined as follows:

> Reliability is the probability that a system or a product will operate properly for a specific period of time in a satisfactory manner under the specified operating conditions. (Blanchard and Fabrycky, 2006)

From the definition, it is easy to see that reliability is a measure of the system's success in providing its functions properly without failure. System reliability has the following four major characteristics:

1. It is a probability. A system becomes unreliable due to failures that occur randomly, which, in turn, also make system reliability a random variable. The probability of system reliability provides a quantitative measure for such a random phenomenon. For example, a reliability of 0.90 for a system to operate for 80 h implies that the system is expected to function properly for at least 80 h, 90 out of 100 times. A probability measures the odds or the fraction/percentage of the number of times that the system will be functional, *not* the percentage of the time that the system is working. An intuitive definition of the reliability is as follows: Suppose there are n totally identical components that are simultaneously subjected to a design operating conditions test; during the interval of time $[0, t]$, $n_f(t)$ components are found to have failed and the remaining $n_s(t)$ survived. At time t, the reliability can be estimated as $R(t) = n_s(t)/n$.

2. Satisfactory performance is specified for system reliability. It defines the criteria at which the system is considered to be functioning properly. These criteria are derived from systems requirement analysis and functional analysis, and must be established to measure reliability. Satisfactory performance may be a particular value to be achieved, or sometimes a fuzzy range, depending on the different types of systems or components involved.

3. System reliability is a function of time. As seen in the definition of system reliability, reliability is defined for a certain system operation time period. If the time period changes, one would expect the value for reliability to change also. It is common sense that one would expect the chance of system failure over an hour to be a lot lower than that over a year! Thus, a system is more reliable for a shorter period of time. Time is one of the most important factors in system reliability; many reliability-related factors are expressed as a time function, such as MTBF.

4. Reliability needs to be considered under the specified operating conditions. These conditions include environmental factors such as the temperature, humidity, vibration, or surrounding locations. These environmental factors specify the normal conditions at which the systems are functional. As mentioned in Chapter 2, almost every system may be considered as an open system as it interacts with the environment regardless, and the environment will have a significant impact on system performance. For example, if you submerge a laptop computer under water, it will probably fail immediately.

System reliability has to be considered in the context of the designed environment; it is an inherent system characteristic. System failures should be distinguished from accidents and damage caused by violation of the specified conditions. This is why product warranties do not cover accidents caused by improper use of systems.

The four elements above are essential when defining system reliability. System reliability is an inherent system characteristic; it starts as a design-independent parameter from the user requirements, along with the design process; eventually, reliability will be translated to systems DDPs and TPMs will be derived in specific and quantitative format, so that reliability of the components can be verified. This translation process requires vigorous mathematical models to measure reliability.

5.2.2 Mathematical Formulation of Reliability

5.2.2.1 Reliability Function

As mentioned above, reliability is a function of time, t, which is a random variable denoting the time to failure. The reliability function at time t can be expressed as a cumulative probability, the probability that the system survives at least time t without any failure:

$$R(t) = P(\mathbf{t} > t) \qquad (5.1)$$

As an assumption for system status, the system is either in a functional condition or a state of failure, so the cumulative probability distribution function of failure $F(t)$ is the complement of $R(t)$, or

$$R(t) + F(t) = 1 \qquad (5.2)$$

So, knowing the distribution of failure, we can derive the reliability by

$$R(t) = 1 - F(t) \qquad (5.3)$$

If the probability distribution function (p.d.f., sometimes called probability mass function, p.m.f.) of failure is given by $f(t)$, then the reliability can be expressed as

$$R(t) = 1 - F(t) = 1 - \int_0^t f(x)\,dx = \int_0^\infty f(x)\,dx - \int_0^t f(x)\,dx = \int_t^\infty f(x)\,dx \qquad (5.4)$$

For example, if the time to failure follows an exponential distribution with parameter λ, then the p.d.f. for failure is

$$f(t) = \lambda e^{-\lambda t} \tag{5.5}$$

The reliability function for time t is

$$R(t) = \int_t^{\infty} \lambda e^{-\lambda x} \mathrm{d}x = e^{-\lambda t} \tag{5.6}$$

One can easily verify Equation 5.6 by using basic integration rules.

5.2.2.2 Failure Rate and Hazard Function

Failure rate is defined in a time interval $[t_1, t_2]$ as the probability that a failure per unit time occurs in the interval, given that no failure has occurred prior to t_1, the beginning of the interval. Thus, the failure rate $\lambda(t)$ can be formally expressed as (Elsayed 1996)

$$\lambda(t_2) = \frac{\int_{t_1}^{t_2} f(t)\mathrm{d}t}{(t_2 - t_1)R(t_1)} \tag{5.7}$$

From Equation 5.4, it may easily be seen that

$$\int_{t_1}^{t_2} f(t)\mathrm{d}t = \int_{t_1}^{\infty} f(t)\mathrm{d}t - \int_{t_2}^{\infty} f(t)\mathrm{d}t = R(t_1) - R(t_2) \tag{5.8}$$

So, Equation 5.7 becomes

$$\lambda(t_2) = \frac{R(t_1) - R(t_2)}{(t_2 - t_1)R(t_1)} \tag{5.9}$$

To generalize Equation 5.9, let $t_1 = t$, $t_2 = t + \Delta t$, then Equation 5.9 becomes

$$\lambda(t + \Delta t) = \frac{R(t) - R(t + \Delta t)}{\nabla t R(t)} \tag{5.10}$$

The instantaneous failure rate can be obtained by taking the limits from Equation 5.10, as

$$\lambda(t) = \lim_{\Delta t \to 0} \frac{R(t) - R(t + \Delta t)}{\nabla t\, R(t)}$$

$$= \frac{1}{R(t)} \lim_{\Delta t \to 0} \frac{R(t) - R(t + \Delta t)}{\nabla t} = \frac{1}{R(t)} \left[-\frac{d}{dt} R(t) \right] \qquad (5.11)$$

and from Equation 5.4, we have

$$\frac{d}{dt} R(t) = -f(t) \qquad (5.12)$$

$f(t)$ is the failure distribution function. So, the instantaneous failure rate is

$$\lambda(t) = \frac{f(t)}{R(t)} \qquad (5.13)$$

For the exponential failure example, $f(t) = \lambda e^{-\lambda t}$ and $R(t) = e^{-\lambda t}$, the instantaneous failure rate according to Equation 5.13 is

$$\lambda(t) = \frac{f(t)}{R(t)} = \frac{\lambda e^{-\lambda t}}{e^{-\lambda t}} = \lambda \qquad (5.14)$$

So, for the exponential failure function, the instantaneous failure rate function is constant over the time. For other types of failure distribution, this might not hold true. For illustration purposes, in this chapter we focus on exponential failure rate function, as exponential failure is commonly found in many applications. For other types of failure functions, one should not assume they have the same characteristics as exponential failure. The specific failure rate function should be derived using Equations 5.10 and 5.11 and the rate is not expected to be constant. Please refer to Appendix I for a comprehensive review for the various distribution functions.

The failure rate λ, generally speaking, is the measure of the number of failures per unit of operation time. The reciprocal of λ is MTBF, denoted as θ:

$$\theta = \frac{1}{\lambda} \qquad (5.15)$$

Here, we use several examples to illustrate the estimation of the failure rate. As the first example, suppose that one manufacturer of an electric component is interested in estimating the mean life of the component. One hundred components are used for a reliability test. It takes 1000 h for all the 100 components to fail under the specified operating conditions. The components are observed and failures in a 200 h interval are recorded, shown in Table 5.2.

The failure rate for each of the 200 h intervals according to Equation 5.13 is given in Table 5.3.

For a constant failure rate, the failure rate can also be estimated by using the following formula:

$$\lambda = \frac{\text{Number of failures}}{\text{Total operating hours}} \tag{5.16}$$

TABLE 5.2

Numbers of Failures Observed in a 200 h Interval

Time Interval (hours)	Failures Observed
0–200	45
201–400	32
401–600	15
601–800	5
801–1000	3
Total failures	100

TABLE 5.3

Failure Rate for Example 1

Time Interval (hours)	Failures Observed	Initial Rate (t)	Failure Rate (per hour)
0–200	45	100	$\dfrac{45}{100 \times 200} = 0.00225$
201–400	32	55	$\dfrac{32}{55 \times 200} = 0.00290$
401–600	15	23	$\dfrac{15}{23 \times 200} = 0.00326$
601–800	5	8	$\dfrac{5}{8 \times 200} = 0.00313$
801–1000	3	3	$\dfrac{3}{3 \times 200} = 0.00500$
Total failures	100		

TABLE 5.4

Observed Failure Time for Example 2

Component Number	Failure Occurrence Time (hours)
1	15
2	19
3	32
4	45
5	61
6	62
7	89
8, 9, 10	All survived 100 h

Let us look at another example which is slightly different from the first example. Suppose that one manufacturer of an electric component is interested in estimating the mean life of the component. Ten components are used for a reliability test of 100 h under the specified operating conditions. During the 100 h, seven failures are observed. Table 5.4 lists the occurrence times for all these failures.

So, based on Equation 5.16, the total number of failures for the 100 h is seven, and total operating hours are the sum of all the ten components' working hours, which is

$$15 + 19 + 32 + 45 + 61 + 62 + 89 + 100 + 100 + 100 = 623$$

so the estimation of the failure rate is

$$\lambda = \frac{7}{623} = 0.01124$$

If there is only one component involved and maintenance actions are performed when this component fails so that it is functional again, the failure rate is estimated by the division of the total number of failures over the total time of the component being functional (total time minus downtime). For example, Figure 5.1 shows a component test for a period of 50 h.

So, the total operating hours $= 50 - 2.5 - 6.5 - 1.6 - 2.9 = 36.5$ h. The failure for this particular component is

$$\lambda = \frac{4}{36.5} = 0.1096 \, / h$$

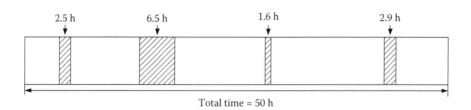

2.5 h   6.5 h   1.6 h   2.9 h

Total time = 50 h

FIGURE 5.1
Failure occurrence for Example 3.

Depending on the different situations in which the test is performed, the appropriate formula should be used to obtain the correct failure rate.

Failure rate, especially instantaneous failure, can be considered as a conditional probability. The failure rate is one of the most important measures for the systems designers, operators, and maintainers, as they can derive the MTBF, or the mean life of components, by taking the reciprocal of the failure rate, expressed in Equation 5.15. MTBF is one common measure for systems reliability due to its simplicity of measurement and its direct relationship to the systems reliability measure.

It is also easily seen that failure rate is a function of time; it varies with different time intervals and different times in the system life cycle. If we plot the system failure rate over time from a system life cycle perspective, it exhibits a so-called "bathtub" curve shape, as illustrated in Figure 5.2.

At the beginning of the system life cycle, the system is being designed, concepts are being explored, and system components are being selected and evaluated. At this stage, because the system is immature, there are many "bugs" that need to be fixed; there are many incompatibilities among components, and many errors are being fixed. The system gradually becomes more reliable along with design effort; thus, the failure rate

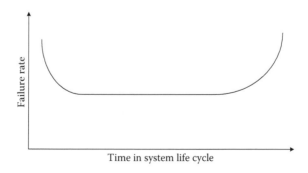

Time in system life cycle

FIGURE 5.2
Illustration of bathtub shape of the failure rate time curve. (Redrawn from Blanchard, B.S. and Fabrycky, W.J., *Systems Engineering and Analysis*, 4th edn, Prentice Hall, Upper Saddle River, NJ, 2006.)

of the system components decreases. This is the typical behavior of the system failure rate in the early life cycle period, as shown in Figure 5.2 in the first segment of the failure rate curve: the decreasing failure rate period, or the "infant mortality region."

Once the system is designed and put into operation, the system achieves its steady-state period in terms of failure rate, and presents a relatively constant failure rate behavior. The system is in its maturity period, as presented in the middle region in Figure 5.2. In this stage, system failure is more of a random phenomenon with steady failure rate, which is expected under normal operating conditions.

When the system approaches the end of its life cycle, it is in its wear-out phase, characterized by its incompatibility with new technology and user needs and its worn-out condition caused by its age, it presents a significantly increasing pattern of failure occurrence, as seen in the last region of the life cycle in Figure 5.2. Failures are no longer solely due to randomness but to deterministic factors mentioned above; it is time to retire the system and start designing a new one and a new bathtub curve will evolve again.

Understanding this characteristic of system failure enables us to make feasible plans for preventive and corrective maintenance activities to prolong system operations and make correct decisions about when to build a new system or to fix the existing one.

5.2.2.3 Reliability with Independent Failure Event

Consider a system with n components, and suppose that each component has an independent failure event (i.e., the occurrence of the failure event does not depend on any other; for a more comprehensive review of independent events, please refer to Appendix I at the end of this book). Components may be connected in different structures or networks within the system configuration; these could be in series, in parallel, or a combination thereof.

5.2.2.3.1 Series Structure

Figure 5.3 illustrates a series structure of components. A series system functions if and only if all of its components are functioning. If any one of the components fails, then the system fails; as seen in Figure 5.3, a failed component will cause the whole path to be broken. Here we use a formal mathematical formulation to define the structure functions, so readers may understand the more complex structure better.

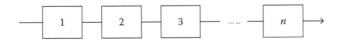

FIGURE 5.3
Series component structure.

Here, we use an indicator variable x_i to denote the whether or not the ith component is functioning:

$$x_i = \begin{cases} 1, & \text{if } i\text{th component is working properly} \\ 0, & \text{if } i\text{th component has failed} \end{cases}$$

Thus, the state vector for all the components is $x = (x_1, x_2, ..., x_n)$. Based on the vector, we can define the system structure function as $\Phi(x)$ such that

$$\Phi(x) = \begin{cases} 1, & \text{if system is working properly when state vector is } x \\ 0, & \text{if system has failed when state vector is } x \end{cases}$$

So, with a series structure, the structure function is given by

$$\Phi(x) = \min(x_1, x_2, ..., x_n) = x_1 x_2 ... x_n = \prod_{i=1}^{n} x_i \qquad (5.17)$$

From Equation 5.17, it is easily seen that $\Phi(x) = 1$ if and only if all the $x_i = 1$, where $i = 1, 2, ..., n$. So, using the structure function, the reliability of the system consisting of n components in a series structure is given by

$$R = (R_1)(R_2)...(R_n) \qquad (5.18)$$

For example, suppose that a system consists of three components A, B, and C in a series structure, failures occurring in the three components are independent, and the time to failure is exponentially distributed, with $\lambda_A = 0.002$ failure/h, $\lambda_B = 0.0025$ failure/h, and $\lambda_C = 0.004$ failure/h, then the reliability for system ABC for a period of 100 h, according to Equations 5.6 and 5.18, is

$$R_{ABC} = R_A R_B R_C = \left(e^{-\lambda_A t}\right)\left(e^{-\lambda_B t}\right)\left(e^{-\lambda_C t}\right) = e^{-(\lambda_A + \lambda_B + \lambda_C)t}$$

$$= e^{-(0.002 + 0.0025 + 0.004)100} = 0.4274$$

If the MTBF is given, then we can use Equation 5.15 to obtain the failure rate by $\lambda = 1/\text{MTBF}$, and Equation 5.18 can be used to obtain the reliability.

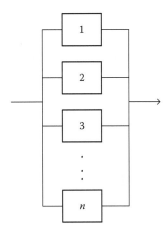

FIGURE 5.4
Parallel component structure.

5.2.2.3.2 Parallel Structure

Figure 5.4 presents a structure of components in parallel.

With a parallel structure, a system fails if and only if all the components fail, or, in other words, a parallel system is functioning if at least one component is functioning. So, for all the components x_i, if at least one $x_i = 1$, then $\Phi(\mathbf{x}) = 1$. Using the same indicator variable and structure function, the structure function of a parallel structure is given by

$$\Phi(\mathbf{x}) = \max(x_1, x_2, \ldots, x_n) = 1 - \prod_{i=1}^{n}(1 - x_i) \qquad (5.19)$$

When $n = 2$, this yields

$$\max(x_1, x_2) = 1 - (1 - x_1)(1 - x_2) = x_1 + x_2 - x_1 x_2$$

So, similarly, the reliability function of a parallel system structure is given by

$$R = 1 - (1 - R_1)(1 - R_2) \ldots (1 - R_n) \qquad (5.20)$$

As an example, suppose that a system consists of three components A, B, and C in a parallel structure, failures occurring in the three components are independent, and the time to failure is exponentially distributed with $\lambda_A = 0.002$ failure/h, $\lambda_B = 0.0025$ failure/h and $\lambda_C = 0.004$ failure/h, then the reliability for system ABC for a period of 100 h is

$$R_{ABC} = 1 - (1 - R_A)(1 - R_B)(1 - R_C) = 1 - \left(1 - e^{-\lambda_A t}\right)\left(1 - e^{-\lambda_B t}\right)\left(1 - e^{-\lambda_C t}\right)$$

$$= 1 - \left(1 - e^{-0.002(100)}\right)\left(1 - e^{-0.0025(100)}\right)\left(1 - e^{-0.004(100)}\right) = 0.9868$$

Some readers may have noticed that with the same components, the parallel structure has a better reliability (0.9868 vs. 0.4274). If we look at the reliability of each component, $R_A = e^{-\lambda_A t} = e^{-0.002(100)} = 0.8187$, $R_B = e^{-\lambda_B t} = e^{-0.0025(100)} = 0.7788$, and $R_C = e^{-\lambda_C t} = e^{-0.004(100)} = 0.6703$, it is obvious that the reliability for the series structure is lower than that of any individual component while the reliability of the parallel structure is higher than that of any individual component. One can prove this proposition easily, since the reliability $0 \le R \le 1$. As a matter of fact, the more components we have in the series structure, the less reliable the system is, and the more components we add to a parallel system, the more reliable it is.

5.2.2.3.3 k-out-of-n Structure and Combined Network

A *k*-out-of-*n* system is functioning if and only if at least *k* components of the *n* total components are functioning. Recall that we defined x_i as a binary function, with $x_i = 1$ if the *i*th component is working, and $x_i = 0$ otherwise. So, the number of working components for the system can be obtained by $\sum_{i=1}^{n} x_i$. Therefore, the k-out-of-n system structure function $\Phi(\mathbf{x})$ can be expressed as

$$\Phi(\mathbf{x}) = \begin{cases} 1, & \sum\limits_{i=1}^{n} x_i \ge k \\ 0, & \sum\limits_{i=1}^{n} x_i < k \end{cases} \tag{5.21}$$

It is easy to see that series and parallel systems are both special cases for the *k*-out-of-*n* structure. The series structure is an *n*-out-of-*n* system and the parallel structure is a 1-out-of-*n* system.

Let us look at the following example: Consider a system consisting of five components, and suppose that the system is functioning if and only if components 1, 2, and 3 all function and at least one of the components 4 and 5 functions. This implies that 1, 2, and 3 are in a series structure and 4 and 5 are placed in a parallel structure. So, the structure function for this particular system is

$$\Phi(\mathbf{x}) = \min(x_1, x_2, x_3) \max(x_4, x_5) = x_1 x_2 x_3 (x_4 + x_5 - x_4 x_5)$$

From the *k*-out-of-*n* structure function, one can easily derive the reliability for the *k*-out-of-*n* system,

$$R_n = P\{\Phi(\mathbf{x}) = 1\} = P\left\{\sum_{i=1}^{n} x_i \geq k\right\} \qquad (5.22)$$

As an example, for a 2-out-of-4 system, the reliability is given by

$$R_4 = P\{\Phi(\mathbf{x}) = 1\} = P\left\{\sum_{i=1}^{4} x_i \geq 2\right\}$$

$$= P\{\mathbf{x} = (1,1,1,1)\} + P\{\mathbf{x} = (1,1,1,0)\} + P\{\mathbf{x} = (1,1,0,1)\}$$

$$+ P\{\mathbf{x} = (1,0,1,1)\} + P\{\mathbf{x} = (0,1,1,1)\} + P\{\mathbf{x} = (1,1,0,0)\}$$

$$+ P\{\mathbf{x} = (1,0,1,0)\} + P\{\mathbf{x} = (0,1,1,0)\}$$

$$+ P\{\mathbf{x} = (1,0,0,1)\} + P\{\mathbf{x} = (0,1,0,1)\} + P\{\mathbf{x} = (0,0,1,1)\}$$

$$= R_1 R_2 R_3 R_4 + R_1 R_2 R_3 (1-R_4) + R_1 R_2 (1-R_3)R_4 + R_1(1-R_2)R_3 R_4$$

$$+ \left(1 - (1-R_1)R_2 R_3 R_4 + R_1 R_2 (1-R_3)(1-R_4) + R_1(1-R_2)R_3(1-R_4)\right.$$

$$+ (1-R_1)R_2 R_3 (1-R_4) + R_1(1-R_2)(1-R_3)R_4$$

$$+ (1-R_1)R_2(1-R_3)R_4 + (1-R_1)(1-R_2)R_3 R_4$$

If all the components are identical, with the same probability of R, Equation 5.22 is given by

$$R_n = P\{\Phi(\mathbf{x}) = 1\} = P\left\{\sum_{i=1}^{n} x_i \geq k\right\} = \sum_{i=k}^{n} \binom{n}{i} R^i (1-R)^{n-i} \qquad (5.23)$$

$\binom{n}{i}$ is a k-combination function, measuring the number of subsets of k elements taken from a set of n elements ($n \geq k$).

$$\binom{n}{i} = \frac{n(n-1)\ldots(n-i+1)}{i(i-1)\ldots 1} = \frac{n!}{i!(n-i)!}$$

Using the concepts above, one can easily solve the reliability for any combined network consisting of series and parallel structures. Let us look at one

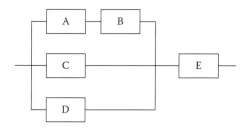

FIGURE 5.5
Sample combined network of components.

example: Suppose that a system consists of five components, A, B, C, D, and E, they are connected in the structure shown in Figure 5.5.

Assuming the failure functions for A, B, C, D, and E are exponentially distributed and the MTBFs for these components are shown as follows:

$$\text{MTBF}_A = 10{,}000\,\text{h}$$

$$\text{MTBF}_B = 8000\,\text{h}$$

$$\text{MTBF}_C = 5000\,\text{h}$$

$$\text{MTBF}_D = 2500\,\text{h}$$

$$\text{MTBF}_E = 20{,}000\,\text{h}$$

What is the probability of the system ABCDE surviving for 5000 h without failure?

From Figure 5.5, we can see that components A and B are in a series structure; A/B, C, and D are connected in a parallel structure; and, finally, ABCD are connected with E in a series structure. The failure rates for these components are

$$\lambda_A = \frac{1}{10{,}000} = 0.0001 \text{ failure/h}$$

$$\lambda_B = \frac{1}{8000} = 0.000125 \text{ failure/h}$$

$$\lambda_C = \frac{1}{5000} = 0.0002 \text{ failure/h}$$

$$\lambda_D = \frac{1}{2500} = 0.0004 \text{ failure/h}$$

$$\lambda_E = \frac{1}{20,000} = 0.00005 \text{ failure/h}$$

The reliability for A and B is

$$R_{AB} = R_A R_B = \left(e^{-\lambda_A t}\right)\left(e^{-\lambda_B t}\right) = e^{-(\lambda_A + \lambda_B)t}$$

$$= e^{-(0.0001 + 0.000125)5000} = 0.32465$$

because

$$R_C = e^{-(\lambda_C)t} = e^{-(0.0002)5000} = 0.36788$$

$$R_D = e^{-(\lambda_D)t} = e^{-(0.0004)5000} = 0.13536$$

The composite reliability for ABCD is

$$R_{ABCD} = 1 - (1 - R_{AB})(1 - R_C)(1 - R_D)$$

$$= 1 - (1 - 0.32465)(1 - 0.36788)(1 - 0.13536) = 0.63088$$

since

$$R_E = e^{-(\lambda_E)t} = e^{-(0.00005)5000} = 0.7788$$

So, the reliability for the overall system ABCDE is given by

$$R_{ABCDE} = R_{ABCD}R_E = (0.63088)(0.7788) = 0.4913$$

This implies that the probability of the system ABCDE surviving for an operating time of 5000 h is about 49.13%, or the system reliability for 5000 h is 49.13%. For a number of operating hours less than 5000, one would expect this reliability to increase.

From the above example, we can see that the general procedure for solving a system reliability problem is quite simple and straightforward; no matter how complex the system structure is, it can always be decomposed into one of the two fundamental structures, series and parallel. So, one would follow these steps:

1. Obtain the reliability value for the individual components for the time t.
2. Start from the most basic structure, and gradually work up to the next level, until the whole system structure is covered. For the above example, we started with the bottom level of the structure, which is A and B, obtaining the reliability of R_{AB}, so A and B may be treated as being equivalent to one component in terms of reliability; then, we address A/B, C and D; they are the next level's basic structure, as they are a three-branched parallel structure; and finally we obtain the system reliability as the overall structure is a large series network between A/B/C/D and E.

Using the above two procedures, one can easily obtain reliability measures for any complex network structures. There are some exercise questions at the end of the chapter; readers may practice applying these procedures and formulas.

The reliability examples we have talked about so far have only considered the case when the first failure occurs; there are certain circumstances in which components can be replaced when one component fails. To simplify the situation, let us assume the replacement happens immediately (i.e., time to replace = 0), or we can imagine a redundant system design: When one component fails, there is a switch to connect a backup component instantly, as shown in Figure 5.6. This type of system is also called a redundant standby network.

In a standby system, the backup component is not put into operation until the preceding component fails. For example, in Figure 5.6, at the beginning, only Component 1 is operative while Components 2 and 3 are standing by. When Component 1 fails, Component 2 is immediately put in use until it

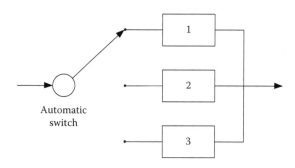

FIGURE 5.6
Three-component redundant standby system.

fails; then, Component 3 becomes operational. When Component 3 fails, the system stops working and this is considered a system failure.

In standby systems, the failures of individual components are not totally independent of each other; this is different from a purely parallel network, in which failures occur independently. In standby structures, failures occur one at a time, while in the parallel network, two parts can fail at the same time.

Assuming that the failure function is still exponential (i.e., the time to fail follows an exponential distribution), to simplify the situation, let us further assume that all the parts are identical. The failures occur one by one, so for an N-component standby system, the system is functional until N failures occur. Put in a formal way, if we denote the random variable $\mathbb{N}_t$ as the number of failure occurring in time t, then we have

$$R(n \text{ standby}) = P(\mathbb{N}_t \leq N) \tag{5.24}$$

It is known that when the interval time between failures follows an exponential distribution with parameter λ, then the probability distribution of the number of failures occurring in any time interval of t follows a Poisson distribution with parameter λt. The Poisson distribution can be defined as follows:

A discrete random variable $\mathbb{N}$ has a Poisson distribution with parameter λ if for $n = 0, 1, 2, \ldots$

$$P(\mathbb{N} = n) = \frac{e^{-\lambda} \lambda^n}{n!} \tag{5.25}$$

So, the number of failures occurring during any time interval t is given by the following formula:

$$P(\mathbb{N}_t = n) = \frac{e^{-\lambda t} (\lambda t)^n}{n!}$$

The reliability for the standby system can be written as

$$R(N \text{ standby}) = P(\mathbb{N}_t \leq N) = \sum_{n=0}^{N} \frac{e^{-\lambda t} (\lambda t)^n}{n!} \tag{5.26}$$

As an example, suppose that a system consists of one component with two other identical ones (three components in total) in standby. Each of the components has an MTBF of 2500 h, distributed exponentially. Determine the system reliability for a period of 100 h.

$$\lambda = \frac{1}{2500} = 0.0004 \text{ failures/h}$$

$$\lambda t = 0.0004(100) = 0.04 \text{ failures}$$

Based on Equation 5.26, the system reliability is

$$R(2 \text{ standby}) = P(\mathbb{N}_t \leq 2)$$

$$= e^{-0.04} + (0.04)e^{-0.04} + \frac{(0.04)^2 e^{-0.04}}{2!} = 0.99997$$

If these three components are configured in a parallel structure, then the reliability is

$$R(3 \text{ parallel}) = 1 - \left(1 - e^{-0.04}\right)^3 = 0.99994$$

which is less than 0.99997. A standby structure provides higher reliability than a parallel structure with the same components. This may be easily seen, as the reliability is a function of time; the standby system uses one part after another fails, so all the parts except the first one have a later start time than parts in a parallel structure. Thus, it is anticipated that those parts in standby systems will last longer.

There are many other situations that can also be modeled as a standby system. For example, when one component fails, it can be replaced by a backup component from the inventory so that the system still functions. If the replacement time is relatively short enough to be ignored, then the reliability of the system can be approximated by treating it as a standby structure. Or, in other words, if we look at all the components together as a whole system, it is as if the overall system MTBF has been prolonged; that is, for an N-component standby system, if each of the components has a failure rate of λ, then the overall system $\text{MTBF} = N(\text{MTBF}) = N/\lambda$, so the system failure rate $\lambda_N = \lambda/N$. However, we cannot use Equation 5.6 to obtain the reliability of the system, because the failure function for the system is no longer exponentially distributed.

5.2.3 Reliability Analysis Tools: FMEA and Faulty Trees

System reliability, as one of the inherent design characteristics, is one of the most important parts of any system's operational requirements, regardless of what type of system it is. Starting from a very high level, requirements regarding system reliability are defined both quantitatively and qualitatively, including

1. Performance and effectiveness factors for system reliability
2. System operational life cycle for measuring reliability
3. Environmental conditions in which the system is expected to be used and maintained (such as temperature, humidity, vibration, radiation, etc.)

The original requirements are derived from users, mission planning, and feasibility analysis. Once the high-level requirements are obtained, lower-level requirements are developed as the system design evolves; system requirements need to be allocated to the system components. System reliability is allocated in the system TPMs and integrated within the functional analysis and functional allocation processes.

When allocating to the lower levels of the system, there is, unfortunately, no template or standard to follow, as every system is different and there may be tens of thousands of parts involved in multiple levels. Most of the allocations utilize a trial-evaluation-modify cycle until a feasible solution is reached. This approach uses a bottom-up procedure as well the top-down process, as different COTS components are considered for selection. Under these circumstances, it is very difficult to arrive at optimum solutions; usually a feasible solution meeting the system requirements and complying with all other design constraints is pursued, and this process is also iterative and often involves users.

Throughout the design, as part of the iterative design and evaluation process, there are many analysis tools that are available to aid the designers to effectively derive the requirements and TPMs for system reliability at different levels of the system structure. For most systems, the reliability requirements are addressed in an empirical manner; with a large volume of requirements and many iterations of analysis-integration-evaluation, one needs to have a practical tool to elicit the structures and relationships required for system reliability, so that the TPMs can be determined. Two of the most commonly used tools are failure mode effect analysis (FMEA) and faulty tree analysis (FTA).

5.2.3.1 *Failure Mode Effect Analysis (FMEA)*

Failure mode effect analysis (FMEA), sometimes called failure mode, effects, and criticality analysis (FMECA), is a commonly used analysis tool for analyzing failures that are associated with system components. It was originally developed by NASA to improve the reliability of hardware design for space programs. Although the original FMEA document is no longer in effect, the FMEA methodology, however, has been well preserved and tested and has evolved. Nowadays, FMEA has become a well-accepted standard for identifying reliability problems in almost any type of systems, ranging from military to domestic and mechanical to computer software design.

Generally speaking, FMEA is a bottom-up inductive approach to analyze the possible component failure modes within the system, classifying them into different categories, severities, and likelihoods, identifying the consequences caused by these failures to develop a proactive approach to prevent them from occurring, and the related maintenance policy for these failures. It is an inductive process, because FMEA starts with detailed specific examples and cases of failure, to gradually derive general propositions regarding system reliability predictions (as opposed to the deductive approach, where the specific examples are derived from the general propositions, as in the faulty tree analysis approach we will discuss in Section 5.2.3.2).

FMEA usually consists of two related but separate analyses; one is FMEA, which investigates the possible failure modes at different system levels (components or subsystems) and their effects on the system if failure occurs; the second is criticality analysis (CA), which quantifies the likelihood of failure occurrence (i.e., failure rate) and ranks the severity of the effects caused by the failures. This ranking is usually accomplished by analyzing historical failure data from similar systems/components and through a team approach, derived in a subjective manner.

To conduct an FMEA analysis, there are some basic requirements that need to be fulfilled first. These requirements include:

1. System structure in schematic form. Without the basic understanding of the system architecture, especially the hardware and software structures, one cannot identify the possible consequences if one or more components fail. This is the starting point of FMEA analysis.

2. System function FFBD. As we stated earlier, the FFBD is the foundation for many analyses; only with the FFBD specified, functions can then be allocated to components, components allocated to hardware, software or humans, and the operational relationships between the components defined, which is necessary to conduct FMEA analysis.

3. Knowledge of systems requirements. System hardware and software architecture is derived based on requirements. At any point in the design, requirements are needed to verify the design decisions; ultimately, this is important for FMEA-related analysis, since FMEA is an inductive approach and everything is assessed on an empirical basis.

4. A comprehensive understanding of the systems components. This includes, but is not limited to, access to current technology, understanding of the COTS items, and knowledge of supply chain operations and structures related to system components.

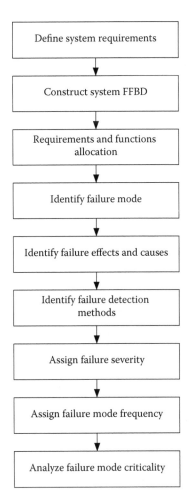

FIGURE 5.7
Typical FMEA work steps.

With the basic sources of information available and preliminary assessment of the system structure, a team approach is applied to develop the FMEA analysis results; the basic steps are illustrated in Figure 5.7.

1. *Define system requirements.* Requirements for system reliability need to be clearly defined, as do the TPMs (such as the MTBF of the system) and the system operating environments. With high-level systems requirements defined and refined at a lower level, the system structures can be identified, via a top-down approach, from system to subsystem level, down to the components and eventually

the hardware and software units to construct the systems. This provides a big picture for system reliability and a starting point to conduct FMEA analysis.

2. *Construct system FFBD.* One thing we have to keep in mind is that FMEA analysis has be based on the system design synthesis and integration results. Ideally, FMEA analysis should be paired with system functional analysis and functional allocation, as FMEA analysis is tied to each system component. To perform an FMEA analysis, the following materials/information are needed from functional analysis:

   a. System functional architecture and mission requirements.

   b. System FFBD.

   c. System operational information for each of the functions, similarly to Figure 4.10; input/output, mechanism, and constraints information are required to identify the failure mode and its effects.

   d. Rules, assumptions, and standards pertaining to components. Understanding the limitations of the current feasible technology and COTS constraints helps us to make predictions of the failure mode more meaningful. The ground rules generally include system mission, phase of the mission, operating time and cycle, derivation of failure mode data (i.e., supplier data, historical log, statistical analysis, subject matter experts' estimates, etc.), and any possible failure detection concepts and methodologies.

3. *Requirements allocation and function allocation*: With the FFBD, the reliability-related TPMs are allocated to the components level; this is parallel to the function allocation process, described in Section 4.3.2. With the requirements allocated to the lower levels, the effects of failure on the system can be specified at a quantitative level.

4. *Identify failure mode.* A failure mode is the manner in which a failure occurs at a system, subsystem, or component level. There are many types of failure mode involved in a single component; to derive a comprehensive analysis of the failure mode, designers should look at the various sources of information, including similar systems, supplier data, historical data for critical incidents and accidents studies, and any information related to environmental impacts. A typical component failure mode should include the following aspects:

   a. Failure to operate in proper time

   b. Intermittent operation

   c. Failure to stop operating at the proper time

   d. Loss of output

   e. Degraded output or reduced operational capability

5. *Identify causes and effects of failure.* The cause of the failure mode is usually an internal process or external influence, or an interaction between these. It is very possible that more than one failure could be caused by one process and a particular type of failure could have multiple causes. Typical causes of a failure include aging and natural wearing out of the materials, defective materials, human error, violation of procedures, damaged components due to environmental effects, and damage due to the storage, handling, and transportation of the system. There are many tools to aid designers in laying out the sources and effects of failures and their relationships; for example, the "fishbone" diagram by Ishikawa, the Swiss model, and the human factors analysis and classification system (HFACS) for human error analysis have been widely used to identify the complex structure of error cause–effect relationships. Failure effect analysis assesses the potential impact of a failure on the components or the overall system. Failures impact systems at different levels, depending on the types of failures and their influences. Generally speaking, there are three levels of failure effect:

   a. *Local effect*: Local effects are those effects that result specifically from the failure mode of the component at the lowest level of the structure itself.

   b. *Parent-level effect*: These are the effects that a failure has on the operation and performance of the functions at the immediate next higher level.

   c. *End-level effect.* These effects are the ones that impact on the operation and functions on the overall system as a whole. A small failure could cause no immediate effect on the system for a short period of time, cause degraded system overall performance, or cause the system to fail with catastrophic effects.

6. *Identify failure detection method.* This section of FMEA identifies the methods by which the occurrence of a failure is detected. These methods include

   a. Human operators

   b. Warning devices (visual or auditory)

   c. Automatic sensing devices

   The detection methods should include the conditions of detection (i.e., normal vs. abnormal system operations) and the times and frequencies of the detection (i.e., periodic maintenance checking to identify signs of potential failure, or diagnosis of failure when symptoms are observed).

7. *Assign failure severity.* After all failure modes and their effects on the system are identified, the level of impact of these failures on the system need to be ranked by assigning an appropriate severity

score. This will enable design teams to prioritize failures based on the "seriousness" of the effect, so that they can be addressed in a very efficient way, especially given the limited resources available. To assign a severity score for each of the failure modes, each failure effect is evaluated in terms of the worst consequences of the failure on the higher-level system, and through iterative team efforts, a quantitative score is assigned to that particular failure. Table 5.5 illustrates a typical severity ranking and scales.

8. *Assign failure mode frequency and probability of detection.* This is the start of the second half of the FMEA analysis, the criticality analysis (CA). It adds additional information about the system failure so that a better design can be achieved by avoiding these failures. The CA part of FMEA enables designers to identify the system reliability- and maintainability-related concerns and address these concerns in the design phase. The first step of CA is to transfer the data collected to determine the failure rate; that is, the frequency of failure occurrence. This rate is often expressed as a probability distribution as the failure occurs in a random manner. It also includes the information about the accuracy of failure detection methods, combining the probability of failure detection (i.e., a correct hit detection, or a false alarm detection) to provide the level of uncertainty of failure occurrence and the probability of the failure being detected.

9. *Analyze failure criticality.* Once the failure rate and detection probability of the failure has been identified, information pertaining to the failure needs to be consolidated to form a criticality assessment of the failure for it to be addressed in the design. Criticality can be assessed quantitatively or qualitatively. For a quantitative assessment, various means or measures can be used to calculate the critical values of the components; for example, the required

TABLE 5.5

Typical FMEA Severity Ranking System

Severity Score	Severity	Potential Failure Effects
1	Minor	No effect on higher system
2–3	Low	Small disruption to system functions; repair will not delay the system mission
4–6	Moderate	May be further classified into low moderate, moderate or high moderate, causing moderate disruption and delay for system functions
7–8	High	Causes high disruption to system functions. Some portion of functions are lost; significant delay in repairing the failure
9–10	Hazard	Potential safety issues, potential whole system mission loss and catastrophic if not fixed

number of backup (redundant) parts for a predetermined reliability level, as expressed in Equation 5.26, identifying the item criticality score (C_m) for each of the items. If the failure rate is λ, the failure mode probability is α, and the failure effect probability is β, then the failure mode criticality score for a time period of t is given by

$$C_m(t) = \alpha\beta\lambda t \tag{5.27}$$

If the item has a number of different failure modes, then the item criticality number is the sum of all the failure mode criticality numbers, given by

$$C_y(t) = \sum C_{mi}(t) = \sum \alpha_i \beta_i \lambda_i t \tag{5.28}$$

Qualitative analysis is used when the failure rate for the item is not available. A typical method used in qualitative analysis is to use the risk priority number (RPN) to rank and identify concerns or risks associated with the components due to the design decisions. The number provides a mean to delineate the more critical aspects of the systems design. The RPN can be determined from:

RPN=(severity rating)×(frequency rating)×(probability of detection rating)

Generally speaking, a component with a high frequency of failure, high impact/severity of failure effect, and difficulty of failure detection usually has a high RPN. Such components should be given high priority in the design consideration.

It is convenient to present a finished FMEA analysis in a tabular format, listing all the required information in different columns. Table 5.6 presents a sample set of FMEA analysis results an automobile.

5.2.3.2 Faulty Tree Analysis (FTA)

A faulty tree analysis, or FTA model, is a graphical method for identifying the different ways in which a particular component/system failure could occur. Compared to the FMEA model, which is considered a "bottom-up" inductive approach, FTA is a deductive approach, using graphical symbols and block diagrams to determine the events and the likelihood (probability) of an undesired failure event occurring. FTA is used widely in reliability analysis where the cause–effect relationships between different events are identified. Figure 5.8 illustrates the basic symbols that an FTA model uses.

TABLE 5.6

Sample FMECA Analysis

Item	Failure Mode	Failure Effects	Severity	Cause	Occurrence	Prevention	Detection	RPN	Criticality
Control unit	Inoperable vehicle	Full vehicle shut down	9	Poor electrical connection/hardware failure/power loss	4	Electrical routing/color coding of cables	Test electrical connections and routing	288	38
ESS cooling system	Fail to cool	Battery failures	10	Poor coolant system routing/increased pressure/poor electrical connection/component failure	7	Proper electrical routing/proper coolant routing/cooling system component monitoring	Coolant temperature sensor/sensor for powered components	140	60
Engine and motor/inverter cooling system	Fail to cool engine, motor, and inverter	Engine, motor, inverter overheating	7	Poor coolant system routing/increased pressure/poor electrical connection/component failure	8	Proper electrical routing/proper coolant routing/cooling system component monitoring	Coolant temperature sensor/current sense for powered components	112	60

| Fuel system | Fail to inject properly | Loss of charge sustaining ability | 9 | Lack of maintenance/no fuel/improper pressure (high & low pressure systems)/ mechanical failure/ improper heat shielding/pump failure/poor electrical connection | 3 | Mechanical integration/ electrical routing | Fuel level sensor/ fuel pressure sensor/air fuel ratio/current sense/engine power | 72 | 25 |

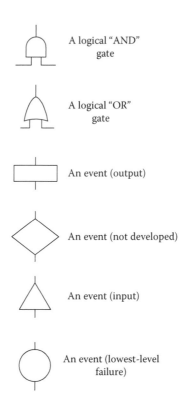

FIGURE 5.8
Some basic FTA constructs and symbols.

FTA models are usually paralleled with functional analysis, providing a concise and orderly description of the different possible events and the combination thereof that could lead to a system/subsystem failure. FTA is commonly used as a design method, based on the analysis of similar systems and historical data, to predict causal relationships in terms of failure occurrences for a particular system configuration. The results of FTA are particularly beneficial for designers to identify any risks involved in the design, and more specifically to

1. Allocate failure probabilities among lower levels of system components
2. Compare different design alternatives in terms of reliability and risks
3. Identify the critical paths for system failure occurrence and provide implications of avoiding certain failures
4. Help to improve system maintenance policy for more efficient performance

Generally speaking, there are four basic steps involved in conducting an FTA analysis:

Step 1: Develop the functional reliability diagram. Develop a functional block diagram for systems reliability, based on the system FFBD model, focusing on the no-go functions and functions of diagnosis and detection. Starting from the system FFBD, following the top-down approach, a tree structure for critical system failure events is identified. This diagram includes information about the structures of the no-go events, what triggers/activates the events, and what the likelihoods and possible consequences of those events are.

Step 2: Construct the faulty tree. Based on the relationships described in the functional diagram, a faulty tree is constructed by using the symbols from Figure 5.8. The faulty tree is based on the functional diagram but is not exactly the same, in the sense that functional models follow the system operational sequences of functions while the FTA tree follows the logical paths of cause–effect failure relationships; it is very possible that, for different operational modes, multiple FTAs may be developed for a single functional path. In constructing the faulty tree, the focus is on the sequence of the failure events for a specific functional scenario or mission profile.

Step 3. Develop the failure probability model. After the FTA is constructed, the next step is to quantify the likelihood of failure occurrence by developing the probability model of the faulty tree. Just as in understanding the models of reliability theory, readers need to familiarize themselves with basic probability and statistics theory. The fundamentals of probability and statistics are reviewed in Appendix I at the end of this book; readers must first engross themselves in these subjects to understand these models. As a matter of fact, in terms of quantitative modeling methodology for systems engineering, probability and statistics are perhaps the most important subjects besides operations research; due to the uncertain and dynamic nature of complex system design, one can hardly find any meaningful solution to a system design problem without addressing its statistical nature. We will be covering more on this subject in later chapters and Appendix I.

The mathematical model of FTA is primarily concerned with predicting the probability of an output failure event with the probabilities of events that cause this output failure. For simplification purposes, we assume all the input failures are independent of each other. Two basic constructs for predicting the output events are the AND-gate and the OR-gate.

5.2.3.2.1 AND-gate

All the input events (E_i, $i = 1, 2, \ldots, n$) attached to the AND-gate must occur in order for the output event (A) above the gate to occur. That is to say, in terms of the probability model, the output event is the *intersection* of all the input events. For example, for the AND-gate illustrated in Figure 5.9, if we know the probability of the input events as $P_1, P_2, \ldots, P_n$, the probability of output failure above the AND-gate can be obtained as $P(A) = P(E_1 \cap E_2 \cap \ldots \cap E_n)$. Since

FIGURE 5.9
AND-gate structure.

all the input events are independent of each other, $P(A)$ is the product of all the input event probabilities, that is, $P(A)=P_1P_2...P_n$.

For example, for a three-branched AND-gate as illustrated in Figure 5.10, if $P_1=0.95$, $P_2=0.90$, and $P_3=0.92$, then

$$P(A) = P_1P_2P_3 = (0.95)(0.90)(0.92) = 0.79$$

5.2.3.2.1 OR-gate

If a failure occurs if one or more of the input events (E_i, $i=1, 2, ..., n$) occurs, then an OR-gate is used for this causal relationship. In terms of the probability model, the OR-gate structure represents the *union* of the input events attached to it. For example, for the OR-gate illustrated in Figure 5.11, if we know the probability of the input events as $P_1, P_2,..., P_n$, the probability of output failure above the OR-gate can be obtained as $P(O) = P(E_1 \cup E_2 \cup ... \cup E_n)$.

FIGURE 5.10
AND-gate FTA example.

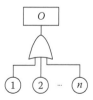

FIGURE 5.11
OR-gate structure.

Since all the events are not mutually exclusive, we cannot simply use the sum of the probability of the events. To solve for $P(O)$, we need to use the concept of the compliment, that is, $P(O) = 1 - P(\bar{O})$, $P(\bar{O})$ is the probability that none of the input events occurs; this means that all the events must not occur together. So we have $P(\bar{O}) = P(\bar{E}_1 \cap \bar{E}_2 \cap ... \cap \bar{E}_n)$; thus we can obtain $P(O)$ by $P(O) = 1 - P(\bar{E}_1 \cap \bar{E}_2 \cap ... \cap \bar{E}_n) = 1 - (1 - P_1)(1 - P_2)...(1 - P_n)$, or

$$P(O) = 1 - \prod_{i=1}^{n}(1 - P_i), \quad i = 1, 2, ..., n$$

For example, for a three-event OR-gate, as illustrated in Figure 5.12, if $P_4 = 0.95$, $P_5 = 0.90$, and $P_6 = 0.92$, then

$$P(B) = 1 - (1 - P_4)(1 - P_5)(1 - P_6)$$

$$= 1 - (1 - 0.95)(1 - 0.90)(1 - 0.92) = 0.9996$$

After talking about AND-gates and OR-gates, some readers may easily see that the calculation of the AND-gate is similar to the series structure and the OR-gate is similar to the parallel structure of the reliability network. This is because the logic for the AND and OR of failure events are the same as the reliability events in the series and parallel structures. Understanding the basic probability model for the AND-gate and OR-gate, we can solve any composite faulty tree structure; we just start from the bottom level and work our way up, until the probabilities for all the events are obtained. Take the example of the FTA in Figure 5.13. If we know $P_1 = 0.60$, $P_2 = 0.75$, $P_3 = 0.90$, $P_4 = 0.95$, and $P_5 = 0.80$, what is the value of $P(C)$?

First, Event A is an OR-gate from Event 1/Event 2, so $P(A) = 1 - (1 - P_1)(1 - P_2) = 1 - (0.40)(0.25) = 0.90$; next, Event B is an AND-gate from Event 3/Event 4, so $P(B) = P_3 P_4 = (0.90)(0.95) = 0.855$; and finally, C is a an OR-gate event from Event A/Event B/Event 5, so $P(C) = 1 - [1 - P(A)][1 - P(B)][1 - P_5] = 1 - (1 - 0.90)(1 - 0.855)(1 - 0.80) = 0.9971$. ∎

Step 4. Identify the critical fault path. With the probability of failure of the system or of a higher-level subsystem, a path analysis can be conducted to

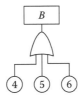

FIGURE 5.12
OR-gate example.

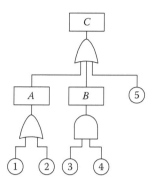

FIGURE 5.13
Composite FTA example.

identify the key causal factors that contribute most significantly to the failure. Certain models can be applied to aid the analysis based on the assumptions made about the faulty tree, such as Bayesian's model, Markov decision model, or simply using a Monte Carlo simulation. For a more comprehensive review of these models, readers can refer to the *Reliability Design Military Handbook* (MIL-HDBK-338B; U.S. Department of Defense 1988).

FTA provides a very intuitive and straightforward method to allow designers to visually perceive the possible ways in which a certain failure can occur. As mentioned before, FTA is a deductive approach; once the bottom-level failures are identified, FTA can easily assist the designers to assess how resistant the system is to various risk sources. FTA is not good at finding the bottom-level initiating faults; that is why it works best when combined with FMEA, which exhaustively locates the failure modes at the bottom level and their local effects. Performing FTA and FMEA together may give a more complete picture of the inherent characteristics of system reliability, thus providing a basis for developing the most efficient and cost-effective system maintenance plans.

5.3 System Maintainability

One of the design objectives is to ensure that the system is operational for the maximum period of time. We have discussed this objective in Section 5.2 about system reliability; a reliable system is certainly our ultimate goal. Unfortunately, failures always occur, no matter how reliable the system is; as Murphy's Law states, "anything that can go wrong will go wrong." Having a high level of reliability and fixing failures quickly when they occur are really the "two blades of the sword"; we need both to improve the level of system availability. In the previous section, we have comprehensively reviewed the system reliability factors, which are the proactive aspects of failures, suggesting how to configure our system so that the inherent system reliability characteristics can be optimized. With system reliability optimized, we now turn our focus to the second aspect, system maintainability, which deals with the measures and methods to manage failures should they occur.

5.3.1 Maintainability Definition

System maintainability measures the ability with which a system is maintained to prevent a failure from occurring in the future and restore the system when a failure does occur. The ultimate goal of system operation is to make the system operational as far as possible; a more realistic measure for this operational capability is system availability. This is because if a system is not available, whether due to failure or routine maintenance, the consequence is similar in the sense that if the system is not operational, it is not generating profits or providing the functions that it is supposed to. So, system reliability and maintainability are two separate but highly related factors concerning the same objective, which is to increase the degree to which the system is available. Reliability is an inherent system characteristic; it deals with the internal quality of the system itself, the better design of the system, and better system reliability. Maintainability, on the other hand, is derived based on the system reliability characteristics; it is a design-dependent parameter that is developed to achieve the highest level of system availability. Although maintainability is inherent to a specific system, one usually cannot specify maintainability until the system requirements on reliability and availability are determined. Maintainability is a design-derived decision, a result of design. As defined in MIL-HDBK-470A, system maintainability is "the relative ease and economy of time and resources with which an item can be retained in, or restored to, a specified condition when maintenance is performed by personnel having specified skill levels, using prescribed procedures and resources, at each prescribed level of maintenance and repair."

Generally speaking, system maintainability can be broken down into two categories, preventive maintenance and corrective maintenance.

1. Preventive maintenance: also called proactive maintenance or scheduled maintenance, this refers to systematic methods of maintenance activities to prolong system life and to retain the system at a better level of performance. These activities include tests, detection, measurements, and periodic component replacements. Preventive maintenance is usually scheduled to be performed in a fixed time interval; its purpose is to avoid or prevent faults from occurring. Preventive maintenance is usually measured in preventive time, or M_{pt}.

2. Corrective maintenance: also called reactive maintenance or unscheduled maintenance. Corrective maintenance is performed when system

failure occurs. Corrective maintenance tasks generally include detecting, testing, isolating, and rectifying system failures to restore the system to its operational conditions. Typical actions of corrective maintenance include initial detection, localization, fault isolation, disassembly of system components, replacing faulty parts, reassembly, adjustment, and verification that system performance has been restored. Corrective maintenance is usually measured in corrective time, or M_{ct}.

5.3.2 Measures of System Maintainability

As an inherent DDP, the effectiveness and efficiency of maintainability is primarily measured using time and cost factors. The goal of maintenance is to perform the tasks in the least amount of time and with the least amount of cost.

5.3.2.1 Mean Corrective Time $\left(\overline{M_{ct}}\right)$

For corrective maintenance, the primary time measurement is the mean corrective time $\overline{M_{ct}}$. Nevertheless, due to the random nature of system failures, the time taken to fix them, M_{ct}, is also a random variable. As a random variable, the distribution function to interpret M_{ct} varies from system to system. Just like any other random variable, the common measures for M_{ct} include the probability distribution function (p.d.f.), cumulative distribution function (c.d.f.), mean, variance, and percentile value. Practically, one can approximate these parameters by observing the M_{ct} sample and analyzing the sample data, assuming each of the observations is individually independently distributed (IID). It has been found that most of the repair times fall into one of the three following distributions (Blanchard and Fabrycky 2006):

1. The normal (or Gaussian) distribution: The normal distribution is most commonly used in systems with relatively straightforward and simple maintenance actions; for example, where system repairs only involve simple removal and replacement actions, and these actions are usually standard and with little variation. Repair times following the normal distribution can be found for most maintenance tasks. Another reason for its popularity is perhaps due to the famous central limit theorem, which states that the mean of a sufficiently large number of independent random variables (or asymptotic independent samples), each with a finite mean and variance, will be approximately normally distributed. This is the reason that the normal distribution is used for such conditions if the true distribution is unknown to us.

2. The exponential distribution: This type of distribution most likely applies to those maintenance activities involving faults with a

constant failure rate. As the failure occurs independently, the constant failure rate will result in a Poisson process (see Appendix I for details) so that the general principles of queuing theory may apply.

3. The lognormal distribution: This is a continuous probability distribution of a random variable whose logarithm is normally distributed. Lognormal distribution has been used commonly in maintenance tasks for large, complex system structures, whose maintenance usually involves performing tasks and activities at different levels, and usually involves a nonstationary failure rate and time duration.

Let us use the normal distribution as an example to illustrate how some typical statistical analysis may be performed. A sample of 60 observations were collected for a maintenance task, as shown in Table 5.7. What is the mean and standard deviation for the task time? And what is the probability that the task time is between 60 and 80 min?

The histogram of the data is presented in Figure 5.14.

The mean corrective time $\overline{M_{ct}}$ is given by

$$\overline{M_{ct}} = \frac{\sum_{i=1}^{60} M_{ct_i}}{60} = 49.71\,(\text{min})$$

And the standard deviation is

$$\sigma = \sqrt{\frac{\sum_{i=1}^{n}\left(M_{ct_i} - \overline{M_{ct}}\right)^2}{n-1}} = 10.17\,(\text{min})$$

To obtain the percentage value, we need to use the standard normal table (a standard normal distribution function is a normal distribution with mean of 0 and variance of 1). First, we need to convert the corrective time normal distribution to a standard normal distribution. If a random variable X is normal distributed with mean μ and standard deviation of σ, or $X \sim N(\mu,\sigma)$, then the random variable $Z = (X-\mu)/\sigma$ follows standard normal distribution, that

TABLE 5.7

Observed M_{ct_i} for a Maintenance Task (min)

60.73	43.95	53.13	49.93	29.78	55.93	48.12	44.64	34.58	60.43
41.02	46.04	60.18	46.35	46.72	32.84	45.08	55.39	21.12	50.96
50.70	45.59	43.70	45.97	56.98	64.13	50.60	40.52	47.50	40.43
49.01	50.96	50.47	55.44	31.95	47.68	51.73	57.66	59.69	32.99
49.74	48.62	53.87	45.31	59.39	58.71	64.60	53.71	25.99	56.63
62.98	58.34	62.75	49.17	56.55	56.90	30.80	62.46	60.00	65.28

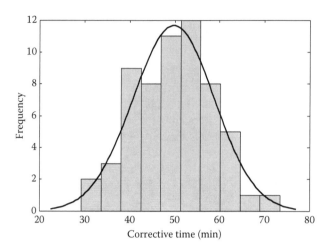

FIGURE 5.14
Histogram of the corrective time sample data.

is, $Z \sim N(0,1)$. So, for our example, $X \sim N(49.71, 10.17)$, we wish to know the percentage between $X_1 = 60$ min and $X_2 = 80$ min, so we have

$$Z_1 = \frac{X_1 - \mu}{\sigma} = \frac{60 - 49.71}{10.17} = 1.012$$

$$Z_2 = \frac{X_2 - \mu}{\sigma} = \frac{80 - 49.71}{10.17} = 2.978$$

Thus, $P(X_1 < X < X_2) = P(Z_1 < Z < Z_2) = P(Z < Z_2) - P(Z < Z_1)$.

From the standard normal table in Appendix II, which presents the cumulative probability of Z, we can obtain $P(Z < Z_2) = P(Z < 2.978) = 0.9986$, and $P(Z < Z1) = P(Z < 1.012) = 0.8438$, so the percentage of corrective time between 60 and 80 min is $P(60 < X < 80) = P(1.012 < Z < 2.978) = 0.9986 - 0.8438 = 0.1548$, or roughly 15.5%. Other statistics of interest, such as the confidence interval (CI), can also be derived:

$$\left[\overline{M_{ct}} - Z_\alpha \left(\frac{\sigma}{\sqrt{n}} \right), \overline{M_{ct}} + Z_\alpha \left(\frac{\sigma}{\sqrt{n}} \right) \right] \tag{5.29}$$

where Z_α is the value obtained from the standard normal table, based on the level of the confidence α. For example, if we desire the 95th percentile value, $Z_\alpha = 1.96$.

The mean corrective time, $\overline{M_{ct}}$, can be estimated by taking the sample mean of an observation. If a system involves multiple elements and each one has a different failure rate λ_i, and each element has a mean corrective time of Mct_i, then the composite overall system mean corrective time is given by

$$\bar{M}_{ct} = \frac{\sum (\lambda_i)(Mct_i)}{\sum \lambda_i} \qquad (5.30)$$

5.3.2.2 Preventive Maintenance Time (M_pt)

Compared to corrective maintenance, preventive time has relatively less variability, as it is usually scheduled at fixed time intervals and the activities involved are very specific and standard. In other words, preventive maintenance activities occur in certain frequencies, or f_{pt}, that is to say, the number of preventive maintenance actions per time period. So, the mean preventive time $\overline{M}_{pt}$ is a function of the frequency, as shown in Equation 5.31.

$$\overline{M}_{pt} = \frac{\sum (f_{pti})(Mpt_i)}{\sum f_{pti}} \qquad (5.31)$$

where Mpt_i is the individual preventive maintenance time for the ith element of the system. For example, a system consists of three elements. The frequency of the scheduled (or preventive) maintenance for Element 1 is once a month, taking 2 h; for Element 2, once in three months, taking 5 h; and for Element 3, once a year, taking 6 h. So we have $Mpt_1 = 2$ h, $Mpt_2 = 5$ h, and $Mpt_3 = 6$ h; $f_{pt1} = 1$ per month, $f_{pt2} = 1/3$ per month and $f_{pt1} = 1/12$ per month. So, the mean preventive time for this system can be obtained as

$$\overline{M}_{pt} = \frac{(1)(2) + \left(\frac{1}{3}\right)(5) + \left(\frac{1}{12}\right)(6)}{1 + \frac{1}{3} + \frac{1}{12}} \approx 2.94\,\text{h}$$

5.3.2.3 Mean Active Maintenance Time $\left(\overline{M}\right)$

With both scheduled (or corrective) maintenance time and unscheduled (preventive) time being defined, we can obtain the mean time required for a piece of maintenance, either scheduled or unscheduled, as both activities cause system unavailability; this is called the mean active maintenance time ($\overline{M}$). This covers only the technical aspects of the maintenance time, assuming that all required tools and parts are available when a maintenance action is required.

$$\bar{M} = \frac{\sum (\lambda_i)(M_{cti}) + \sum (f_{pti})(M_{pti})}{\sum (\lambda_i + f_{pti})} \tag{5.32}$$

5.3.2.4 Mean Down Time (MDT)

Sometimes, delays in fixing the system are caused by nontechnical factors. For example, when a system breaks down, we find out that the replacement part is not in stock; we need to order it and it takes some lead time to arrive. This type of delay is called logistic delay, and time taken due to logistic delay is logistic delay time (LDT). Besides LDT, there are also periods of administrative delay time (ADT). ADT is referred to as the time delay for administrative reasons, such as supervisor approval, board review, organizational structure flow, and so forth. Neither LDT nor ADT are technical factors for maintenance but they both produce similar effects on maintenance efficiency, preventing the system from being restored on time, and they inevitably happen, as logistics and administration are two key components of system operations. Considering LDT and ADT gives us a more realistic picture of system maintenance requirements; thus, a more realistic measure of the maintenance time is mean down time (MDT), given by Equation 5.33:

$$MDT = \bar{M} + LDT + ADT \tag{5.33}$$

With these time factors from different scopes defined, we can now look at the different measures of system availability.

5.3.3 System Availability

Simply put, system availability is the portion of time in which a system is in its operational or functional state under the specified environmental conditions. System availability is highly related to system reliability. As we learned in Section 5.2, reliability is one of the system's inherent characteristics; as reliability increases, it is obvious that the system will become more available. However, availability is not just reliability; as seen in the previous sections, it includes factors that are not covered by system reliability. Reliability only addresses system failures caused by breakdowns; failures occur randomly and maintenance activities are primarily corrective. Availability may also be increased by making strategic plans of preventive maintenance activities, by regularly testing and replacing parts before they fail to prolong the time between failures occurring. So, based on different perspectives, there are three different measures for availability (Blanchard and Fabrycky 2006):

1. Inherent availability (A_i). This is "the probability that a system, when used under stated conditions or design specified ideal environment, will operate satisfactorily at any point in time, as required." A_i excludes preventive maintenance, logistics, and administrative delays; it is only concerned with random failure-induced maintenance actions. It primarily reflects the quality of the system; the higher the reliability (larger MTBF), the shorter time required to fix failures (smaller Mct), the higher inherent availability. A_i can be expressed as in Equation 5.34:

$$A_i = \frac{\text{MTBF}}{\text{MTBF} + \overline{M_{ct}}} \qquad (5.34)$$

2. Achieved availability (A_a). This is the probability that a system will operate or function in a satisfactory manner in the ideal supporting environment. Compared to A_i, achieved availability considers both corrective and preventive maintenance activities; it is a more practical measure than A_i, since preventive maintenance activities will help to avoid failures from occurring. A_a can be expressed as in Equation 5.35:

$$A_a = \frac{\text{MTBM}}{\text{MTBM} + \overline{M}} \qquad (5.35)$$

MBTM is the mean time between maintenance; it is the measure of maintenance time considering both corrective and preventive maintenance activities. MBTM is given by Equation 5.36:

$$\text{MTBM} = \frac{1}{\lambda + f_{pt}} = \frac{1}{\dfrac{1}{\text{MTBM}_u} + \dfrac{1}{\text{MTBM}_s}} \qquad (5.36)$$

3. Operational availability (A_o). This is the probability that the system will operate in a satisfactory manner in the actual operational environment. The actual delays within the system consist of both technical aspects (corrective and preventive maintenance) and nontechnical factors (logistical and administrative delays). Operational availability gives the most realistic and practical measure for system availability, as it considers all the aspects of the system delay factors and reflects the efficiency of the

maintenance at the organizational level. A_o can be expressed as in Equation 5.37:

$$A_o = \frac{MTBM}{MTBM + MDT} \tag{5.37}$$

For most system designs, availability is a more realistic measure for the overall efficiency, considering system reliability and maintainability together. As mentioned earlier, reliability is a measure of dealing with random failures; it depends on the quality of the design, and once the design is finalized, reliability cannot be directly controlled. System maintainability, on the other hand, offers full control for the system designers to improve the degree of availability by providing well-planned maintenance strategies. These strategies are determined with the system reliability characteristics in mind, as there is a trade-off relationship between reliability and maintainability. To achieve a higher availability, a system with better reliability may require less frequent maintenance actions—both preventive and corrective—and vice versa. Understanding the trade-off relationships between reliability and maintainability will help us to create a more efficient system maintainability plan, both in terms of cost and time.

5.3.4 System Design for Maintainability

As one of the key design considerations and design-dependent measures, maintainability should be considered in the early planning of the design phase, starting from the conceptual design stage. One thing to keep in mind is that, as with other design parameters and TPMs, it is difficult to design a hard and fast maintainability plan, due to the dynamic nature of system design process. With changing requirements, design for maintainability should also be flexible and evolve continuously. Such design is an iterative process, evolving with the test and evaluation processes. It primarily includes five major activities:

1. Derive requirements from the systems requirements. Maintainability is a DDP; it is design derived, based on the mission requirements for the systems availability and reliability profile. Analytical modeling and experiments are necessary to aid in the translation process.

2. Define resources and constraints for system maintainability. These resources and constraints cover the whole spectrum of system availability, including support facilities, tools and equipment, personnel skills levels and training requirements, and management style/policies. These factors will all play a role in determining the maintainability policy for the system.

3. Define the maintenance level. Maintenance levels need to be specified for the system after the system functional structure is designed

and physical models are configured. These levels include the nature of maintenance tasks and detailed information for each of them, both corrective and preventive (i.e., who, where, when, and how these tasks are performed). Sometimes, analysis models such as task analysis can be applied to aid in deriving this information.

4. Maintenance function identification and allocation. As stated earlier, the functional structure of maintenance needs to be identified. This structure is based on the system functional architecture, with consideration of no-go functions, further expanding them into system maintenance functions. These maintenance functions are allocated in a top-down process to lower levels; trade-off studies and decision-making models are sometimes necessary to balance the requirements concerned with different aspects, such as system life cycle cost, reliability, usability, and supportability.

5. Establish the maintenance program management plan. As part of the system engineering management plan (SEMP), the factors described above are organized into one management document to guide system maintenance activities throughout the system life cycle. The major sections of the maintenance program plan should include

   a. Maintenance requirements and TPM objectives

   b. System maintenance functional structure and relationships with other system functions

   c. Maintenance organization and personnel structure and their requirements

   d. Logistics, supply, facility, and tools and equipment support for maintainability

   e. Job training and documentation requirements for maintenance personnel

   f. Test, evaluation, and demonstration methods, and models/data related to system maintainability

Many standards, such as MIL-STD-1472D (Human Engineering Design Criteria for Military Systems and Facilities), MIL-STD-470B (Maintainability Program Requirements for Systems and Equipment), MIL-STD-471A (Maintainability Verification/Demonstration/Evaluation), MIL-HDBK-472 (Maintainability Prediction), and DOD-HDBK-791 (Maintainability Design Techniques) (U.S. Department of Defense 1966, 1973, 1988, 1989a, 1989b), provide good sources for design guidelines for maintenance issues. Although primarily focused on military systems, most of the standards are very general and universally valid for most other types of systems.

Many standards and published guidelines provide some general recommendations for the selection of components and personnel for system design.

General guidelines for components selection:

1. Use standardized components and materials. These are easier to find, quicker to replace, and, most importantly, due to standardized production and the existence of large suppliers for these components, are most likely less expensive to procure.
2. Limit the need for special tools and equipment. This is based on similar reasons to (1), to minimize the time and cost involved for maintenance tasks.
3. Design for the consideration of ease of maintenance. This includes modular parts to minimize the impact to other components, separate control adjustability, the use of self-diagnosis and self-detection to rapidly identify failures, provisions to preclude errors in the installation phase, the provision of easy accessibility to avoid obstruction of the items to be serviced, ensuring access to spaces for test equipment and tools, making the most frequently serviced components the most accessible, avoidance of short-life components, especially for critical system items, and using proper labeling and identification for effective failure identification.

In terms of maintenance personnel and key human factors issues, the general design should consider the following factors:

1. Human performance consideration
2. Human machine interface/usability
3. Maintainer skills and training programs
4. Environmental conditions (noise, vibration, humidity, temperature, etc.)
5. Design simplicity
6. Safety

These may be found in any human factors text, and need to be tailored to each individual system design.

System maintainability is a DDP; the issues we have discussed here underscore the need for a thorough, methodical design process. The key here is planning and requirements driven, as any neglect of small issues in the early stages may cause a catastrophe in the later phases, as the more design details are involved, the larger the scope of the design becomes. Proper modeling and analysis are needed to balance the constraints and conflicts between different requirements; they are often highly related and support each other. System maintainability is highly correlated to system reliability, and its functions are supported by system supportability and logistical factors, which we will discuss next.

5.4 System Supportability

To prolong system operational time, the system needs to be reliable, and, more importantly, a well-defined maintenance policy to complement the reliability design is necessary to make sure the system and its components are well maintained while preventing failures from occurring in the most effective and efficient manner. Maintenance requires logistical support, including facilities, personnel, tools, equipment, and spare parts. These support functions are performed on a continuous basis, throughout the system life cycle and beyond. To ensure an effective maintenance performance, the necessary support infrastructure needs to be in place and operate efficiently. In the current social and economic environment, it is essential to consider the support functions within the context of the global supply chain, as this has become an integral component for the operations of all businesses and organizations. We cannot discuss system support functions without addressing supply-chain-related logistics. In the next section, we will first define system supportability based on supply chain management.

5.4.1 Definition of Supportability

System supportability refers to the ease and economy of design, installation, and implementation of the support infrastructure that enable effective and efficient system maintenance and support of the system throughout its life cycle. The goal of system supportability is to develop a cohesive support infrastructure that is highly responsive to demand from system maintenance activities, and that is efficient in terms of time and cost with minimum impact to other system functions. Supportability is an inherent system characteristic; it is a derived DDP, developed for a specific system configuration.

The basic elements for system supportability include:

1. *Maintenance support requirements.* One of the important elements of system supportability is to define a clear set of requirements based on the requirements for reliability, maintainability, and availability. Requirements should define the goals for support functions, infrastructure, activities, organizations, and hardware/software that are involved in system support activities at the system level, iteratively refined and decomposed to lower levels. These requirements are defined starting from the conceptual design stage.

2. *Support personnel.* This category addresses the personnel required to perform support functions, including system users, maintainers, and logistical/supply chain management personnel. Support functions cover a wide range of activities throughout the system life cycle, from initial system installation and system-sustaining support all the way to system retirement.

3. *Training and training support.* System support should address training for system operators and maintainers throughout the system life cycle, and support for implementing the training, including documentation, training materials, and the necessary training resources. This training covers the initial training program for new personnel, daily on-the-job training, and training program assessment (i.e., feedback, data analysis, and improvement).

4. *Inventory and supply support.* As an integral part of system logistics and supply chain management, inventory control and management plays an important role; the quantities and qualities of the inventory items have a significant impact on the overall supply chain operation. Common inventory items include spare parts, repair parts, consumables, special supplies, and supporting supplies.

5. *Support tools and equipment.* This category of support elements includes the tools and equipment that are necessary for carrying out support functions. Tools include those that are required for performing maintenance activities; testing, measuring, and diagnosing the system; and calibration equipment. These tools should be maintained and kept in an operational state whenever they are needed. They also include any designated computer hardware and software that will perform the support functions.

6. *Packaging, handling, storage, and transportation.* A large part of system support functions involves the flow of material from one location to another. From the procurement of the parts to their final destination, support items need to be properly packaged, handled, stored, and transported. This category of system support addresses supplier relationships, the global supply chain infrastructure, and resources for an effective and efficient material flow.

7. *Facilities.* This category includes all the facilities that are necessary to support all the scheduled and unscheduled maintenance activities at different levels: user site level, depot level, and central headquarters level. These facilities may include the buildings, laboratories, vehicles, and any other fixed or mobile units that will house the support functions.

8. *Data, documentation, and analysis.* Data may include all the technical information concerning the system configuration and procedures that involve system installation, operation and maintenance instruction, procurement, and modifications. This data may be in quantitative TPM format, or in graphic format, as in blueprints or schematic drawings, or in electronic format in information systems, such as CAD/CATIA data, databases (e.g., the enterprise resource planning [ERP] database in supply chain management). This data is collected on an ongoing basis throughout the system life cycle and should be well maintained for documentation and analysis to improve the support functions.

5.4.2 Supply Chain and System Supportability

Nowadays, one cannot talk about supportability without talking about the supply chain, as every business organization is part of at least one supply chain, and it is not uncommon to see that many organizations are part of multiple supply chains. A supply chain is a sequence of organizations, people, information, resources, and activities that are involved in producing and/or delivering a product or service. A product in the supply chain starts with the raw materials; through a sequence of processes in various facilities (e.g., warehouses, factories, distribution centers, retail stores, and offices), it evolves to its final form and is delivered to its users. A typical supply chain is illustrated in Figure 5.15.

Sometimes, a supply chain is also referred as a value chain; as the material progresses through the chain, value is added to the materials. Increasing the value-added activity efficiencies and minimizing the non-value-added activities are the key concepts of supply chain management (SCM). This is the process of planning, implementing, and controlling the operations for more efficient supply chain operations. Within the supply chain organizations, management and operations have different responsibilities for ensuring an effective and efficient supply chain. From the management perspective at higher levels, the strategic responsibilities include

1. Aligning the supply chain strategies with the system mission and overall organizational strategic planning. Any decisions on outsourcing and procurement should be based on the overall system mission requirements.

2. Supply chain structure configuration: determining the number and locations of the suppliers, warehouses, distribution centers, and support facilities.

3. Information technology determination: selecting the application of information technology to manage supply chain information, including data collection, information processing and sharing, inventory status checking, and event tracking. Technology and methods such as ERP may be applied to integrate internal and external operations across the entire organization, facilitating the flow of information between different support functions.

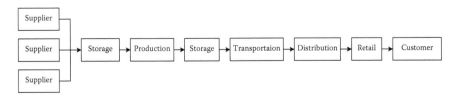

FIGURE 5.15
A typical supply chain structure.

4. Capacity planning of the supply chain: based on the system require-ments, determining the long-term and short-term capacity needs, and the level of flexibility required to accommodate the risks involved.

5. Risk reduction and management: identifying the potential sources of risks and making appropriate decisions so that these risks can be measured and controlled to a minimum level.

The commonly used measures for supply chain factors within support-ability are (Blanchard and Fabrycky 2006)

1. Capacity. To what extent can the support functions can be accomplished, and what is the capability and ability of handling large volumes and uncertain transactions? Together with reliability and flexibility, long-term capacity is planned to cope with the large degree of uncertainty involved in the supply chain, as many unexpected events may occur within the process. With a well-defined supply chain capacity, an organization can fulfill a need or service with high probability. The capacity of the supply chain measures its capability of dealing with uncertain demands and fluctuation of transactions; it refers to the combined effects of the demand pattern and cost functions that are involved in the supply chain.

2. Reliability and availability. The availability of the supply chain mea-sures the inherent capability of being readily available for smooth transactions whenever required. A more realistic measure for this capability is to use operational availability (A_o) to measure overall organizational efficiency.

3. Quality of the supply chain. The quality measure of the supply chain consists of the following subcategories:

   a. Response time: since all the activities and transactions take time within the supply chain, with many administrative/transporta-tion delays and the handling of lead time, it is desired that the response time should be as short as possible. The response time is a composite function of many other related factors, includ-ing purchasing and order fulfillment, inventory management, and supplier flexibility/cooperation. The response time is one of the primary measures of supply chain management (SCM) effectiveness.

   b. Total processing time/cycle time: The total time to fulfill a need, from the identification of the need for a product until it is deliv-ered in the required quantities and quality to the customer and service is completed, should be as short as possible. This has a much wider scope than the response time for a certain event. Modern technologies, such as e-business, electronic bar codes, and radio frequency identification (RFID) are often applied to fast tracking order statuses and sharing information quickly among partners.

c. Total cost. This includes all the cost factors of processing transactions from the sources of the materials to the end users, including product/service cost, transportation cost, inventory cost, and costs involved in risks, such as returns and defects. Total cost should be minimized, and is often traded off with other supply chain measures, such as response time and quality.

An effective supply chain design to support system supportability relies on the integration of all factors within the supply chain; these factors include trust among partners, effective communication, fast information flow, visibility and transparency of the supply chain, management capability of handling uncertain events, and appropriate performance measure metrics.

The key to SCM is to support system maintenance functions by having the highest quality of parts in the shortest period time with minimum cost involved. It is a trade-off between the various cost factors (i.e., holding cost, shortage cost, procurement cost, etc.) and the demand rate. Here, we use a simple economic order quantity (EOQ) model to illustrate how to determine the proper quantity to minimize the total cost of the transaction.

5.4.3 Inventory Management: EOQ Model

Inventory management answers two fundamental questions: (1) When should an order be placed? (2) How many units should each order have? The factors that regulate these two questions are the various costs that are involved in inventory management. To make clear assumptions for the EOQ model, the following costs are considered.

5.4.3.1 Ordering and Setup Cost

For most orders, there is a fixed cost factor involved, regardless of the size of the order; for example, the cost of labor to set up the order (cost of communication, paper, billing process, etc.) and, sometimes, a flat-rate transportation cost. The order and setup cost is assumed to be fixed for each order placed and denoted as K.

5.4.3.2 Unit Purchasing Cost

This is simply the variable cost (or price) for each unit of the product purchased. This cost sometimes includes the shipping cost if that cost depends on the quantity ordered. The unit purchasing cost is denoted as p.

5.4.3.3 Unit Holding Cost

This is the cost of holding one unit of inventory for one time period. The holding cost usually includes the storage cost, insurance cost, taxes on the inventory, and costs due to unexpected losses such as theft, spoilage, and damage. The holding cost is denoted as h/unit/time period.

Basic assumptions for the deterministic EOQ model include

1. Constant demand rate: to simplify the problem, it is assumed that demand is deterministic and occurs at a constant rate, denoted as D units/time period.

2. Shortage cost is ignored.

3. Although a lead time may be involved, as seen in Figure 5.16, the lead time does not really have an impact on the overall total cost for the simple EOQ model. For simplicity of calculation, we assume that the order arrives immediately. So, we have identical cycles (the large triangle) for the EOQ model. We just need to look at each *cycle* to derive the total cost per period.

Here is the summary of all the symbols used in the deterministic EOQ model:

TC: total cost per time period

Q: quantities ordered each time (this is the variable we are trying to determine)

D: demand rate (number of units consumed per time period)

T: number of periods in each ordering cycle

K: ordering and setup cost per order

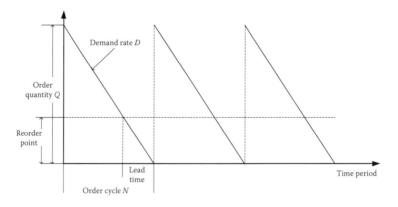

FIGURE 5.16
Illustration of EOQ model.

p: unit cost (price)

h: unit holding cost (per period)

We need to determine the value of Q that minimizes the total cost TC(Q). Obviously, we have the following formula for TC(Q):

$$TC(Q) = \text{Setup cost/time} + \text{purchasing cost/time} + \text{holding cost/time}$$

or

$$TC(Q) = SC + PC + HC$$

For every order placed (at the beginning of each cycle), there is a fixed setup cost K involved; that is to say, for every cycle time T, there is a cost of K incurred, so we have

$$SC = \frac{K}{T}$$

And because the constant demand rate $D = Q/T$ (the slope of the line), then

$$SC = \frac{K}{Q/D} = \frac{KD}{Q}$$

The purchasing cost per period is obtained by

$$PC = p\left(\frac{Q}{T}\right) = pD$$

and for the holding cost, we need to obtain the average inventory level $\bar{I}(T)$ for each cycle, given by

$$\bar{I}(T) = \frac{\int_0^T I(t)\,dt}{T}$$

For the deterministic EOQ model, the integral of $I(t)$ equals the area of the triangle in Figure 5.16, which is $(QT)/2$, so $\bar{I}(T) = Q/2$. Thus, the average holding cost per period is

$$HC = h\bar{I}(T) = h\frac{Q}{2}$$

So, combining the ordering costs, purchasing cost and holding cost, we can obtain

$$TC(Q) = \frac{KD}{Q} + pD + h\frac{Q}{2}$$

To find the value of Q that minimizes TC(Q), we let the first-order derivative TC'(Q)=0 (the necessary condition of the local minimum), or

$$TC'(Q) = -\frac{KD}{Q^2} + \frac{h}{2} = 0$$

So, solving this equation, we obtain the economy of quantity as

$$Q^* = \sqrt{\frac{2KD}{h}} \tag{5.38}$$

And, to verify this is truly the minimum value, we obtain the second-order derivative of TC(Q) as

$$TC''(Q) = 2\frac{KD}{Q^3} > 0$$

So, we know Q^* truly is a minimum value. (For more on the minimization and maximization of the functions, please refer to Chapter 7 for a more in-depth review.) The overall relationship of the EOQ model can be illustrated in Figure 5.17.

As an example of how EOQ is utilized, we assume that a support facility uses 600 parts per year. Each time an order for parts is made, there is a fixed cost of $7.50 incurred. Each part costs $2, and the holding costs $0.10/part/year. Assuming the demand rate is constant and the shortage cost is ignored, what is the EOQ? How many orders will be placed each year? And what is the length of each order cycle?

From the problem, we know that $K=7.5$, $p=2$, $h=0.1$ and $D=600$. So, using Equation 5.36, we can obtain the EOQ amount by

$$Q^* = \sqrt{\frac{2KD}{h}} = \sqrt{\frac{2(7.5)(600)}{0.1}} = 300$$

Hence, the EOQ amount is 300 and the number of cycles per year is given by

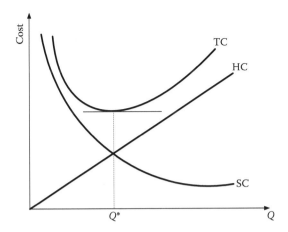

FIGURE 5.17
Relationship of various costs in EOQ model.

$$\frac{D}{Q^*} = \frac{600}{300} = 2\left(\frac{\text{orders}}{\text{year}}\right)$$

and the cycle length is $Q^*/D = 0.5$ year, or half a year. ■

The deterministic EOQ model with a constant demand rate is the simplest case of an EOQ model. More advanced models will consider lead time effects, when quantity discounts are allowed, with a continuous production rate involved (economic production quantity or EPQ model), with back orders allowed, and the shortage cost included. For a more in-depth review of these variations of the EOQ model, readers can refer to Winston (1994), chapter 16.

5.5 Human Factors and Usability Engineering

The parameters we have discussed so far primarily pertain to system hardware and software components; little has been said about the human elements. However, the human components are extremely essential to system success, and are sometimes a determining factor, as every system is eventually used by humans; even in so-called unmanned systems, such as autonomous aerial vehicles, there is still a need for the human-in-the-loop (HILP) ground control stations to monitor the status of the vehicle, and take over the controls if necessary. Understanding the needs of human elements, including system operators and maintainers, is of utmost importance to accomplish the system mission. The subject that addresses human elements in system design is called human factors engineering.

5.5.1 Definition of Human Factors

Human factors engineering, according to Chapanis (1996), is not the same as human factors. Human factors is a "body of information about human abilities, human limitations, and human characteristics that is relevant to design," while human factors engineering is the "application of human factors information to the design of the tools, machines, systems, tasks, jobs and environments for safe, comfortable and effective human use." Based on these definitions, human factors is an applied science discipline while human factors engineering refers to engineering.

As the foundation of human factors engineering, human factors study humans, utilize knowledge discovered from biology, physiology, psychology, and life sciences, and derive the information that is relevant to the interaction between human and engineered systems. This section is not intended to give a comprehensive review of the human factors body of knowledge, as human factors cover a wide range of topics that exceeds the scope of this book. There are many excellent references available for a more in-depth review, such as Wickens et al.'s (2003) text on human factors engineering.

Generally speaking, human factors studies the following subjects:

1. *Human visual sensory system*: Over 90% of information is perceived by the visual system. Human factors studies the human eyeball system and optic nerves, including the lens and the visual receptor system, and investigates the effects of visual stimulus (light) on human visual reception, such as the location of the stimulus, acuity, sensitivity, color, adaption, and differential wavelength sensitivities. From the study of the human visual system, we can obtain the advantages and disadvantages or limitations of human vision, such as contrast sensitivity, color sensation, and night vision. These understandings have significant impact on designing for human visual information processing; for example, designs to facilitate visual search and detection, and provide comfort and signal discrimination.

2. *Auditory, tactile, and vestibular system*: As the second most used sensory channel, the human auditory system responds to sound stimuli. Human factors studies the physical properties of the sound, understanding the nature of the sound, its measurements (amplitude and frequency), envelope information, and sound location. The human receptor of the sound stimulus is the ear and vestibular system. The experience of human hearing is investigated to understand the relationship between loudness and pitch, and masking effects

of different sound sources. This provides implications for design-ing sound systems for human users, including alarms, speech com-munication and recognition, managing and controlling noises, and providing hearing protection if the noise is above certain danger thresholds. Other senses, including the tactile and haptic senses, and the proprioception and kinesthesis channels, are also important for certain types of user interaction with systems.

3. *Cognition*: The basic mechanisms by which humans perceive, think, and remember things are the focus of the study of cognitive psy-chology. The core mechanism describing human cognition is a top-down, linearly ordered process: the information processing model, as shown in Figure 5.18.

   From the information processing model, human factors study the selective attention pertaining to different sensory channels, the three perceptual processes (bottom-up feature analysis, utilization, and top-down processing), and investigate the effects of association of stimuli and contextual information, short-term memory capacity limitation (e.g., Miller's 7 ± 2 chunks models) and long-term memory mechanisms (forgetting and retrieving information). This informa-tion is essential for the design of better systems to aid human situ-ational awareness and easy learning and recalling of knowledge and procedures.

4. *Human decision making*: Decision making is at the latter stages of the information processing model. After perceiving what is pres-ent and understanding what it means, humans need to decide on a course of action to respond to the information perceived. Human decision making and problem solving is a highest-level human

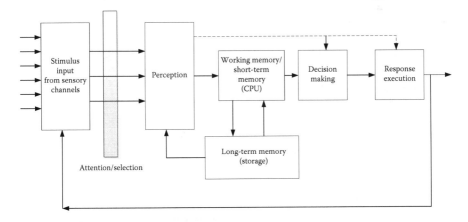

FIGURE 5.18
Information processing model. (Redrawn from Wickens, C.D. et al., *An Introduction to Human Factors Engineering*, Prentice Hall, Upper Saddle River, NJ, 2003.)

cognitive behavior; it involves information processing from multiple sensory channels and complex processes involving short-term and long-term memory. Human factors is concerned with different decision-making models that can capture human decision-making activities, such as normative decision-making models, descriptive decision-making models, and, sometimes, heuristics and biases to simplify decision-making problems, as our information processing models pose significant limits to decision-making capabilities concerning complex problems. Human factors focuses on task design, decision support system (DSS), visual aids, and displays to facilitate a more rational decision-making process.

5. *Motor skills and control*: Human control is the last stage of the information processing model to execute a response based on the results of decision making. The primary psychological measures of the effectiveness of human control is the accuracy of that control and the response time. One factor impacting the response time is the complexity involved in the decision; there are many empirical models, such as the Hick–Hyman model to address the relationship between the response time and the number of alternatives. Design features to facilitate user control and motor skills include the visibility of the stimulus, the physical feel of the control and feedback, size, and labeling. There are many empirical models to address different types of control mechanism. For example, a well-known model for positioning control devices is Fitts's model, or Fitts's law, which explains the relationships between movement time and difficulty of the movement (A = amplitude of the movement and W = size of the target), or $MT = a + b \log_2(2A/W)$. The device characteristics, such as direct/indirect control, control gain, control order, time delay, closed loop/open loop, and stability all play a role in determining the human control performance.

6. *Anthropometry*: Anthropometry is the study of human body dimensions, to match the physical dimensions of the system and workplaces to human users. Humans have a large degree of variability, in terms of age, gender, race, and occupation. The most useful tool for addressing such variability is the use of statistics. Readers should review the materials in Appendix I to familiarize themselves with statistical concepts, to understand the models involved in this book, particularly within the human factors field. A typical application of the statistics applied in anthropometry is to derive a percentile value for a particular body dimension. By using the normal (Gaussian) distribution, especially the normalized standard normal table (Appendix II), we can obtain a percentile value by $X = \mu + Z_\alpha \sigma$, where Z_α is the Z-value for a certain α-level, μ is the mean, and σ is the standard error for the body dimension. Most anthropometry data

is static, but when movement is involved, necessary adjustments are needed. When using anthropometry data, we need to (1) first determine the user population for the system and then (2) determine the relevant body dimensions; (3) determine the percentile value used for this design based on the design requirements, and calculate the value based on the data; and finally (4) make the necessary adjustments to accommodate for the dynamic work environment. Areas of application of anthropometry may involve upper and lower limits for special user groups, adjustability design, posture and normal line of sight, components arrangement, and workplace design (work surface height and inclination, etc.) with environmental conditions in mind.

7. *Biomechanics*: Awkward body posture is not the only factor to cause injury to the human body; sometimes, forces resulting from improper exertion may also cause severe body damage. Biomechanics studies the physics involved in the physical work of humans, trying to understand the impact of external forces on different body components. First, we need to understand the human musculoskeletal system, the muscles and bones of the human body, and the biomechanical models of humans performing physical tasks, such as the application of Newton's law on human joints and muscles. This helps designers to understand different ways that the body may be injured by external forces, such as lower back pain problems. There are many standards and regulations for man material handling jobs, such as those of the National Institute for Occupational Safety and Health (NOISH), which published a lifting guide for recommended weight limits (RWL) in 1991, based on three kinds of criteria: biomechanical, physiological, and psychophysical. RWL are a product of several multipliers, including the horizontal, vertical, distance, asymmetric, coupling, and frequency multipliers. Many guidelines were published based on this limit, such as manual material handling guidelines, seated work and chair design, and proper hand-tool design to prevent cumulative trauma disorders (CTDs).

8. *Work physiology*: For humans to perform a physical task, they need enough energy to support muscular activities. Physiology studies how human physiological systems work together to meet the energy requirements for human activities, both physical and mental. The central topic of work physiology is the study of muscle structure and metabolism involved in muscular activities, including aerobic and anaerobic metabolism, and the circulatory and respiratory systems (i.e., heart, blood vessels, and lungs), including blood function/flow and lung structure/capacity. A fundamental measure of work physiology is to calculate the energy cost for the work, measured in calories per minute. The workload of activities

can be measured in terms of oxygen consumption, heart rate, blood pressure and minute ventilation, and sometimes through subjective surveys and questionnaires. The main goal of the study of work physiology is to avoid body fatigue, in both the short and long term, as such fatigue, if not properly controlled, will lead to stress and long-term permanent body damage.

Since the 1940s, human factors engineering has demonstrated its value in systems design processes. Traditionally, human factors engineers and professionals are not involved in systems design until some types of system prototypes are developed, as the role of human factors professionals is primarily thought of as back-end verification and evaluation. When systems become large and complex, and many problems are found in the later stages that are extremely difficult and costly to be implemented, there is a need for concurrent integration of human factors engineering into the design process. Instead of only involving human factors professionals in the later stages of the design, the design team should include all the relevant stakeholders and players from the very beginning, incorporating all requirements at the conceptual stage to avoid the difficulties of unnecessary late changes. Almost every system needs some human factors support; as mentioned above, a system has be used, operated, and maintained by human users. Human factors professionals do not work in an isolated way, but rather team up with other designers and engineers, bridging the gap between the system technical specifications and the intuitive and straightforward user interaction with the system. For many decades, many successful stories have shown the value that human factors has offered to system design. Human factors is consulted at almost every stage of the design. There are many great texts presenting various techniques that human factors professionals apply in systems engineering; by no means do we intend to repeat these techniques in great detail here. Since this book is primarily about systems engineering, we just give readers a brief overview of the subject and present the three most commonly used human factors models in systems engineering application; that is to say, work system design, anthropometry and ergonomics design, and usability engineering in user-centered interaction.

5.5.2 Work System Design

The quality of the workplace determines the efficiency of the work performed in that place. Human are not machines, and it is true that most humans work to earn a living; however, other aspects beyond that basic purpose are also important for humans' work performance and sometimes their safety and well-being in the workplace; these factors include emotions, motivations, self-esteem, and the need for socialization. The quality of work life can be measured by several factors; these factors include the physical working conditions and work compensation (Stevenson 2009).

1. *Working conditions.* Physical working conditions play a significant role in humans' safety and thus have a great impact on their productivity and work performance. These factors include temperature, humidity, ventilation, illumination, noise, and vibration. There are well-defined condition limits for these factors, usually specified in government regulations and standards; for example, OSHA standards. Besides these factors, there are other regulating factors such as work time and work breaks, which also have a significant impact on humans' health. Appropriate shift length and break frequency will not only provide human operators with time to rest from fatigue and boredom, but also give a sense of freedom and control over one's work. Occupational safety measures are mandatory to ensure workers' safety and prevent accidents from happening, primarily through job design and workplace housekeeping, to make the job safer, to make humans aware of unsafe actions, and, most importantly, to eliminate potential hazards that may cause injuries.

2. *Work compensation.* Work compensation is an important factor in motivating human workers to be productive and efficient. Appropriate compensation attracts the best people to work for the employer and the best compensation keeps competent employees. Different organizations use different ways to compensate employees. The most commonly used approaches are time-based, output-based, and knowledge-based systems. Time-based systems compensate employees based on the hours they spend on the job. It is the most widely used compensation system overall; the method is straightforward and easy to manage. It is suitable for work that is difficult to put an incentive on, such as office work and administration. When incentives are desirable, output-based systems should be used to compensate work based on the amount of output produced; this ties the compensation directly to the efforts, making it possible to earn more if one is performing well. However, incentive compensation makes it difficult for management to predict the cost of the production, is difficult to implement together with the time-based system, and, sometimes, since it is more flexible than the time-based system, it also increases scheduling problems. The knowledge-based compensation system is used to reward employees who have higher skills. With more systems becoming complex and more advanced technology being involved, skillful employees who are capable of multiple tasks are more valuable. Knowledge-based compensation systems reward people with more skills, encouraging them to undergo training and education to acquire more skills to be more competitive.

Here we present a simple example of the work-time measurement models that can be used in work system design to develop a time standard (Stevenson 2009). The time standard involves three time components, the observed time

(OT), the normal time (NT), and the standard time (ST). Imagine that, for a particular task, we observe a sample of task times, x_i, $i = 1, 2, \ldots, N$, where N is the number of observations. The OT is simply the average of the time samples collected, as shown in Equation 5.39:

$$OT = \frac{\sum_{i=1}^{N} x_i}{N} \tag{5.39}$$

The normal time, or NT, is the observed time adjusted for human performance; it is computed by multiplying the OT by a performance rating factor (PR), as shown in Equation 5.40:

$$NT = OT \times PR \tag{5.40}$$

Performance factors capture the variability of the time spent by various people, due to various reasons, to account for slowness at a rate deviated from the norm due to this variability. A normal PR is 100%; a PR of 90% indicates a pace that is 90% of the norm.

Standard time (ST) accounts for more realistic situations beyond NT, such as personal delays (i.e., restroom breaks, phone calls, drinks and snacks, etc.) and other unavoidable delays (machine failure, supervisor checking, material handling lead time, etc.). An allowance factor (AF) is assigned to represent these delay factors, as shown in Equation 5.41:

$$ST = NT \times AF \tag{5.41}$$

The AF can be based either on the job time itself or the total time worked (i.e., a work day). If the AF is based on job time, then the allowance is computed as $AF = 1 + A$, where A is the allowance percentage based on the job time; if the AF is based on the whole work time, then $AF = 1/(1 - A)$, where A is the allowance percentage based on the work day. Let us use an example to illustrate the difference between these two cases:

Suppose $A = 0.10$; the allowance factor for 10% of the job time is $AF = 1 + A = 1 + 0.10 = 1.10$ or 110%; the allowance factor for 10% of the work time is $AF = 1/(1 - A) = 1/(1 - 0.10) = 1/0.90 = 1.11$.

5.5.3 Application of Anthropometric Data

One of the most important applications of human factors engineering in systems engineering is to design proper tools, equipment, and workplaces, to fit the physical dimensions of the design to the physical requirements and constraints of human users. A good source for the design comes from quantitative anthropometric data. Anthropometry, originating from the Greek words "anthropos" (meaning "man") and "metron" (meaning "measure"), is a scientific discipline that studies and measures human body dimensions. As mentioned earlier, humans have a large number of variabilities; these arise from different sources, such as age, gender, race, occupation, and generational variability. To account for these variabilities, statistics have to be applied to anthropometry data.

In anthropometry, the following terms are used for a unified and standard theme for measurements of human body dimensions (Wickens et al. 2003).

Height: A straight-line, point-to-point vertical measurement

Breadth: A straight-line, point-to-point horizontal measurement across the body

Depth: A straight-line, point-to-point horizontal measurement running fore–aft through the body

Distance: A straight-line, point-to-point measurement between body landmarks

Circumference: A closed measurement following a body contour (not circular)

Curvature: A point-to-point measurement following body contours (neither circular nor closed)

A large amount of anthropometry data has been compiled since the 1980s by a group of researchers and organizations. For example, a survey of personnel under the age of 40 was completed by the U.S. Army in 1989, looking at several body measurements of men and women; NASA compiled anthropometry data and guidelines for the design of space systems in the 1990s. Some private organizations also conducted their own studies and surveys for their own system design; this data is also available for purchase.

Table 5.8 summarizes some of the anthropometry data collected in the United States (Chengalur et al. 2004)

When dealing with anthropometry data, the Gaussian (normal) distribution is commonly used to derive the percentile values. To investigate body dimensions, they can be modeled as random variable (RV) x due to their variability between individual humans. If we assume that x is normally distributed with mean μ and variance σ^2, then we can convert the random variable x into a standard normal random variable (with mean of 0 and variance of 1) by using Equation 5.42 (Figure 5. 19):

$$Z = \frac{x - \mu}{\sigma}$$

(5.42)

TABLE 5.8

U.S. Anthropometry Data (in inches)

Measurement	Males 50th Percentile	Males ±1 S.D.	Females 50th Percentile	Females ±1 S.D.	Population Percentile 50/50 Male/Female 5th	50th	95th
Standing							
1. Forward functional reach							
a. Include body depth at shoulder	32.5	1.9	29.2	1.5	27.2	30.7	35.0
	(31.2)	(2.2)	(28.1)	(1.7)	(25.7)	(29.5)	(34.1)
b. Acromial process to functional pinch	26.9	1.7	24.6	1.3	22.6	25.6	29.3
c. Abdominal extension	(24.4)	(3.5)	(23.8)	(2.6)	(19.1)	(24.1)	(29.3)
2. Abdominal extension depth	9.1	0.8	8.2	0.8	7.1	8.7	10.2
3. Waist height	41.9	2.1	40.0	2.0	37.4	40.9	44.7
	(41.3)	(2.1)	(38.8)	(2.2)	(35.8)	(39.9)	(44.5)
4. Tibial height	17.9	1.1	16.5	0.9	15.3	17.2	19.4
5. Knuckle height	29.7	1.6	28.0	1.6	25.9	28.8	31.9
6. Elbow height	43.5	1.8	40.4	1.4	38.0	42.0	45.8
	(45.1)	(2.5)	(42.2)	(2.7)	(38.5)	(43.6)	(48.6)
7. Shoulder height	56.6	2.4	51.9	2.7	48.4	54.4	59.7
	(57.6)	(3.1)	(56.3)	(2.6)	(49.8)	(55.3)	(61.6)
8. Eye height	64.7	2.4	59.6	2.2	56.8	62.1	67.8
9. Stature	68.7	2.6	63.8	2.4	60.8	66.2	72.0
	(69.9)	(2.6)	(64.8)	(2.8)	(61.1)	(67.1)	(74.3)
10. Functional overhead reach	82.5	3.3	78.4	3.4	74.0	80.5	86.9

Seated							
11. Thigh clearance height	5.8	0.6	4.9	0.5	4.3	5.3	6.5
12. Elbow rest height	9.5	1.3	9.1	1.2	7.3	9.3	11.4
13. Midshoulder height	24.5	1.2	22.8	1.0	21.4	23.6	26.1
14. Eye height	31.0	1.4	29.0	1.2	27.4	29.9	32.8
15. Sitting height normal	34.1	1.5	32.2	1.6	32.0	34.6	37.4
16. Functional overhead reach	50.6	3.3	47.2	2.6	43.6	48.7	54.8
17. Knee height	21.3	1.1	20.1	1.0	18.7	20.7	22.7
18. Popliteal height	17.2	1.0	16.2	0.7	15.1	16.6	18.4
19. Leg length	41.4	1.9	39.6	1.7	37.3	40.5	43.9
20. Upper-leg length	23.4	1.1	22.6	1.0	21.1	23.0	24.9
21. Buttocks-to-popliteal length	19.2	1.0	18.9	1.2	17.2	19.1	20.9
22. Elbow-to-fist length	14.2	0.9	12.7	1.1	12.6	14.5	16.2
	(14.6)	(1.2)	(13.0)	(1.2)	(11.4)	(13.8)	(16.2)
23. Upper-arm length	14.5	0.7	13.4	0.4	12.9	13.8	15.5
	(14.6)	(1.0)	(13.3)	(0.8)	(12.1)	(13.8)	(16.0)
24. Shoulder breadth	17.9	0.8	15.4	0.8	14.3	16.7	18.8
25. Hip breadth	14.0	0.9	15.0	1.0	12.8	14.5	16.3
Foot							
26. Foot length	10.5	0.5	9.5	0.4	8.9	10.0	11.2
27. Foot breadth	3.9	0.2	3.5	0.2	3.2	3.7	4.2
Hand							
28. Hand thickness, metacarpal III	1.3	0.1	1.1	0.1	1.0	1.2	1.4
29. Hand length	7.5	0.4	7.2	0.4	6.7	7.4	8.0
30. Digit 2 length	3.0	0.3	2.7	0.3	2.3	2.8	3.3
31. Hand breadth	3.4	0.2	3.0	0.2	2.8	3.2	3.6

(Continued)

TABLE 5.8 (Continued)
U.S. Anthropometry Data

Measurement	Males 50th Percentile	±1 S.D.	Females 50th Percentile	±1 S.D.	Population Percentile 50/50 Male/Female 5th	50th	95th
32. Digit 1 length	5.0	0.4	4.4	0.4	3.8	4.7	5.6
33. Breadth of digit 1 interphalangeal joint	0.9	0.05	0.8	0.05	0.7	0.8	1.0
34. Breadth of digit 3 interphalangeal joint	0.7	0.05	0.6	0.04	0.6	0.7	0.8
35. Grip breadth, inside diameter	1.9	0.2	1.7	0.1	1.5	1.8	2.2
36. Hand spread, digit 1 to digit 2, first phalangeal joint	4.9	0.9	3.9	0.7	3.0	4.3	6.1
37. Hand spread, digit 1 to digit 2, second phalangeal joint	4.1	0.7	3.2	0.7	2.3	3.6	5.0
Head							
38. Head breadth	6.0	0.2	5.7	0.2	5.4	5.9	6.3
39. Interpupillary breadth	2.4	0.2	2.3	0.2	2.1	2.4	2.6
40. Biocular breadth	3.6	0.2	3.6	0.2	3.3	3.6	3.9
Other Measurement							
41. Flexion-extension, range of motion of wrist (°)	134	19	141	15	108	138	166
42. Ulnar-radical range of motion of wrist (°)	60	13	67	14	41	63	87
43. Weight (kg)	183.4	33.2	146.3	30.7	105.3	164.1	226.8

Source: Chengalur, S.N., et al., *Kodak's Ergonomics Design for People at Work.* Hoboken, NJ: Wiley, 2004. With permission.
Note: The data is taken primarily from the military studies, where several thousands of people were studied. Numbers in parentheses are from industrial studies where 50–100 women and 100–150 men were studied. All measurements are in inches unless otherwise stated.

Then, by using the standard normal table from Appendix II, we can obtain the value of Z for any percentile value of Z; we can easily obtain the value of *x* by applying Equation 5.43:

$$x = \mu + Z\sigma \qquad (5.43)$$

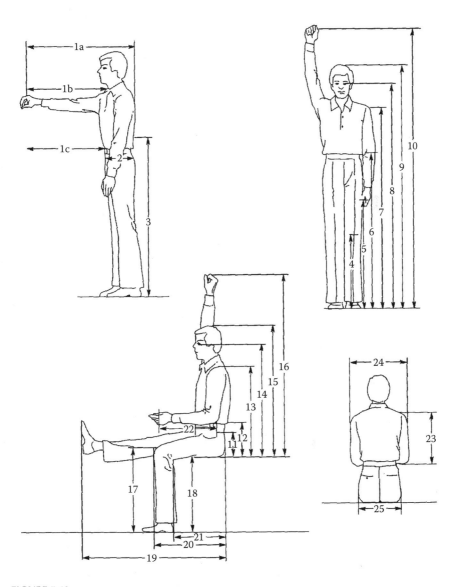

FIGURE 5.19
Standard postures for measuring anthropometric data. (From Chengalur, S.N., et al., *Kodak's Ergonomics Design for People at Work*. Hoboken, NJ: Wiley, 2004. With permission.)

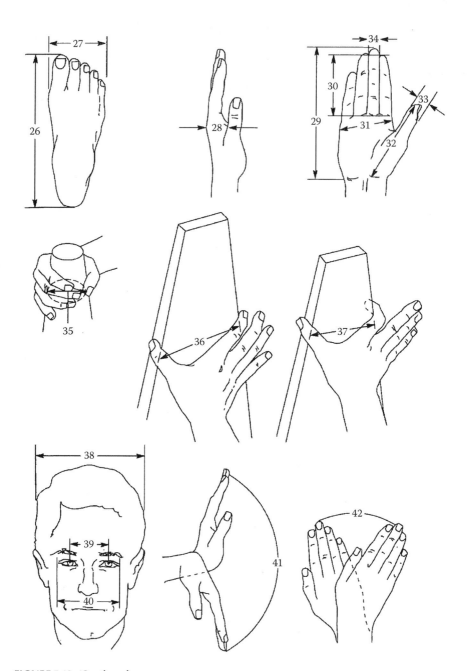

FIGURE 5.19 (Continued)
Standard postures for measuring anthropometric data. (From Chengalur, S.N., et al., *Kodak's Ergonomics Design for People at Work*. Hoboken, NJ: Wiley, 2004. With permission.)

Let us give a simple example here. Assume we are interested to know the 75th percentile value for a body dimension x. We have measured a sample of x and estimated that x has a mean value of 25.5 in. and a standard deviation of 3.6 in. From the standard normal table, we know the Z-value for a 75th percentile value is approximately 0.674 (for readers who are not familiar with the standard normal table, please refer to Appendices I and II of this book for a brief review of the normal distribution and a quick reference standard normal table). Using Equation 5.43, we can derive the percentile value for x as $x = \mu + Z\sigma = 25.5 + (0.674)(3.6) = 27.93$ in.

By using these steps we can derive any percentile value of our choice. However, a general rule of thumb of applying percentile values in the design is to use the 5–95 range. For example, NASA's 1995 design of space system guidelines chose a range from a 5th-percentile Japanese woman to a 95th-percentile American man as the data for inclusion. Whether to use the 5th or 95th percentile depends on the nature of the design, or in other words, the lower or upper limit for the design.

Lower limit refers to the physical size of the system, not the human user per se. The lower limit implies that the system cannot be smaller, otherwise it would be unusable by a larger person. An example of the lower limit would be the height of a doorway, or the sitting weight capacity strength of a chair. In the case of the lower limit, the high percentile value (i.e., the 95th percentile) is used to determine this limit.

Upper limit, on the other hand, refers to the maximum value that system cannot exceed, otherwise a small user would have difficulties using the system. An example of the upper limit is the weight of a tool to be carried by a human, which has to be below a certain level (the upper limit) so that the smallest user is able to carry it. Usually a low percentile value (i.e., the 5th percentile) is used to set the upper limit.

Knowing the difference between the lower limit and upper limit enables the designers to specify the appropriate levels of design features to meet user needs. A typical design that includes anthropometry data usually involves the following steps (Wickens et al. 2003):

1. Determine the intended user population. Based on the design requirements, find out who will be using the system and their workplace, identify the variability factors involved for the target user groups, including the gender, age, race, and occupational characteristics.

2. Determine the relative body dimensions that are involved in the design. Find out the main body dimensions for the intended use of the systems. For example, a chair design primarily is concerned with height while seated, hip breadth, and leg length, while a control panel primarily involves arm reach and finger size.

3. Determine the appropriate percentile value for the selected body dimensions to account for the variability involved. A general rule of

thumb for consideration is to use the "design for extreme" approach first, that is, considering the data for the individuals at the extremes such as the lower or upper limits mentioned above; if considering these individuals will not meet the requirements, then the next step is to "design for adjustable range," in which the design can be adjusted within a range so that different users will be accommodated; of course, this would require more sophistication in the design. If there are some constraints to or a lack of feasibility in making the design adjustable, a third approach would use the "design for the average," which uses the 50th percentile value to accommodate a majority of the users involved. For example, many big department store designs use this percentile to plan the checkout counter to accommodate a large group of customers, since the variability of the customers is very large and, also, it is impossible to make the counter adjustable. Once the target value has been determined, the percentile value will be calculated using the appropriate data sources and equations above.

4. Make the necessary modifications for the calculated percentile value. Much of the anthropometry data is measured in a very ideal situation, with standard posture and minimum clothing involved, which is not realistic for most systems designs. A necessary adjustment for the percentile value is necessary to account for clothing, protection devices, and dynamic body movement for user tasks.

5. Evaluate the design by testing it using a prototype or simulation. Before finalizing the design using the calculated anthropometry data, the design needs to be verified by using the prototype or simulation to make sure it truly meets the requirements. Many design software packages, such as CATIA, allow designers to visualize and simulate the physical fit of the users in the system configuration and workplace, inspecting the interaction from different perspectives, simulating various scenarios to identify any potential difficulties and problems with the design. Any problems found can then be addressed before the design is finalized.

5.5.4 Usability Engineering

Another area in which human factors engineering is applied extensively in the context of systems engineering is design for usability. Usability plays an important role in our daily lives. It makes our interactions with any interface easier to understand and to operate. For a human–system interface, it is similar to "user-friendliness," but it is not a single, one-dimensional property of a user interface. According to Nielsen (1994), usability is a quality issue of a system interface; it carries two separate but related meanings.

First, usability is the assessment how friendly an interface is, and second, it also refers to the methods and models to improve the ease of use during the design process.

Regarding the measurement of usability, it is typically measured by having a number of representative users interact with the system to perform a specified and predetermined task. Nielsen (1993) proposed a detailed methodology for interface testing; his methodology includes testing goals and plan development, obtaining testing users, choosing experimenters, the ethical aspects of the study of human subjects, developing testing tasks, performing tests, and measurement. For the design to improve usability, there is no set template to follow as every system has its unique design features; one needs to tailor usability principles to accommodate different types of systems. Nielsen (1993) summarized five main elements or principles for usability design:

- *Learnability*: Learnability refers to how easy the system is for users to learn the functionality and how easy it is to accomplish basic tasks for the first time. Learnability implies a good match between users' expectations and experience to facilitate learning to use the system.

- *Efficiency*: Efficiency means that once users have learned the system, how efficiently can they perform tasks in terms of time and errors?

- *Memorability*: Memorability aspects of usability refers to the level of retention of learned tasks performance, especially when users come back to use the system after a period of not using it; how easily can they recall what they learned about their interaction with the system to reestablish proficiency?

- *Errors*: A system with good usability should enable the users to minimize the possibilities of making errors, in terms of how many errors users make, how severe those errors are, and how easily users can recover from those errors.

- *Satisfaction*: Satisfaction refers to the look and feel of the interface to the user; in other words, how pleasant and attractive is it to use the design? Satisfaction is usually measured subjectively, using methods such as surveys, questionnaires, interviews, and focus groups.

As part of the system requirements, requirements concerning system usability are collected in the early stages of the design, as, following the top-down process, these requirements are gradually translated into design specifications, similar to other types of systems requirements, as mentioned in Chapter 3. The functionalities of user interaction, such as menus and controls, are determined through an iterative process, from concepts to components, by using the various levels of prototyping and evaluation, just as for the rest of the system requirements.

Usability engineering contributes to system engineering mainly in the area of the testing phases. Design requirements concerning usability vary a great deal among different groups of users; this makes an empirical approach more appropriate when specifying usability issues. Ever since the concept of "usability" was introduced into interface design, many researchers have completed particular research and experimentation on usability evaluation. Typically, user testing is driven by scenario-based tasks that users need to perform. For example, the following list illustrates a sample usability testing scenario list for an online voting system interface design:

- Scenario 1: Vote registration
- Scenario 2: Reading and understanding a specific item on a ballot
- Scenario 3: Use of tools and navigation through different levels of the system
- Scenario 4: Review and modify voting choices
- Scenario 5: Exit the system and confirm the vote

In a typical user test, the experiment includes three phases in a session: the planning phase, the testing phase, and the reporting phase. During the planning phase the testing procedure will be explained to the subjects using a set of training scenarios. The postevaluation questionnaires to which they are supposed to respond are also explained, if there are any after the test. The postevaluation questionnaire deals with users' general impression of the system, usage of terminology, information content, and information structure. During the testing, problems and feedback from the users will be recorded. In the posttesting session, users are given the opportunity to provide feedback and opinions regarding the problems that they have faced during the test. This session also serves as an opportunity for the observer to clarify any doubts that they might have had during the test with regard to the observations made. In the reporting stage, inherent problems and inconsistencies, according to postevaluation questionnaires, interviews, and expert discussions, are identified. Problems are usually identified using standard statistical methods, such as descriptive statistics (mean or standard deviation, for example) or analysis of variance (ANOVA) if multiple designs are being compared.

In addition to usability testing, other testing methods that have been used in usability evaluation are heuristic evaluation, cognitive walk-through, and competitive evaluation. The idea of "heuristics" came from the fact that interface design principles are fairly broad and would apply to many types of user interfaces (Nielsen 1993). Implementing usability testing could be costly and time consuming; heuristic evaluation and testing could be a faster and more economical way of obtaining initial results. Heuristic evaluation is done by having experts evaluate an interface and form an opinion about what is good and bad about it. After years of experience, heuristic evaluation has become a systematic inspection of a user interface design. Evaluation is usually performed by a small set of expert evaluators using recognized usability principles. Nielsen (1993) gave ten general heuristic evaluation criteria:

1. *Visibility of system status*: The system interface should always keep users informed about what is going on, providing a visible status for the system functions through an appropriate feedback format within a reasonable time frame.

2. *Match between system and the real world*: The system should speak the users' language, with terms, words, phrases, and concepts familiar to the user, matching users' mental models and expectations with system functions. It should follow real-world conventions, presenting information in a natural and logical order and format.

3. *User control and freedom*: When encountering problems and difficulties, there should be little difficulty for users to leave the current, undesired state easily with a clearly marked "exit", without having to go through an extended number of tedious steps; the system should support the undoing and redoing of the most recent user actions and functions.

4. *Consistency and standards*: The style of the interaction throughout the system interface levels should be consistent to minimize unnecessary learning and confusion between different levels of interactions.

5. *Error prevention*: A good usable interface is always designed in a way that prevents a problem from occurring in the first place; this should include eliminating error-prone conditions or checking for those conditions and present users with options before they make an error. This feature is important for all levels of user groups, especially when the system complexity increases.

6. *Recognition rather than recall*: A good design should minimize the user's memory load by making objects and actions visible and easily understandable. The user should be presented with ample information for them to choose from, not have to remember lots of information and retrieve them from memory. A good analogy for this principle is multiple-choice questions versus essay questions in a test; a multiple-choice question provides more recognition while an essay question primarily relies on one's recall ability.

7. *Flexibility and efficiency of use*: A good interface should provide flexibility for different users to suit their needs for efficiency. For example, while a novice user might need lots of detailed tutorials for them to learn to use the system efficiently, an experienced user might want to skip those tutorials and get straight to the tasks, even

using some accelerators/shortcuts to speed up the tasks. A good system interface should have this flexibility to cater for both inexperienced and experienced users and allow users to tailor their actions.

8. *Aesthetic and minimalist design*: A good usability interface should be attractive. This requires that every piece of information in the design should be relevant, minimizing clutter, maximizing visibility, and complying with the psychological principles of user visual comfort.

9. *Help users recognize, diagnose, and recover from errors*: Error messages are inevitable in every system. When an error occurs, the system should provide the error message in plain language (no codes), indicate precisely the nature of the problem, and constructively suggest a solution or directions for solving the problem.

10. *Help and documentation*: A good system should provide help and documentation. Such information should be easy to search and written from users' perspectives. This documentation should be available at all times and not be too large in size.

There are many studies have shown that both user testing and heuristic analysis are needed in systems design. These two methods have different strengths; the best evaluation of a user interface comes from applying multiple evaluations.

It is believed that the difference in nature of these two techniques would make them appropriate for different testing purposes. Most of the time, heuristic analysis finds more problems than user testing because it provides more freedom to explore the interface, while user testing needs a well-developed test bed and a more controlled environment (Rogers et al. 2011). Typically, in the earlier design stages, the interface is often not fully developed. Heuristic analysis would be able to project potential usability problems, a quality that user testing lacks. Feedback from heuristic analysis can be used to create a design standard for the rest of the interface. After design improvements are made following the initial heuristic analysis, thorough user testing is required, as user testing and heuristic analysis find very different types of problems. User testing is able to assess the usability issues most pertinent to users much more directly, without bothering with basic problems. Feedback from user testing can be used to fine-tune the interface, which is typically done in the later stages of the design process. User testing may also detect potential new usability problems that are the direct result of the design improvement. In other words, both user testing and heuristic analysis are needed for usability in system design. To reap the optimal benefits, it is believed that both user testing and heuristic analysis should be used in different stages of the user interface design process. We believe that heuristic analysis should be implemented in the early stages of the development process, while user testing should be conducted at a later stage.

5.6 Summary

Systems design is driven by requirements; however, requirements cannot design systems. They need to be translated into quantitative design parameters, so that the requirements can be materialized and the system can be brought into being. Besides the unique functionalities that each system will have for its own purposes, there are some common parameters that most systems will probably possess, including system reliability, maintainability, supportability, and usability. In this chapter, we reviewed these common system design parameters, giving a detailed definition for each of them, and described some of the models for them.

Regarding system reliability, we defined its elements, and then the mathematical modeling of reliability was introduced. The basic component structures for system reliability are series and parallel relationships. Any complex system configuration can be decomposed by using a combination of these two basic structures. Some examples were given to illustrate the procedure for working out reliability (the probability of success for a given period of time), with exponential failures used as the failure distribution functions. Practical methods and tools for design for reliability in the system life cycle were also discussed; these models included FMEA and FTA. These two methods address the occurrence of system failures from two different perspectives; FMEA looks at the bottom-up level of component failures, and induces the possible effects caused by one or more such failures; FTA, on the other hand, looks at system failures from another perspective, specifying the possible causes of a particular failure occurring. These two methods are complimentary to each other; to obtain a complete picture of system failures and their effects, FMEA and FTA need to be combined together.

System maintainability is a DDP to ensure that the system is in operational status for the maximum period. Maintainability refers to the ease and economy of system maintainability activities, and is usually measured in terms of time and cost. We defined a number of related terms for system maintainability, including MTBF (unscheduled maintenance), mean corrective maintenance time, mean time between scheduled maintenance, mean preventive maintenance time, mean active maintenance time, and MDT. Based on these terms, we defined system availability at different levels, such as inherent availability, operational availability, and achieved availability. Practical issues of design for maintainability were discussed for efficient and effective maintenance planning for maintainability in the system life cycle.

System supportability is a DDP that supports system maintainability activities. Supply-chain-based system supportability was defined, and its main factors and design principles were discussed in the text. We then introduced the simplest linear EOQ model, derived the optimal order quantity formula, and illustrated the application of the model via some numerical examples.

Human factors engineering has also had significant impact in large, complex systems design. Human factors is the study of human information, including human characteristics and limitations, while human factors engineering is the application of human factors in systems design, to make the interaction between humans and systems safer, more efficient and enjoyable. Human factors engineering covers a wide range of topics; we briefly reviewed each topic so that readers would become familiar with the concepts, and then focused on three main areas in which human factors engineering is mostly applied within system engineering: work system design, anthropometry, and usability engineering. In work system design, we introduced the work-time measurement model; in anthropometry, the calculation of percentile values based on the normal distribution was discussed and some examples were given to illustrate how to apply the model in the design. Some general design principles for the proper use of anthropometry data were presented; for usability design, although the nature and style of the interaction varies a great deal between different systems, there are some fundamental concepts for measuring usability that can be applied to all types of systems; they can be summarized by the five elements of usability given by Nielson (1994). The application of usability engineering in systems design focuses mostly on two areas; at the front end, the translation of usability requirements to interface design specification, and at the back end of the design, usability engineering is primarily concerned with measurement of user performance to verify the fulfillment of requirements. Two different evaluation methods, user testing and heuristic evaluation, were compared in terms of effectiveness. It was found that these two methods are complimentary in nature and should be combined together to give more comprehensive evaluation results. Heuristics evaluation uses experts and gives more freedom to them to evaluate the interface according to basic usability heuristics, while user testing utilizes real users, usually giving them specific task scenarios to perform, and the results are narrower in scope but more in-depth. By combining the two methods in the design life cycle and using different methods at different design stages, one can obtain a more complete picture of interaction usability issues in a very efficient way.

PROBLEMS

1. What does DDP mean? Referring to one of your favorite systems, give three examples of DDPs for that system.

2. What does TPM mean? Why are TPMs important in systems design? What is the relationship between DDPs and TPMs? Referring to one of your favorite systems, give three examples of TPMs for that system.

3. Define reliability and list the major elements of systems reliability.

4. Define failure rate. Estimate the failure rate for the following set of system components:

   Suppose we test ten components in a reliability test for 200 h simultaneously. During the test, six failures occurred, as shown in the following table.

Component No.	Failure Occurrence Time (hours)
1	10
2	12
3	56
4	89
5	110
6	155
7, 8, 9, 10	All survived 200 h

5. Referring to problem 4, what is the probability that a component will work for at least 250 h without any failure if the failure is distributed exponentially?

6. A system has a MTBF of 2500 h. It has survived 1000 h without any failure. What is the probability that this system will survive another 1000 h without any failure if the failure is distributed exponentially?

7. A system consists of four components connected in a series structure. The individual component reliabilities are

   Component A = 0.95

   Component B = 0.90

   Component C = 0.99

   Component D = 0.98

   Determine the overall system reliability.

8. A system ABCD consists of four components, connected in a series structure. The MTBFs for these components are

   Component A: MTBF = 1000 h

   Component B: MTBF = 2000 h

   Component C: MTBF = 2000 h

Component D: MTBF = 2500 h

Assuming all the failures are independently distributed exponentially, show that the failure for system ABCD overall is also distributed exponentially. Estimate the MTBF for system ABCD.

9. A system consists of four components connected in a parallel structure. The individual components reliabilities are

Component A = 0.95

Component B = 0.90

Component C = 0.99

Component D = 0.98

Determine the overall system reliability.

10. Compare the results of Problems 7 and 9. What conclusion can you draw?

11. Estimate the reliability for the following system:

Component	Reliability
A	0.95
B	0.96
C	0.98
D	0.92
E	0.88
F	0.90

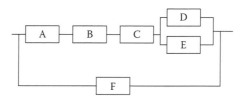

12. A system is running with three other identical parts in standby mode. Each part has an MTBF of 4000 h, distributed exponentially. Determine the system reliability for a period of 1000 h.

13. Estimate the MTBF for Problem 12.

14. Define FMEA and FTA. What are the differences between these two methods?

15. The following diagram shows the faulty tree of a system design. Given that $P_D = 0.10$, $P_E = 0.30$, $P_F = 0.40$, $P_G = 0.10$, what is the value of P_A?

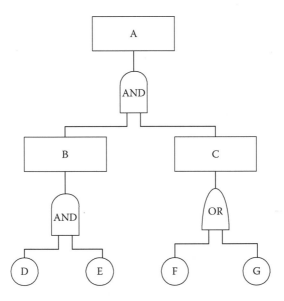

16. Define system maintainability. Explain the relationship between reliability and maintainability.

17. What is the difference between MTBF and MTBM?

18. The following table illustrates a sample of maintenance task time observations.

74.66	75.40	47.18	59.63	59.87
57.36	55.94	66.00	43.45	51.22
62.26	69.15	66.61	53.03	59.51
53.97	46.62	63.77	70.50	68.93
56.73	40.51	57.30	55.07	54.89
35.85	69.60	52.93	61.05	64.62
43.85	53.27	50.87	59.96	42.15
51.67	61.65	42.87	57.15	52.57
60.49	53.82	43.75	64.07	67.35
54.10	68.07	55.83	34.05	57.83

a. Calculate the $\overline{M_{ct}}$ and standard deviation for this sample.

b. Find out the percentage of corrective time between 40 and 50 min by using the standard normal distribution table in Appendix II.

19. Giving the following information, estimate $\overline{M}$, MTBM, $\overline{M_{ct}}$, A_i, A_o, and A_a.

    Total operation time $= 20,000$ h

    Number of scheduled and unscheduled maintenance occurrences $= 50$

    $\lambda = 0.0025$

    $\overline{M_{pt}} = 5$ h

20. What the main activities involved for the design of maintainability?

21. To facilitate an appropriate design for maintainability, what are the general guidelines and principles for selecting parts and personnel?

22. Define the supply chain and the main factors that are involved in supply chain management. Why is the supply chain important in supportability design?

23. Determine the economic order quantity (EOQ) for spare parts, when the cost per unit is $5, the cost of setting up the order, including all the fixed costs, is $100, and the cost to hold an item is $1 per unit per year. It is estimated that the annual demand of the part is $1000.

24. Define human factors and briefly explain the role of human factors engineering in the systems design process.

25. What are the major factors for work system design?

26. Explain how anthropometry data is utilized in the design. What are the lower limit and upper limit in the context of anthropometry data? Give an example for each of the limits.

27. A certain design dimension has a mean value of 20 in. and standard deviation of 5 in.; find the 75th percentile value for this dimension using the standard normal distribution table in Appendix II.

28. Following the above question, what is the percentage for a value of 22.5 in.?

29. What is usability engineering? List the major elements of system usability.

30. Compare usability testing and heuristic evaluation. Under what circumstances should each method be used to evaluate the usability of the system?

6

Decision-Making Models in Systems Engineering

Humans make decisions every day. These decisions range from ones with longer-term impacts, such as whether to take a job after college or go to graduate school to everyday decisions such as what to eat for lunch. Similarly, in systems design, decisions are being made constantly on a daily basis. This is because

1. The resources involved in systems design are always limited. One of the most important objectives and challenges for systems engineers is to utilize limited resources in the most effective and efficient way; this goal is achieved only through a well-thought-out decision-making process.

2. The systems engineering design process is itself an iterative decision-making process. Systems engineers make decisions in a continuous basis; from the development of requirements, selection of design concepts, choice of appropriate components, and optimization of system parameters to the final system installation and operation, design decisions are being made at different levels. Due to the complex nature of the designs, these design decisions usually involve multiple criteria and, sometimes, high-risk stakes and uncertainties, which will have significant impact on the quality, cost, and safety of the designed system.

With a high degree of uncertainty, multiple criteria, and, sometimes, conflicting decision goals involved, systems engineers strive to find the best solution for these types of complex decision-making problems within a relatively short period of time. To simplify these problems, proper models are necessary to articulate and formulate the decisions. In this chapter, we will review some of the commonly used models for solving decision-making problems in the systems engineering context. These models are largely quantitative in nature, providing a rigorous foundation for systems engineers to address the decision-making problems at different levels. On completion of this chapter, readers should have a good understanding of the decision-making problems within the systems context in general, know how to identify decision objectives, criteria, and factors, formulate problems into solvable mathematical models, and obtain feasible solutions. More specifically, the following concepts and topics will be covered in this chapter:

- A rigorous definition of decision-making problems and the characteristics/elements of a decision-making problem
- Review of deterministic decision-making models
- Illustration of the difference between decision risks and uncertainty; review of decision-making models under risks and uncertainty
- Review of commonly used decision-making models including utility theory and decision tree models
- Review of the decision-making problems that involve multiple criteria; more specifically, ranking of the decision criteria by analytical hierarchical process (AHP) methods

At the end of the chapter, a brief summary is given to help readers to organize the concepts discussed in the text, providing a general framework for considering decision problems in complex systems design.

6.1 Elements of Decision Making

Decision-making is a high-level human cognitive process to select a preferred action among several alternatives; it is a process in which "a person must select one option from a number of alternatives" with "some amount of information available" and under the influence of context uncertainty (Wickens et al. 2003). Some of the unique elements of systems engineering decision-making tasks include the following:

1. It has a goal to achieve. All decision-making tasks are carried out to achieve a specific goal. The goal is usually associated with a certain degree of optimization, that is, to maximize overall revenue or to minimize the cost of the completion of a project.
2. It is a selection process. There is usually more than one approach to achieve the goal; these different approaches are called decision alternatives. Such alternatives are usually complementary to each other, that is to say, one alternative is better in one aspect but worse in another. If an alternative, A, is better in every single aspect than a second alternative, B, it is said that Alternative A dominates Alternative B; a dominating decision makes the problem easier to solve. In most cases, it is not always easy to obtain dominating alternatives. Given a set of candidate alternatives that are complementary, a decision maker's job is to select the best alternative that

serves the decision goals, usually by using some types of decision aids and models.

3. Realistically, any decision-making tasks are subject to some time constraints; decision makers have a limited time window to make that decision. However, for most systems design decision-making problems, time is usually not an issue; system engineers should have sufficient time to put together all the information, develop decision models, and solve the problem. Any decisions regarding systems designs should not be rushed. So, for decision making in systems engineering, we usually assume that the decision time is not a factor, and there is no time limit to reach a decision. Decision making under time pressure is a slightly different problem and has been studied extensively in psychology and human factors; for example, Rasmussen's skill, rule, knowledge (SRK) model of behavior and Klein's recognition-primed decision (RPD) model. These models differ from the decision-making models in systems engineering in the sense that they are mainly qualitative and empirical in nature, involving many heuristics.

4. Decision making usually involves uncertainty. Uncertainty can be characterized as the unknown probability or likelihood of a possible outcome of a decision or the lack of complete information on which to base a decision. Uncertainty plays a significant part in modeling the decision-making process in systems engineering and is usually modeled by the use of probability and statistics theories. Readers should be familiar with the fundamental concepts of probability and statistics to understand the concepts and models of this chapter. For a preliminary study and review, please refer to Appendix I of this book.

Regardless of the nature of the problem, the decision-making process usually consists of a sequence of six activities, described as follows:

1. Identify the problem and outline the goals to be achieved. These goals serve as criteria against which to compare different alternatives for solving the problem.

2. Gather the data that is necessary to solve the problem.

3. Based on the data collected, generate alternatives to achieve the goals. An alternative is a candidate solution for the problem.

4. Articulate the problem by using the appropriate model, integrating all the data and decision criteria together to evaluate the alternatives.

5. Evaluate the alternatives using the model. Based on the decision criteria, the best alternative is selected.

6. Consider practical implications and limitations and execute the decision actions.

In the following sections, we will introduce the most commonly used decision-making models in systems engineering, starting from the decision-making models under risks and uncertainty, followed by utility theory, the decision tree, and the decision-making models for multiple criteria, including the AHP model. Most of these models originate from the normative models based on probability and statistics, which can be found in most operations research texts.

6.2 Decision-Making Models under Risks and Uncertainty

To explicitly explain the procedure for solving decision-making problems, let us denote the following:

m: numbers of alternatives

n: numbers of possible future events that could happen

A_i: the ith alternative, where $i = 1, 2, ..., m$

F_j: the jth future event, where $j = 1, 2,..., n$

P_j: the probability that the jth future event will occur in the future, where $j = 1, 2, ..., n$

E_{ij}: the payoff if the ith alternative is selected and jth future event occurs.

Based on the above notation, a decision-making problem can be formulated by using the decision evaluation matrix (see Figure 6.1).

For decisions under certainty—that is to say, the future event is certain, meaning that there is only one future possible and E_i is the payoff if the ith alternative is selected—the solution to the problem is very straightforward; that is to select the alternative that can maximize the payoff (in terms of the cost, E_i would be negative, so the maximization of the payoff would still apply), as shown in Equation 6.1:

	Probability of the future	$(P_j), j = 1, 2, ..., n$
	Possible future events	$F_j, j = 1, 2, ..., n$
Alternatives: $A_i, i = 1, 2, ..., m$		E_{ij}

FIGURE 6.1
Decision evaluation matrix. (Redrawn from Blanchard, B. S. and Fabrycky, W. J., *Systems Engineering and Analysis*, 4th edn, Prentice Hall, Upper Saddle River, NJ, 2006.)

$$\max_i [E_i] \quad \text{for} \quad i = 1, 2, \ldots, n \qquad (6.1)$$

6.2.1 Decision Making under Uncertainty

For decision making under uncertainty, it is assumed that all future events are possible and there is no information about the likelihood (probability) of the future. Five possible decision criteria can be used to make a decision: Laplace, maximax, maximin, Hurwicz, and minimax regret. To illustrate these criteria, let us consider the following example.

A company is considering building a retail store in a new location to increase sales revenue. The company can choose not to build the store, but to use other stores to fulfill demand, or build the store in three different sizes, small, medium, or large. There are three possible levels of future demand for this new location, which can be quantified as low demand, medium demand, and high demand. For different levels of demand, a certain alternative yields a unique payoff. For example, if a small store is built, and if the demand for the new market is small, an annual equivalent payoff value of $10,000 will be obtained. The annual equivalent payoff value for all the alternatives are presented in the decision evaluation matrix shown in Table 6.1 (in thousands of dollars).

Before we proceed to solve the problem by using different decision criteria, the first thing to do is to examine the alternatives, to see if any dominated alternative exists. If the payoff of Alternative A is superior (better) than that of Alternative B for all future events, then Alternative A dominates Alternative B, or Alternative B is dominated by Alternative A. Mathematically, a dominated alternative can be defined as follows:

An alternative A_i is dominated by an alternative $A_{i'}$ if for all possible future events, $F_j, j = 1, 2, \ldots, n, E_{ij} \leq E_{i'j}$.

A dominated alternative should be eliminated from the pool of alternatives. For the above example, it is easy to see that A_1 (do nothing) is dominated by

TABLE 6.1

Decision Evaluation Matrix for Store Size Example (in thousands of dollars)

	Possible Future		
	Low Demand	Medium Demand	High Demand
A_1: Do nothing	5	6	7
A_2: Build small	10	12	12
A_3: Build medium	7	15	15
A_4: Build large	−3	5	20

TABLE 6.2

Modified Decision Evaluation Matrix for Store
Size Example (in thousands of dollars)

	Possible Future		
	Low Demand	Medium Demand	High Demand
A_2: Build small	10	12	12
A_3: Build medium	7	15	15
A_4: Build large	−3	5	20

A_2 (build small) or A_3 (build large), so A_1 should be eliminated from consideration. The problem thus can be simplified to Table 6.2. A further examination finds that there is no alternative in Table 6.2 that is dominated by other alternatives.

6.2.1.1 Laplace Criterion

In the absence of the probability of future events, it is not possible to assume one event is more likely than another, but rather we assume that they are all equally possible. This assumption is based on the principle of insufficient reason (principle of indifference), or the Laplace principle, which states that if n possibilities are indistinguishable from each other except for their names, then each possibility should be assigned an equal probability of $1/n$. As Laplace (1902) indicates,

> The theory of chance consists in reducing all the events of the same kind to a certain number of cases equally possible, that is to say, to such as we may be equally undecided about in regard to their existence, and in determining the number of cases favorable to the event whose probability is sought. The ratio of this number to that of all the cases possible is the measure of this probability, which is thus simply a fraction whose numerator is the number of favorable cases and whose denominator is the number of all the cases possible.

Based on the Laplace principle, the probability of the occurrence of future events for the above example is 1/3. Or, in other words, the best alternative to be selected is the one that has the largest arithmetic average. This is illustrated in Table 6.3.

From the results, it is apparent that A_3 (build medium) is the best choice.

6.2.1.2 Maximax Criterion

The maximax criterion selects the alternative with the largest payoff value out of the best payoff values for each alternative. For each alternative, determine

TABLE 6.3

Laplace Criterion (in thousand dollars)

	Low Demand	Medium Demand	High Demand	Laplace Criterion
A_2: Build small	10	12	12	$(10+12+12) \div 3 = 11.33$
A_3: Build medium	7	15	15	$(7+15+15) \div 3 = 12.33$
A_4: Build large	−3	5	20	$(-3+5+20) \div 3 = 7.33$

the best payoff (or largest value); the maximax criterion chooses the alternative with the best of the best outcomes. For this reason, the maximax criterion is also called the "optimistic" criterion, as it only looks at the best-case scenario for each alternative. Mathematically, the maximax chooses the alternative A_i with the largest value of $\max_j E_{ij}$, or $\max_i \{\max_j E_{ij}\}$. Considering the example Table 6.2 presents, the maximum payoff value for each alternative is shown in Table 6.4.

From Table 6.4, the maximum number is 20; thus, the selection based on the maximax criterion is A_4 (build a large store).

6.2.1.3 Maximin Criterion

Contrary to the optimistic maximax criterion, the maximin is the pessimistic criterion that selects the best of the worst-case scenarios for each of the alternatives. The maximin chooses the alternative A_i with the largest value of $\min_j E_{ij}$, or $\max_i \{\min_j E_{ij}\}$. Consider the example Table 6.2 presents; the minimum payoff value (worst-case scenario) for each alternative is shown in Table 6.5.

TABLE 6.4

Payout according to the Maximax Criterion (in thousand dollars)

Alternative	$\max_j E_{ij}$
A_2: Build small	12
A_3: Build medium	15
A_4: Build large	20

TABLE 6.5

Payout according to the Maximin Criterion (in thousand dollars)

Alternative	$\min_j E_{ij}$
A_2: Build small	10
A_3: Build medium	7
A_4: Build large	−3

From Table 6.5, the maximum number is 10; thus, the selection based on the maximin criterion is A_2 (build a small store).

6.2.1.4 Hurwicz Criterion

The maximax and maximin criteria above present the two extreme ends of decision making. But, realistically, a decision maker is usually neither extremely optimistic nor pessimistic. Most of the times, a decision maker compromises between the maximax and maximin criteria. The Hurwicz criterion adds certain levels of realism into the formula, by calculating the weighted average between the maximax and maximin using a coefficient of optimism, indicated by α, with $0 \le \alpha \le 1$. The Hurwicz criterion is represented by Equation 6.2:

$$H_i = \alpha[\max_j E_{ij}] + (1-\alpha)[\min_j E_{ij}], \quad i = 1,2,\ldots,m \quad (6.2)$$

So, a decision maker would select the alternative with the largest value of H_i, or $\max_i H_i$. α is called the coefficient of optimism and $1-\alpha$ is the coefficient of pessimism. One can easily see that if $\alpha=1$, the Hurwicz criterion becomes a maximax criterion and if $\alpha=0$, then it becomes a pure maximin criterion. So we can say that the maximax and maximin criteria are two special cases for the Hurwicz criterion. The selection of α is largely subjective; it depends on the decision maker's personal feelings regarding optimism and pessimism. A different α will change the results of the decision. As an example, using the matrix from Table 6.2 and an α of 0.8, the results are presented in Table 6.6. Clearly the choice would be A_4, as it has the largest payoff value of $15,400.

6.2.1.5 Minimax Regret

The minimax regret criterion uses the concept of opportunity cost to arrive at a decision. An opportunity cost is also called a regret; a regret for an alternative for a certain future event is the difference between the payoff for that alternative and the best payoff for that future. For example, for the future event of low demand, the best payoff is $10,000 dollars, which is alternative

TABLE 6.6

Example of the Hurwicz Criterion with
α = 0.8 (in thousand dollars)

	$\max_j E_{ij}$	$\min_j E_{ij}$	Hurwicz with α=0.8
A_2	12	10	$0.8(12)+(1-0.8)(10)=11.6$
A_3	15	7	$0.8(15)+(1-0.8)(7)=13.4$
A_4	20	−3	$0.8(20)+(1-0.8)(-3)=15.4$

TABLE 6.7

Regret Matrix

	Low Demand	Medium Demand	High Demand	Maximum Regret
A_2: Build small	0	3	8	8
A_3: Build medium	3	0	5	5
A_4: Build large	13	10	0	13

A_2, so the regret for A_2 is 0, the regret for A_3 is 3, which is the difference between the payoff for A_3 (7) and the best payoff (10), and similarly, the regret for A_4 is 13. The regrets for all the alternatives are listed in Table 6.7.

The minimax regret criterion selects the alternative with the best of the "worst" regrets, or the minimum of the maximum regret. For the above example, the minimax criterion recommends alternative A_3 (build medium).

The minimax criterion is similar to the maximin, as it also looks at the best of the worst-case scenarios. However, since the minimax criterion applies to the regret rather than to the payoff value itself, it is believed not to be as pessimistic as the maximin approach.

6.3 Decision Making under Risks

Risks, in the context of systems engineering, lie between the two extreme cases of certainty and uncertainty. With risks, future events are still random but we have knowledge of the explicit probability of the occurrence of future events; in other words, information about P_i, $i = 1, 2, ..., n$ is available for future F_i, $i = 1, 2, ..., n$, and $\sum_{i=1}^{n} P_i = 1$.

The most commonly used criterion for decision making under risks is the expected value criterion. The expected value criterion chooses the alternative that yields the maximum expected payoff value. The expected value of an alternative is calculated as the sum of the products of the payoff value and its corresponding probability, or mathematically

$$\text{EX}(A_i) = \sum_{j=1}^{n} P_j E_{ij} \quad i = 1, 2, ..., m \tag{6.3}$$

For the example from Table 6.2, if we know that $P_1 = 0.25$, $P_2 = 0.45$, and $P_3 = 0.30$, then the expected value for A_2 is

$$\text{EX}(A_2) = 0.25(10) + 0.45(12) + 0.30(12) = 11.50 \text{ (thousand dollars)}$$

TABLE 6.8

Expected Value Criterion

Probability	0.25	0.45	0.30	
Demand	Low	Medium	High	Expected Value
A_2: Build small	10	12	12	$0.25(10)+0.45(12)+0.30(12)=11.50$
A_3: Build medium	7	15	15	$0.25(7)+0.45(15)+0.30(15)=13.0$
A_4: Build large	−3	5	20	$0.25(−3)+0.45(5)+0.30(20)=7.50$

The expected values for all the alternatives are presented in Table 6.8. Clearly, from Table 6.8, the expected value criterion suggests A_3 would be the best choice.

Using the expected value for decision-making risks seems rational, as it selects the maximum expected payoff value. However, this criterion may not capture the true behavior of humans when making decisions under risks, as different people might have different preferences regarding the values of the risks. To illustrate, let us assume two lottery alternatives between which a decision maker must choose: Lottery 1 (L1): There is a 100% chance (certainty) of receiving $100; Lottery 2 (L2): There is a 50% chance of receiving $300 and a 50% chance of receiving nothing. Let us apply the expected value criterion to these two alternatives. Clearly, the expected values are $EX(L1)=\$100$ and $EX(L2)=0.5(300)+0.5(0)=\$150$. The expected value criterion recommends us to choose L2 over L1 as it has a larger expected value (150 > 100), but, in the real world, many people are not willing to risk a 50% chance of receiving nothing, but would rather play safe, walking away with a guaranteed $100. The decision is made based on people's individual perceptions of the risk, and how much the risk is worth to each individual person; the perception of risks varies among different individuals, which cannot be interpreted accurately by the expected value. We introduce a more comprehensive model to address individual decision making under risks, the Von Neumann-Morgenstern Utility Theory, in Section 6.4.

6.4 Utility Theory

John Von Neumann and Oskar Morgenstern (1944) developed a model of expected utility (VNM utility) to replace the expected value criterion to represent individual preferences regarding risks. The utility, defined by Von Neumann and Morgenstern, is a quantitative measure of decision preferences of choices associated with risks. To define the utility functions, let us first revisit the decision problems involved with risks.

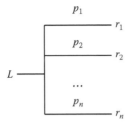

FIGURE 6.2
Graphical representation of a lottery.

In VNM utility theory, a decision option is denoted as a *lottery*, as an analogy to gambling options. A lottery is a situation where an individual (or agent) will receive one reward r_i with a probability of p_i, $(i=1, 2, ..., n)$, from n mutually exclusive outcomes. A lottery is denoted as $L(p_1, r_1; p_2, r_2; ...; p_n, r_n)$ and can be easily represented by a tree structure, with the tree branches representing the possible outcomes, as shown in Figure 6.2.

For the lottery examples mentioned above, in Lottery 1 (L_1), one receives $100 with certainty:

$$L_1 \underline{\qquad 1 \qquad} \$100$$

And in Lottery 2 (L_2), there is a 50% chance of receiving $300 and a 50% chance of receiving nothing:

$$L_2 \begin{cases} \underline{\quad 0.5 \quad} \ \$300 \\ \underline{\quad 0.5 \quad} \ \$0 \end{cases}$$

In the above examples, as we have mentioned, although L_2 has a higher expected value, in reality most people would choose L_1 instead, simply because L_2 involves risks while L_1 has no risks at all. Utility theory is intended to incorporate the effects of risks in choosing between lotteries, assuming we only can choose one lottery, not both. If L_1 is preferred over L_2, we denote this choice as $L_1 p L_2$; or we say that L_1 has a higher utility than L_2.

If we are allowed to change the probability for lottery L_2—say we will increase the probability of receiving the $300 from 0.5 to 0.8—this results in a new lottery, L_3.

$$L_3 \begin{cases} \underline{\quad 0.8 \quad} \ \$300 \\ \underline{\quad 0.2 \quad} \ \$0 \end{cases}$$

With this increased probability, most people would now be willing to choose L_3, if they believe L_3 is indifferent from L_1 as the risk becomes lower, then we say that L_1 and L_3 are equivalent, or $L_1 i L_3$. This implies that L_1 has the same utility as L_3.

To address the effects of risks on lottery decision making, we use utility to replace the monetary value to reflect the decision makers' feelings about the risks involved in lotteries. With the preference and indifference of lotteries defined, we can now give a formal definition of utility as follows (Winston 1994, 2005):

> A utility of a reward r_i, denoted as $u(r_i)$, is the value of the probability q_i, that makes the following two lotteries equivalent:

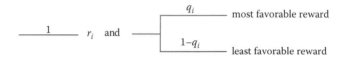

The utility $u(r_i)$ can be thought as a number q_i where a decision maker feels indifferent between receiving r_i for certain and the risk q_i between the most favorable and least favorable rewards. This utility is usually expected to be different between different decision makers; it is a subjective preference that an individual feels about the value of the risks with respect to the amount to be received. It is a common practice that makes the following assumptions about the most favorable and least favorable rewards:

$$u\left(\text{most favorable reward}\right) = 1$$

$$u\left(\text{least favorable reward}\right) = 0$$

r_i is a reward value between the most favorable and least favorable rewards, $u(r_i) = q_i$, and $0 \leq q_i \leq 1$.

When comparing the lotteries, we use utility to replace the actual monetary reward to represent the decision maker's preference concerning the combined effects of the monetary value and its associated risks. The utility function allows decision makers to tie their personal attitudes to the rewards and risks to alternatives (lotteries); thus, the decision made is more realistic and individual using the expected value criterion. Let us look at the following example to see how the utility function is derived and how we use it to choose between different lotteries. Consider three lotteries as follows:

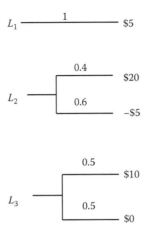

If we apply the expected value criterion, we will find that all three lotteries have the same expected value. Readers can easily verify that all the lotteries have the same expected value of $5. So, under the expected value criterion, decision makers should feel indifferent between these lotteries. But, in reality, different decision makers will have different preferences concerning them. Let us see how the utility model can be applied to rank these three lotteries. For all three lotteries, it is easy to see that the most desirable reward is $20, and the least desirable outcome is −$5. For all other rewards (r_i=$0, $5, $10), a decision maker is asked to determine the probability p_i so that the decision maker feels indifferent between the two lotteries.

Suppose that, for $0, the decision maker is indifferent between

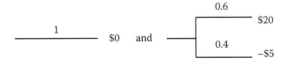

And for $5, the decision maker is indifferent between the following two lotteries:

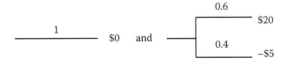

and for $r_i = \$10$, the decision maker is indifferent between the following two lotteries:

Using the above definition of indifferent lotteries, the decision maker can transform the original lotteries L_1, L_2, and L_3 to their equivalent lotteries L'_1, L'_2, and L'_3, shown in the following, such that they only contain the most favorable and the least favorable outcomes:

$$L'_1 \quad \begin{cases} 0.8 & \$20 \\ 0.2 & -\$5 \end{cases} \quad \equiv \quad L_1 \quad \xrightarrow{\ 1\ } \ \$5$$

$$L'_2 \quad \begin{cases} 0.4 & \$20 \\ 0.6 & -\$5 \end{cases} \quad \equiv \quad L_2$$

$$L'_3 \quad \begin{cases} 0.5 \begin{cases} 0.9 & \$20 \\ 0.1 & -\$5 \end{cases} \\ 0.5 \begin{cases} 0.6 & \$20 \\ 0.4 & -\$5 \end{cases} \end{cases} \quad \equiv \quad L_3 \quad \begin{cases} 0.5 & \$10 \\ 0.5 & \$0 \end{cases}$$

L'_3 is called a *compound lottery* because it consists of multiple lotteries within L'_3. A compound lottery such as L'_3 can be reduced to a *simple lottery* by combining the probabilities of the same reward. For example, for $20, the probability is $0.5(0.9) + 0.5(0.6) = 0.75$, and for −$5, the combined probability is $(0.5)(0.1) + (0.5)(0.4) = 0.25$, so L'_3 is reduced to the following simple format:

$$L'_3 \quad \begin{cases} 0.75 & \$20 \\ 0.25 & -\$5 \end{cases}$$

From the new lotteries L'_1, L'_2, and L'_3, since they all contain either the most desirable or least desirable outcomes, one can easily make a decision by comparing the probabilities of receiving the most desirable rewards; since

$0.8 > 0.75 > 0.4$, we can obtain the preferences for these three lotteries as $L_1' p L_3' p L_2'$. This implies that, for this decision maker, $L_1 p L_3 p L_2$.

Another way to rank these three lotteries is to use the utility function. According to the definition of the utility, we have $u(\$20) = 1$ and $u(-\$5) = 0$, and, by definition, we can obtain the utility values for other rewards since the decision maker is indifferent between

So, $u(\$0) = 0.6$. Similarly, we can obtain the utility function values for the remaining rewards: $u(\$5) = 0.8$ and $u(\$10) = 0.9$.

With all the utility function values known, we can calculate the expected utility for each of the lotteries, which will give the same ranking results. The expected utility can be defined as the following:

For a given lottery $L:(p_1, r_1; p_2, r_2; ...; p_n, r_n)$, the expected utility, written as $E(U \text{ for } L)$, is

$$E(U \text{ for } L) = \sum_{j=1}^{n} p_j u(r_j) \tag{6.4}$$

Using expected utility, we can easily choose between two lotteries by the following criteria:

$L_1 p L_2$ if and only if $(\leftrightarrow)$ $E(U \text{ for } L_1) > E(U \text{ for } L_2)$

$L_1 i L_2$ if and only if $(\leftrightarrow)$ $E(U \text{ for } L_1) = E(U \text{ for } L_2)$

$L_2 p L_1$ if and only if $(\leftrightarrow)$ $E(U \text{ for } L_1) < E(U \text{ for } L_2)$

Thus, for our example above, we have

$$E(U \text{ for } L_1) = (1) u(\$5) = 0.8$$

$$E(U \text{ for } L_2) = (0.4) u(\$20) + (0.6) u(-\$5) = (0.4)(1) + 0.6(0) = 0.4$$

$$E(U \text{ for } L_3) = (0.5) u(\$10) + (0.5) u(\$0) = 0.5(0.9) + 0.5(0.6) = 0.75$$

So, based on the expected utility criteria, we can rank these three lotteries as $L_1pL_3pL_2$, since $0.8 > 0.75 > 0.4$, which gives us the same results as using the composite lotteries.

To use expected utility theory, Von Neumann and Morgenstern proposed and proved the following four axioms as prerequisites.

Axiom 6.1: Complete Ordering Axiom

For any arbitrary two lotteries, L_1 and L_2, exactly one of the possible must be true: (1) L_1pL_2, (2) L_1iL_2, or (3) L_2pL_1.

Axiom 6.2: Transitivity Axiom

If L_1pL_2 and L_2pL_3, then L_1pL_3.

Axiom 6.3: Continuity Axiom

If L_1pL_2 and L_2pL_3, then there exists some value of p, $0 < p < 1$, such that L_2iL_2' where

Axiom 6.4: Independence Axiom

The independent axiom implies that for the same decision-maker, two lotteries mixed with a third one will remain the same preference order as when the two lotteries are compared independently of the third lottery. Suppose that we L_1pL_2 and for any $0 < c < 1$ and comparing with another lottery L_3, the same preference will hold, that is,

These four axioms will define a rational decision maker; expected utility theory applies if and only if these four axioms hold true.

By defining the utility function, we can derive information about a decision maker's attitude to risks; he/she is either a risk-seeking, risk-neutral, or risk-averse person. This information can be obtained by looking at the *certainty equivalent* (CE) and *risk premium* (RP) values.

A CE of a lottery *L*, denoted as CE(*L*), is the reward *r* such that the decision maker is indifferent between the lottery *L* and receiving a certain reward *r*. Then CE(*L*)=*r*. For example, if a decision maker feels indifferent about the following two lotteries

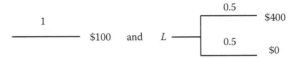

then CE(*L*)=$100.

Knowing the value of CE(*L*), we now can define the RP of a lottery, denoted as RP(*L*), which is given by RP(*L*)=EV(*L*)−CE(*L*), where EV(*L*) is the expected value of the lottery's outcomes. Using the above example, EV(*L*)=0.5($400)+ 0.5($0)=$200, so RP(*L*)=EV(*L*)−CE(*L*)=$200−$100=$100.

Using RP(*L*), we can assess a decision maker's attitude to risk. For any lottery *L* that involves a risk (i.e., any lottery that has more than one possible outcome, sometimes called a *nondegenerate* lottery), a decision maker is

1. *Risk averse* if and only if RP(*L*)>0. A risk-averse decision maker avoids risk; he/she will prefer the lottery with more certainty unless risks are adequately compensated.
2. *Risk neutral* if and only if RP(*L*)=0. A risk-neutral decision maker is insensitive to risk; he/she is completely indifferent to the risk involved in a lottery and only concerned about the expected value of the rewards.
3. *Risk seeking* if and only if RP(*L*)<0. A risk-seeking decision maker is attracted to risks. He/she will prefer a lottery with a lower expected return on rewards but with greater risks involved.

For the above example, since the decision maker's RP(*L*)=$100>$0, the decision maker is considered a risk-averse person. Another way to differentiate attitudes to risk is to look at the shape of the utility function $u(x)$; if $u(x)$ is differentiable (meaning $u(x)$ is continuous and the derivative function exists), then we can have an alternative definition for risk attitudes:

1. *Risk averse* if and only if the second-order derivative $u''(x)<0$ or $u(x)$ is strictly concave
2. *Risk neutral* if and only if $u''(x)=0$ or $u(x)$ is a linear function
3. *Risk seeking* if and only if $u''(x)>0$ or $u(x)$ is strictly convex

To prove that function $u(x)$ is strictly concave (convex), we can use the sign of the second-order derivative $u''(x)<0$ ($u''(x)>0$ for convexity), or we can use the definition of a concave (convex) function; that is, for any arbitrary value x_1 and x_2 and positive number $0<\alpha<1$, strict concavity is defined as

$$u\big[ax_1+(1-a)x_2\big] > au(x_1)+(1-a)u(x_2)$$

and strict convexity implies

$$u\big[ax_1+(1-a)x_2\big] < au(x_1)+(1-a)u(x_2).$$

Consider the following lottery:

$$
L \quad
\begin{array}{l}
\xrightarrow{\quad p \quad} x_1 \\
\xrightarrow{\quad 1-p \quad} x_2
\end{array}
\qquad \text{(Assume } x_1 < x_2)
$$

The relationship between $u(x)$ and RP(L) is illustrated in Figure 6.3.

To illustrate how to use this definition, assume that we have a decision maker who has an asset of \$50,000; his utility function is $u(x)=x^{1/3}$, and every year, he has two options to choose from:

1. L_1: For a small insurance fee of \$200, the decision maker is guaranteed to have no monetary loss in that year.
2. L_2: Without buying the insurance, there is a 10% chance that the decision maker will lose \$40,000 in that year.

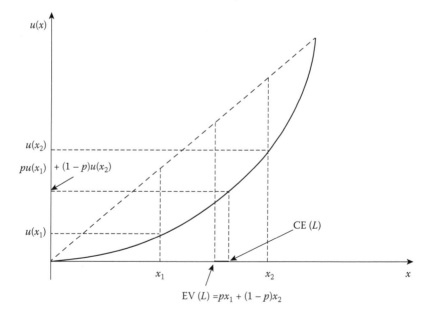

FIGURE 6.3
Illustration of the relationship between $u(x)$ and RP(L).

Should the decision maker buy the insurance or not? Is this person risk seeking, risk neutral, or risk averse?

Solution: The decision maker has the following two lotteries to choose from

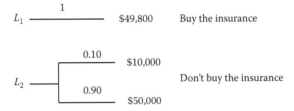

We can compare the expected utilities for these two lotteries:

$$u(\$49,800) = 36.79$$

$$u(\$10,000) = 21.54$$

$$u(\$50,000) = 36.84$$

So, we have

$$E(U \text{ for } L_1) = 36.79$$

and

$$E(U \text{ for } L_2) = 0.10(21.54) + 0.90(36.84) = 35.31$$

Since $E(U \text{ for } L_2) < E(U \text{ for } L_1)$, the decision maker should take Option 1, which is to buy the insurance.

To determine whether this decision maker is risk averse or risk seeking, let us look at the nondegenerate lottery L_2, since $E(U \text{ for } L_2) = 35.31$, so

$$CE(L_2) = (35.31)^3 = \$44024.37$$

and

$$EV(L_2) = 0.10(\$10,000) + 0.90(\$50,000)$$

$$= \$46,000$$

So

$$RP(L_2) = EV(L_2) - CE(L_2)$$

$$= \$46,000 - \$44024.37$$

$$= \$1975.63 > 0$$

which implies that this decision maker is a risk-averse person. This result can also be derived by using the differentiation condition of the utility function

$$u''(x) = -\frac{2}{9}x^{-\frac{5}{3}}$$

We will leave this to readers to solve for practice, and we have also provided some exercise questions in the problem section at the end of the chapter.

Although some criticism of utility theory has been posed since the 1950s (Weaver 1963; March 1978) regarding violations of the fundamental assumptions and axioms in the real world, utility theory provides a rigorous and comprehensive approach for decision makers to incorporate risks into decision-making problems. It is still applied nowadays, through extensions and adaptations, in a wide range of applications.

6.5 Decision Tree

There are many circumstances in which a decision maker needs to make a sequence of decisions at different time points, and the outcome and the chances of obtaining certain outcomes of the following decisions are dependent on what decision is made at a previous point in time. For example, a person can decide now either to pursue a doctorate directly or simply a master's degree, and later on he/she will decide what career to pursue, an engineering job in industry or a teaching job in college. Obviously, the probability of obtaining a certain type of job will depend on the prior decision made (e.g., getting a doctorate will help with obtaining a teaching job more than a master's degree). When dealing with decision-making problems involving multistage and conditional probabilities, a *decision tree* is a very useful tool to help the decision maker to formulate the complex problem into a tree structure, so that a complex problem can be decomposed into a series of smaller decision problems. It is a graphical (schematic) representation of all possible alternatives and the chances (probability) and the consequences of choosing each alternative.

To illustrate how to use a decision tree to solve a complex decision problem, let us start with an example. Company X from Daytona Beach is trying to determine whether or not to build a factory in the city of Guangzhou in China to sell a newly designed product to the local market. Company X has estimated that it will cost $500,000 to build the factory, and if the market is successful, a $750,000 profit will be obtained; if the market turns out to be a failure, a $250,000 profit is expected. Currently, the company believes that there is a 45% chance that the product will be successful in the local market. As an alternative, Company X can hire a local research team at a price of $20,000 to conduct an analysis on the potential market for this new product. If the market survey results are favorable, then it is believed that the chance of local market success would increase to 90%. If the market survey results are not favorable, then the chance of local market success is only 15%. Prior to determining whether or not to hire the research team, the company believes that there is a 50% chance the research will turn out to be favorable. Let us use a decision tree to determine Company X's course of actions to maximize the expected net profit.

To formulate the decision problem using a decision tree, we first need to define what the *decisions* and the *events* are for the problem. A decision in the decision tree, denoted by a square node, represents a point in the decision process or a time at which the decision maker has to make a decision. The decision fork represents the possible actions, and only one action can be chosen at a time. For example, the company has to decide whether or not to hire the research team:

An event in the decision tree, denoted by a circular node, represents all the possible outcomes that will occur when a decision action is taken. Each branch of the event fork represents a possible outcome, usually expressed in terms of the reward/revenue with the chances (probability) of the event. At the event fork, there is no need to pick one branch; all outcomes are possible (although, in the future, only one event can happen). A special case of the event is the termination branch, when there are no further events or actions possible; for example, when Company X achieves local market success, this would be a terminal branch for this problem. A termination branch is represented by a bar line. An example of an event for the above scenario is that, when investing in the local market, there could be market success or market failure:

To construct the decision tree, we first need to work out the logical sequence of decisions and events and lay them out in a tree diagram from left to right. For Company X, the first decision to make is to determine whether or not to hire the research team before deciding to invest in the local market. Knowing the decision sequence, the decision tree is constructed as shown in Figure 6.4.

For each of the termination events, we first need to identify the net profit. For example, for the termination event of (hiring the research team, and when the report is favorable, choosing to build the factory and it turns out to be a market success), the final net profit is $750,000 (market success) – $500,000 (cost of the building the factory) – $20,000 (cost of hiring the local research team) = $230,000. To determine the decision that maximizes the expected net profit, we need to work from the right, starting from the termination events, calculating the expected net profit. The rules of the calculation and selections that we need to keep in mind are:

1. For each of the event forks (circular), an expected value is calculated. Since there are possible outcomes for a particular event, each outcome is associated with one branch that has a likelihood of occurring (the probability).
2. For each of the decision forks (square), a decision has to be made. The branch of the decision fork that has the highest expected value is chosen (if there is a tie, break the tie arbitrarily).

The above two steps are performed from right to left, until we reach the beginning of the decision tree. By tracking back, a sequence of decisions that maximize the total expected value can be obtained.

Let us walk through this process together:

1. For the event of building the factory after obtaining a favorable report, the expected value is $0.9(230,000) + (0.1)(-270,000) = 180,000$.
2. For the event of building the factory after obtaining an unfavorable report, the expected value is $(0.15)(230,000) + (0.85)(-270,000) = -195,000$.
3. For event of building the factory without research, the expected value is $(0.45)(250,000) + (0.55)(-250,000) = -25,000$.

With the expected value for each event calculated, we now can evaluate all the decision forks.

1. Decision after favorable report: Since the expected value for "Build the factory" (180,000) is larger than "Don't build" (-20,000), we choose the decision "Build the factory," and its expected value is that of the chosen branch, 180,000.
2. Decision after unfavorable report: Since the expected value for "Build the factory" (-195,000) is less than "Don't build" (-20,000), we

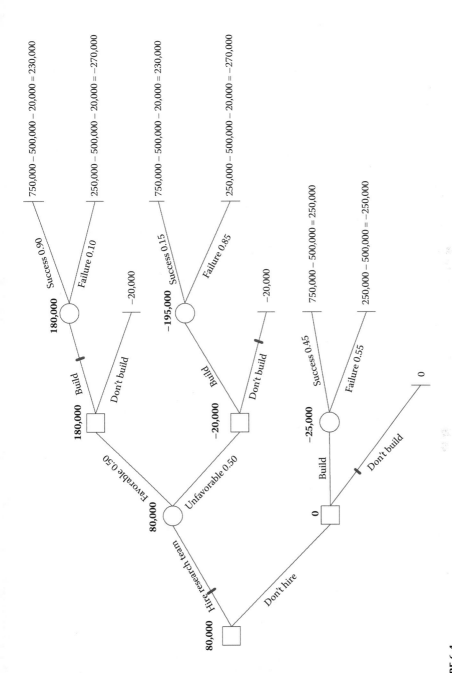

FIGURE 6.4
Decision tree for Company X example.

choose the decision "Don't build," and its expected value is that of the chosen branch, −20,000.

3. Decision without hiring the research team: the expected value for "Build the factory" (−25,000) is less than "Don't build" (0), so we choose the decision "Don't build."

Moving to the next level on the left, we can calculate the expected value for hiring the research team, which is $(0.5)(180,000) + (0.5)(-20,000) = 80,000$; this is greater than the expected value of not hiring the team (0), so, at the beginning of the decision tree, we choose to hire the research team. By tracing back the decision tree to the terminal event, we now have the optimal sequence of decisions determined, which is to hire the research team; if the result of the research is favorable, then build the factory; otherwise, do not build the factory. This would obtain an optimal expected net profit of $80,000. The whole process is illustrated in Figure 6.4.

The above example uses the expected value (risk-neutral) model; the result would be different if the decision maker had a different attitude to risk. For example, let us suppose the decision maker is a risk-seeking person, with a utility function as in Figure 6.5. Would the result change?

The basic procedure for incorporating the utility model is basically the same; one would just need to replace the expected value with the expected utility and try to maximize the expected utility value, instead of the expected

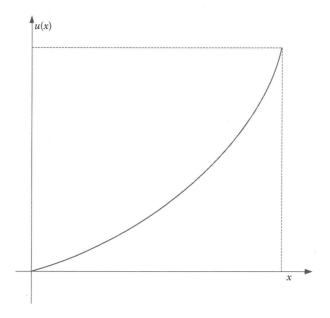

FIGURE 6.5
Illustration of a risk-seeking utility function.

value. There are some exercises for readers to practice in the problems section at the end of the chapter.

6.5.1 Expected Value of Sample Information

The expected value of sample information (EVSI), is the increase of the expected value (or the expected utility value if the expected utility is being used to make decisions) that one can obtain from a sample of information before making that decision. The additional information obtained from the sample can help to reduce the level of uncertainty, thus further making the outcome of the decision more definite. Taking the above example, the sample information obtained from research conducted locally will help to increase the chances of market success from 45% to 90% if the research result is favorable, and will help the company to know that the chances of failure are between 55% and 85% if the result is unfavorable. With this decreased level of uncertainty, the decision maker is more informed, and thus a better decision can be made to increase the expected value.

To calculate the EVSI, we assume that the research costs nothing; the expected value with free research is called the expected value with sample information (EVWSI). For the above example, EVWSI = $80,000 + $20,000 = $100,000. The expected value without any sample information (EVWOI) is the expected value without the research from the above decision tree analysis result, that is, EVWOI = $0. EVSI is the difference between EVWSI and EVWOI, or EVSI = EVSWI − EVWOI, so in this case, EVSI = 100,000 − 0 = 100,000. In the above example, the cost of the research ($20,000) is far less than EVSI, which makes hiring the research team more desirable. That is why the decision is to choose to hire the team.

6.5.2 Expected Value of Perfect Information

In decision theory, perfect information implies that a decision maker has perfect knowledge of the situation before he/she decides what to do. With perfect information, a decision maker would always make the correct decision to maximize profit. In the above example, the company would choose to build if they knew that market success would result, and choose not to build if perfect information of market failure was obtained. To achieve the expected value with perfect information (EVWPI), we use the following tree (Figure 6.6) to indicate the decisions made for different information.

So we can obtain

$$EVWPI = 0.45(250,000) + 0.55(0) = 112,500$$

The expected value of perfect information (EVPI) can obtained by EVPI = EVWPI − EVWOI. For the above example, EVPI = 112,500 − 0 = 112,500.

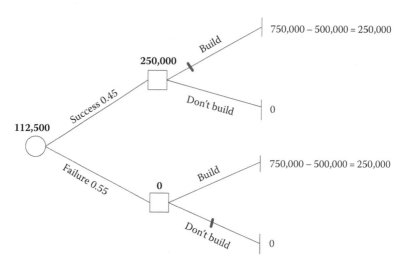

FIGURE 6.6
Decision tree with perfect information.

Generally speaking, EVPI should be greater than EVSI. EVPI is often considered as an upper bound for the value of sample information (or research or testing) for a decision-making problem. It can be used to justify the value of the provision of such a service or information.

6.6 Decision Making for Multiple Criteria: AHP Model

Most of the decision making in our daily lives and engineering problems typically involves more than one criterion and these criteria often conflict when used to evaluate the decision alternatives. For example, when deciding what equipment to buy, cost (or price) is one of the main criteria, and other decision criteria are often in conflict with cost. It is desirable to obtain the equipment costing the least amount of money; meanwhile, we want the equipment to be of high quality, the most reliable, and the safest. These criteria do not quite agree with each other. The trade-off between these conflicting criteria requires us to solve the decision-making problem with a structured approach, not by intuition, to discover the optimal choice, especially when the stakes are high.

Multiple-criteria decision making is a subdiscipline of operations research that has been studied explicitly and extensively. Many models and methods have been developed for articulating and solving these types of problems. Typical methods include mathematical programming, multiattribute utility theory, and ranking of criteria by using methods such as the analytical hierarchy process (AHP). We will introduce mathematical programming models in the next chapter on system optimization models. In this chapter, we will first introduce the AHP model.

To illustrate the application of the AHP model, let us first look at the following example: Company Y is trying to determine where to build their next manufacturing facility. Currently, there are three locations for consideration: Locations A, B, and C. To determine the best location, Company Y has identified five major criteria for consideration:

1. Labor cost (LC).
2. Local government incentives (LG).
3. Local labor skills (LS).
4. Tax rate (TR).
5. Closeness to resources (CR): It is desirable to have the facility built near the location of the raw materials; the closer they are, the better.

The above decision-making problem is illustrated in Figure 6.7. There are several challenges in solving this example decision-making problem. First, each of the criteria listed above is not equally important in site selection; some criteria have greater values than others. For a more accurate comparison of the alternatives, these five criteria need to be given an order of priority. Second, for each of the criteria, the three candidate locations have different performances. For example, one location is the closest to the raw materials, which would save shipping costs, but the company would have difficulty finding qualified labor with sufficient skills there. For each of the five criteria, we need to rank the three locations. A final rank for the decision alternatives may be obtained by the weighted sums of the priority ranking scores. More specifically, if the weightings for all the criteria are $w_i, i = 1, 2, 3,$

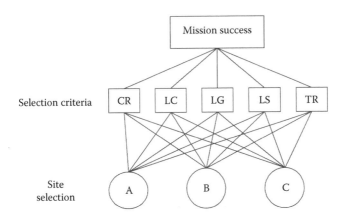

FIGURE 6.7
Hierarchical structure of the site selection problem.

4, 5, and for each of the criteria $i = 1, 2, 3, 4, 5$, each of the three candidates $j = 1, 2, 3$ has a performance of p_{ij}, then the final ranking for the location alternative is $S_j = \sum_{i=1}^{5} w_i p_{ij}$.

When comparing multiple criteria in the decision-making process, it is quite challenging for humans to give absolute judgments over a group of alternatives together, but it has been found that humans find it easier to compare two alternatives at a time, as this is the comparison that occurs in short-term memory most often. Inspired by the psychology of human judgment, the AHP method, first developed by Saaty in the 1970s, is a well-defined quantitative technique that facilitates the structuring of a complex multiattribute problem, and provides an objective methodology for deciding between a set of alternatives to solve that problem. Its application has been reported in numerous fields, such as transportation planning, portfolio selection, corporate planning, and marketing. What AHP does is to decompose a complex decision-making problem into smaller units, turning the comparison of multiple criteria into smaller pairwise comparisons, thus making it easier for decision makers to make judgments.

According to Saaty (2008), the AHP method includes the following steps:

- Define the problem and mission objectives.
- Identify the criteria for making decisions and possible decision alternatives to be considered.
- Develop a hierarchical structure of the decision problem in terms of the overall objective, starting from the top with the decision goals, through the intermediate levels of the decision criteria, all the way to the bottom level with the decision alternatives.
- Determine the relative priorities of criteria that express their relative importance in relation to the element at the higher level, on a pairwise basis; construct a set of pairwise comparison metrics for all the criteria and all the alternatives with regard to each criterion. The pairwise comparison score is based on Table 6.9.

TABLE 6.9

Pairwise Comparison Scale (i to j) for AHP Preferences

Verbal Judgment	Numerical Rating[a]
Extremely preferred	9
Very strongly preferred	7
Strongly preferred	5
Moderately preferred	3
Equally preferred	1

[a] 2, 4, 6, 8 are the intermediate values

- Apply the AHP algorithms (which will be explained in Section 6.6.1) to calculate the overall rating of the decision alternatives.
- Check the consistency of the decision maker's comparisons.
- Continue this process of weighing and adding until the final priorities of the alternatives in the bottommost level are obtained.

6.6.1 AHP Algorithms

Step 1: The original pairwise comparison is given by the following matrix A for n factors

$$
A = \begin{bmatrix}
1 & a_{12} & \cdots & a_{1n} \\
a_{21} & 1 & \cdots & a_{2n} \\
\cdots & \cdots & \cdots & \cdots \\
a_{n1} & a_{n2} & \cdots & 1
\end{bmatrix}
$$

where $a_{ij} = 1/a_{ji}$; a_{ij} is derived by using Table 6.9.

Step 2: To derive the relative weighting for each of the factors, first, matrix A is normalized by dividing each element of the matrix by the sum of its column, that is,

$$
a'_{ij} = \frac{a_{ij}}{\sum_{i=1}^{n} a_{ij}}, \quad i = 1,2,\ldots n; \quad j = 1,2,\ldots n
$$

Step 3: The weighting of each factor can be obtained by taking the mean of each role of the new normalized matrix A',

$$
w_i = \frac{\sum_{j=1}^{n} a'_{ij}}{n}, \quad i = 1,2,\ldots n
$$

Step 4: Consistency check: To ensure a valid comparison, we need to check the consistency of the pairwise matrix. Although the decision maker is rational, decision makers may sometimes make inconsistent pairwise comparisons that violate the transitivity property (which implies that for three factors i,j,k, the pairwise comparison results among them should satisfy $a_{ij}a_{jk} = a_{ik}$. So, for example, if $a_{ij} = 3$, and $a_{jk} = 2$, then a_{ik} is expected to be 6). AHP tolerates some degree of inconsistency, as absolute consistency is hard to achieve (also, the comparison score is bounded by the upper limit of 9). However, severe inconsistency might cause the decision-making results to become invalid; for example, if a decision maker prefers $A > B$, $B > A$, and $C > A$, this would not be a valid comparison. For each of the pairwise comparisons that involve more than three factors, its consistency needs to be checked to validate the results of the comparison before it can be used to make decisions. For this purpose, a simple four-step procedure can be used to calculate the *consistency index* (CI). Let A denote the original pairwise comparison matrix and w denote our estimate of weightings.

$$
A = \begin{bmatrix}
1 & a_{12} & \cdots & a_{1n} \\
a_{21} & 1 & \cdots & a_{2n} \\
\cdots & \cdots & \cdots & \cdots \\
a_{n1} & a_{n2} & \cdots & 1
\end{bmatrix}
$$

and

$$
\mathrm{w} = [w_1, w_2, \ldots, w_n]
$$

1: Compute $A\mathrm{w}^T$.
2: Compute

$$
R = \frac{1}{n} \sum_{i=1}^{n} \frac{i\text{th entry in } A\mathrm{w}^T}{i\text{th entry in } \mathrm{w}^T}
$$

3: Compute CI as follows:

$$
\mathrm{CI} = \frac{R - n}{n - 1}
$$

4: Compare CI to the random index (RI). Compute CI/RI. RI is the *random index* that is dependent on n, as illustrated in Table 6.10.

An RI is developed based on the methods developed by Saaty (1980). The RI was developed as the benchmark to determine the magnitude of deviation

TABLE 6.10

Random Index Table

Factors number	1	2	3	4	5	6	7	8	9	10
Random index (RI)	0	0	0.58	0.90	1.12	1.24	1.32	1.41	1.45	1.49

Source: Data from Saaty, T. *The Analytic Hierarchy Process*. 1980. New York: McGraw-Hill.

from consistent results. The RI is developed by using a completely random pairwise comparison, that is to say, for an $n \times n$ matrix, a_{ij} is taken randomly from the set of values $(1/9, 1/8, 1/7, ..., 7, 8, 9)$. An average RI is calculated for this random matrix, which only depends on the size of the matrix, as shown in Table 6.10.

If the CI is sufficiently small, the subject matter experts' comparisons are probably consistent enough to give a useful estimate of the weightings for the objective. Usually, if CI/RI < 0.1, the degree of consistency is satisfactory; otherwise, serious inconsistencies may exist. If the ratio CI/RI is closer to 1, indicating that the pairwise comparison matrix is closer to a random one, this, in turn, implies inconsistency.

Let us illustrate the AHP method by solving the decision-making problem of Company Y. It has five attributes (labor cost [LC], local government incentives [LG], local labor skills [LS], tax rate [TR], and closeness to resources [CR]) for selection of the facility site and three alternative locations (A, B, and C).

SOLUTION:

Step 1: Company Y uses expert opinions to gather information on the relative importance of these attributes based on a pairwise comparison, as shown in Table 6.11.

Or

$$ A = \begin{bmatrix} 1 & 3 & \frac{1}{2} & 6 & \frac{1}{5} \\ \frac{1}{3} & 1 & \frac{1}{6} & 2 & \frac{1}{9} \\ 2 & 6 & 1 & 8 & \frac{1}{3} \\ \frac{1}{3} & \frac{1}{2} & \frac{1}{8} & 1 & \frac{1}{9} \\ 5 & 9 & 3 & 9 & 1 \end{bmatrix} $$

TABLE 6.11

Matrix of Pairwise Comparisons

	LC	LG	LS	TR	CR
LC	1.0	3.0	1/2	6.0	1/5
LG	1/3	1.0	1/6	2.0	1/9
LS	2.0	6.0	1.0	8.0	1/3
TR	1/6	1/2	1/8	1.0	1/9
CR	5.0	9.0	3.0	9.0	1.0

Step 2: To derive the relative weightings for each of the factor, first, the pairwise comparison matrix A is normalized by dividing each element of the matrix A by the corresponding sum of its column.

	LC	LG	LS	TR	CR
LC	1.0	3.0	1/2	6.0	1/5
LG	1/3	1.0	1/6	2.0	1/9
LS	2.0	6.0	1.0	8.0	1/3
TR	1/6	1/2	1/8	1.0	1/9
CR	5.0	9.0	3.0	9.0	1.0
Sum	8.500	19.500	4.792	26.000	1.755

The normalized matrix is

$$A' = \begin{bmatrix} 0.118 & 0.154 & 0.104 & 0.231 & 0.114 \\ 0.039 & 0.051 & 0.035 & 0.077 & 0.063 \\ 0.235 & 0.308 & 0.209 & 0.308 & 0.190 \\ 0.020 & 0.026 & 0.026 & 0.038 & 0.063 \\ 0.588 & 0.462 & 0.626 & 0.346 & 0.570 \end{bmatrix}$$

To illustrate how an element in A' is derived, for example: $a'_{51} = 0.588 = a_{51} / 8.500 = 5.000 / 8.500$. Readers can verify all the other elements in A'. Table 6.12 shows the row sum and column sum for the normalized matrix.

Step 3: The weighting of each factor can be obtained by taking the mean of each row of the new normalized matrix A'. For example, the weighting for CR = 0.721/5 = 0.144. The calculated weightings for all the factors are shown in Table 6.13.

TABLE 6.12

Normalized Matrix with Row Sums and Column Sums

	LC	LG	LS	TR	CR	Sum
LC	0.118	0.154	0.104	0.231	0.114	0.721
LG	0.039	0.051	0.035	0.077	0.063	0.265
LS	0.235	0.308	0.209	0.308	0.190	1.249
TR	0.020	0.026	0.026	0.038	0.063	0.173
CR	0.588	0.462	0.626	0.346	0.570	2.592
Sum	1.000	1.000	1.000	1.000	1.000	5.000

TABLE 6.13

Normalized Matrix with Row Sums, Column Sums and Calculated Weightings

	LC	LG	LS	TR	CR	Sum	Weighting
LC	0.118	0.154	0.104	0.231	0.114	0.721	*0.144*
LG	0.039	0.051	0.035	0.077	0.063	0.265	*0.053*
LS	0.235	0.308	0.209	0.308	0.190	1.249	*0.250*
TR	0.020	0.026	0.026	0.038	0.063	0.173	*0.035*
CR	0.588	0.462	0.626	0.346	0.570	2.592	*0.518*
Sum	1.000	1.000	1.000	1.000	1.000	5.000	*1.000*

From Table 6.13, it can be easily seen that CR has the highest weighting (0.518) and TR is ranked as the least important factor (0.035). The final weighting for all the factors is

$$w = [0.144, 0.053, 0.250, 0.035, 0.518]$$

Step 4: Check of consistency:

1. Compute Aw^T

$$Aw^T = \begin{bmatrix} 1 & 3 & \frac{1}{2} & 6 & \frac{1}{5} \\ \frac{1}{3} & 1 & \frac{1}{6} & 2 & \frac{1}{9} \\ 2 & 6 & 1 & 8 & \frac{1}{3} \\ \frac{1}{3} & \frac{1}{2} & \frac{1}{8} & 1 & \frac{1}{9} \\ 5 & 9 & 3 & 9 & 1 \end{bmatrix} \begin{bmatrix} 0.144 \\ 0.053 \\ 0.250 \\ 0.035 \\ 0.518 \end{bmatrix} = \begin{bmatrix} 0.742 \\ 0.270 \\ 1.309 \\ 0.174 \\ 2.780 \end{bmatrix}$$

2. Compute

$$R = \frac{1}{n} \sum_{i=1}^{n} \frac{i\text{th entry in } Aw^T}{i\text{th entry in } w^T}$$

So

$$R = \frac{1}{5} \left[\frac{0.742}{0.144} + \frac{0.270}{0.053} + \frac{1.309}{0.250} + \frac{0.174}{0.035} + \frac{2.780}{0.518} \right]$$

$$= \frac{1}{5}(25.830) = 5.166$$

3. Compute the CI:

$$CI = \frac{R-n}{n-1} = \frac{5.166-5}{5-1} = 0.042$$

4. Compare CI to the RI. With $n=5$, the RI value $=1.12$ according to Table 6.10.

$$\frac{CI}{RI} = \frac{0.042}{1.12} = 0.037 < 0.1$$

Since CI/RI < 0.1, we can conclude that the degree of consistency is satisfactory, which, in turn, implies that the weightings are valid.

Now we have the weightings for all the decision factors, next, we need to determine how well each of the alternatives scores on each factor. To determine the "scores", we need to use pairwise comparisons in a similar way to that in which we derived the weightings for the factors.

For LC, suppose the decision maker has obtained the following pairwise comparison matrix:

LC	A	B	C
A	1	1/3	2
B	3	1	4
C	1/2	1/4	1

The normalized matrix is

$$A_{\text{norm\_LC}} = \begin{bmatrix} 0.222 & 0.210 & 0.286 \\ 0.667 & 0.632 & 0.571 \\ 0.111 & 0.158 & 0.143 \end{bmatrix}$$

Taking the average of each row yields the score of how well each alternative scores with respect to the LC factor:

$$A \text{ score for LC} = \frac{0.222+0.210+0.286}{3} = 0.239$$

$$B \text{ score for LC} = \frac{0.667+0.632+0.571}{3} = 0.623$$

$$C \text{ score for LC} = \frac{0.111+0.158+0.143}{3} = 0.137$$

Applying the procedures for checking consistency, we find that

$$\frac{CI}{RI} = 0.016 < 0.1$$

So, the decision maker's comparison exhibits consistency. For LG, the pairwise comparison matrix is

LG	A	B	C
A	1	2	2
B	1/2	1	1
C	1/2	1	1

The normalized matrix is

$$A_{norm\_LG} = \begin{bmatrix} 0.500 & 0.500 & 0.500 \\ 0.250 & 0.250 & 0.250 \\ 0.250 & 0.250 & 0.250 \end{bmatrix}$$

Taking the average of each row yields the score of how well each alternative scores with respect to the LG factor:

$$A \text{ score for LG} = \frac{0.500 + 0.500 + 0.500}{3} = 0.500$$

$$B \text{ score for LG} = \frac{0.250 + 0.250 + 0.250}{3} = 0.250$$

$$C \text{ score for LG} = \frac{0.250 + 0.250 + 0.250}{3} = 0.250$$

The decision maker has made a perfectly consistent decision, since all three columns are identical, which implies that a perfect transitivity relationship has been met.

For LS, the pairwise comparison matrix is

LS	A	B	C
A	1	1/5	1/2
B	5	1	3
C	2	1/3	1

The normalized matrix is

$$A_{norm\_LS} = \begin{bmatrix} 0.125 & 0.130 & 0.111 \\ 0.625 & 0.652 & 0.667 \\ 0.250 & 0.217 & 0.222 \end{bmatrix}$$

Taking the average of each row yields the score of how well each alternative scores with respect to the LS factor:

$$A \text{ score for LS} = \frac{0.125 + 0.130 + 0.111}{3} = 0.122$$

$$B \text{ score for LS} = \frac{0.625 + 0.652 + 0.667}{3} = 0.648$$

$$C \text{ score for LS} = \frac{0.250 + 0.217 + 0.222}{3} = 0.230$$

Applying the procedures for checking consistency, we find that

$$\frac{CI}{RI} = 0.003 < 0.1$$

So, the decision maker's comparison exhibits consistency for comparing the alternatives with regard to LS.

Following a similar approach we have the pairwise comparison matrices for TR and CR, as follows:

TR	A	B	C
A	1	1/3	1/7
B	3	1	1/2
C	7	2	1

$$A_{norm\_TR} = \begin{bmatrix} 0.091 & 0.100 & 0.087 \\ 0.273 & 0.300 & 0.304 \\ 0.636 & 0.600 & 0.609 \end{bmatrix}$$

So

$$A \text{ score for TR} = \frac{0.091 + 0.100 + 0.087}{3} = 0.093$$

$$B \text{ score for TR} = \frac{0.273 + 0.300 + 0.304}{3} = 0.292$$

$$C \text{ score for TR} = \frac{0.636 + 0.600 + 0.609}{3} = 0.615$$

Applying the procedures for checking consistency, we find that

$$\frac{CI}{RI} = 0.002 < 0.1$$

which implies consistency.

CR	A	B	C
A	1	1/2	1/2
B	2	1	2
C	2	1/2	1

$$A_{\text{norm\_CR}} = \begin{bmatrix} 0.200 & 0.250 & 0.143 \\ 0.400 & 0.500 & 0.571 \\ 0.400 & 0.250 & 0.286 \end{bmatrix}$$

So

$$A \text{ score for CR} = \frac{0.200 + 0.250 + 0.143}{3} = 0.198$$

$$B \text{ score for CR} = \frac{0.400 + 0.500 + 0.571}{3} = 0.490$$

$$C \text{ score for CR} = \frac{0.400 + 0.250 + 0.286}{3} = 0.312$$

And for the consistency check, we have

$$\frac{CI}{RI} = 0.046 < 0.1$$

This also implies consistency.

Once we obtain all the scores for the alternatives with regard to the five factors, we can then rank the alternatives by calculating the weighted sum for each of the alternatives. The weightings for all the factors and scores for the alternatives are listed in Table 6.14.

TABLE 6.14

Weighted Sum Calculation for Decision Alternatives

	LC	LG	LS	TR	CR	Weighted Sum
Factor weighting	0.144	0.053	0.250	0.035	0.518	
A	0.239	0.500	0.122	0.093	0.198	**0.197**
B	0.623	0.250	0.648	0.292	0.490	**0.529**
C	0.137	0.250	0.230	0.615	0.312	**0.274**

To illustrate how Table 6.14 is derived, the weighted sum is computed as follows:

Alternative A overall score

$$= 0.144(0.239) + 0.053(0.500) + 0.250(0.122)$$
$$+ 0.035(0.093) + 0.518(0.198) = 0.197$$

Alternative B overall score

$$= 0.144(0.623) + 0.053(0.250) + 0.250(0.648)$$
$$+ 0.035(0.292) + 0.518(0.490) = 0.529$$

Alternative C overall score

$$= 0.144(0.137) + 0.053(0.250) + 0.250(0.230)$$
$$+ 0.035(0.615) + 0.518(0.312) = 0.274$$

Thus, the company would choose Site B as their new manufacturing site.

The AHP algorithm involves lots of matrix operations, which can be carried out easily using spreadsheet software such as Microsoft Excel. Taking LS as an example, the Excel procedure is illustrated in Figure 6.8.

AHP applies a hierarchical structure to assess decision-making problems for multiple decision criteria. The basic assumption of this hierarchical structure is that the attributes are homogeneous in terms of all the objectives. Another issue that needs to be mentioned here is that, for an effective application of AHP, it is important that the hierarchical structure includes only criteria that are independent, not redundant and additive ones. This ensures a valid comparison and high consistency of all the criteria. For dependent cases, the AHP framework should be modified using the feedback and supermatrix approaches (Rangone 1996).

	A	B	C	D	E	F	G
1							
2		LS	A	B	C		
3		A	1.000	0.200	0.500		
4		B	5.000	1.000	3.000		
5		C	2.000	0.333	1.000		
6		Column Sum	8.000	1.533	4.500		
7							
8		A_norm	A	B	C	Row Average	
9		A	0.125	0.130	0.111	0.122	
10		B	0.625	0.652	0.667	0.648	
11		C	0.250	0.217	0.222	0.230	
12							
13							
14		AW	0.367		AW/W	3.001317957	
15			1.948			3.007145076	
16			0.690			3.00262697	
17							
18					R	3.003696668	
19					CI	0.001848334	
20					RI	0.58	
21					CI/RI	0.003186783	
22							

	A	B	C	D	E	F
1						
2		LS	A	B	C	
3		A	1	=1/5	0.5	
4		B	=1/D3	1	3	
5		C	=1/E3	=1/E4	1	
6		Column Sum	=SUM(C3:C5)	=SUM(D3:D5)	=SUM(E3:E5)	
7						
8		A_norm	A	B	C	Row Average
9		A	=C3/$C$6	=D3/$D$6	=E3/$E$6	=AVERAGE(C9:E9)
10		B	=C4/$C$6	=D4/$D$6	=E4/$E$6	=AVERAGE(C10:E10)
11		C	=C5/$C$6	=D5/$D$6	=E5/$E$6	=AVERAGE(C11:E11)
12						
13						
14		AW	=MMULT(C3:E5,F9:F11)		AW/W	=C14/F9
15			=MMULT(C3:E5,F9:F11)			=C15/F10
16			=MMULT(C3:E5,F9:F11)			=C16/F11
17						
18					R	=SUM(F14:F16)/3
19					CI	=(F18-3)/2
20					RI	0.58
21					CI/RI	=F19/F20
22						

FIGURE 6.8
Illustration of AHP computation using Microsoft Excel (top image shows the AHP results and bottom image shows the Excel formula used to obtain the results).

6.7 Summary

In systems engineering design, resources are always limited, which requires system designers to constantly make decisions. Decision making is a process in which a decision maker must select one from a set of alternatives to achieve a specific goal. It is a selection process with some amount of information and usually involves some degrees of uncertainty. A decision-making problem can be solved by applying a six-step process: identify the problem, gather data, generate the alternatives, create an appropriate model based on the decision problem, compare the alternatives using the model, and finally execute the decision action.

In this chapter, we reviewed several decision-making models that are commonly applied in systems engineering, including decision making under risks and uncertainty. "Risks" implies that the likelihood of random future events can be estimated by probabilities, and "uncertainty" means that no knowledge of the future event's probability can be obtained. On decision making under uncertainty, the Laplace, maximax, maximin, Hurwicz and minimax regret criteria were described; on decision making under risks, we introduced the expected value criterion.

The expected value criterion is considered risk neutral since the risks (probability) and expected value are linearly related; however, this might not capture the true behavior of humans. Different decision makers have different attitudes to risks; some may be risk averse and some may like taking risks (risk seeking). Von Neumann and Morgenstern's utility theory provides a rigorous measure to incorporate attitudes to risk into the decision-making process. A decision maker's attitude to risk can be obtained by having him/her choose the appropriate level of chance between receiving the best and worst possible payoff values. Using this concept, the utility of the decision maker with regard to a certain payoff value r_i, denoted as $u(r_i)$, is the value of the probability q_i, that makes the following two lotteries equivalent:

$$\underline{\qquad 1 \qquad} r_i \quad \text{and} \quad \left\{ \begin{array}{l} \underline{\qquad q_i \qquad} \text{ most favorable reward} \\ \\ \underline{\qquad 1-q_i \qquad} \text{ least favorable reward} \end{array} \right.$$

By defining the utility function, we use the expected utility value instead of the expected value to solve the decision-making problems that are involved with risks. Using the expected utility, we can easily choose between two lotteries by the following criteria:

$$L_1 p L_2 \quad \text{if and only if} \, (\leftrightarrow) \quad E\big(U \text{ for } L_1\big) > E\big(U \text{ for } L_2\big)$$

$$L_1 i L_2 E \quad \text{if and only if} (\leftrightarrow) \quad (U \text{ for } L_1) = E(U \text{ for } L_2)$$

$$L_2 p L_1 \quad \text{if and only if} \quad E(U \text{ for } L_1) < E(U \text{ for } L_2)$$

The decision maker's attitude to the risks can be assessed by looking at the risk premium (RP) value of the decision maker's utility function. For a non-degenerate lottery, a decision maker is risk averse if and only if RP is positive, risk neutral if RP is zero, and risk seeking if RP is negative. Another way to derive this is to use the shape of the utility function; a strict concave utility function implies that the decision maker is risk averse, a strict linear function implies risk neutral, and a strict convex function implies risk seeking.

A decision tree is used to solve a decision-making problem with a sequence of decisions involved. There are two main constructs for developing a decision tree; the decision node and the event node. Once the decision tree is constructed, the expected value is calculated for each possible outcome and the decision is made by tracing the tree from the end events all the way back to the beginning of the decision tree; the optimal course of decision actions can thereby be obtained. The EVSI and EVPI values may be computed to evaluate feasibility when there is prior knowledge of a decision outcome.

In decision-making problems that involve multiple criteria/factors, the AHP method can be applied to rank the criteria and alternatives. AHP uses pairwise comparison between decision factors, and a four-step algorithm to develop normalized weightings for all the factors. To achieve valid comparison results, the consistency of the pairwise comparisons need to be checked to ensure the transitivity principle is satisfied. A CI can be computed for the pairwise comparison matrix and compared with an RI; if the ratio is less than 0.1, then the comparison can considered consistent, thus valid. An Excel template was presented at the end of the section to help readers to implement the AHP procedure efficiently.

PROBLEMS

1. What is decision making? Why is decision making is important in systems engineering?
2. What are the elements of decision making? Give an example of decision making and specify the elements of your example.
3. What are the basic steps involved in making a decision?
4. Discuss why using models is important when solving decision-making problems.
5. A company is trying determine which project of five potential options to bid on to maximize profits. The payoff matrix is illustrated as follows (in thousands of dollars). The probabilities of the possible future events are shown above each future F_i in parentheses.

	(0.10) F_1	(0.30) F_2	(0.15) F_3	(0.20) F_4	(0.25) F_5
A_1	10	40	50	−20	30
A_2	0	60	40	25	15
A_3	−20	30	30	45	20
A_4	20	50	10	50	35
A_5	−15	20	35	15	10

From this payoff matrix, find

    a. Which alternative is dominated, so can be eliminated from the list?

    b. Which alternative should be chosen if we use the expected value criterion?

6. Based on Problem 5, if the probabilities for all five futures are unknown, which alternative should be chosen based on

    a. The maximax criterion

    b. The maximin criterion

    c. The Laplace criterion

    d. The minimax regret criterion

    e. The Hurwicz criterion, if $\alpha = 0.20$

7. What is utility? Why is utility theory used to assess human decision-making behavior?

8. If a utility function is

    a. $u(x) = lnx$

    b. $u(x) = 2x + 3$

    c. $u(x) = x^3$

how would we choose between the following two lotteries?

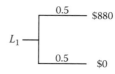

and

L_2
    0.9 —— $500
    0.1 —— −$100

9. Calculate the RP for 8a, b, and c. Do these values imply a risk-seeking, risk-neutral or risk-averse decision maker?

10. The human factors and systems (HFS) department is trying to determine which of two printers to purchase. Both machines will satisfy the department's needs. Machine 1 costs $1500 and has a warranty contract, which, for an annual fee of $120, covers all the repairs. Machine 2 costs $2000, and its annual maintenance cost is a random variable. At the present time, the HFS department believes there is a 30% chance that the annual maintenance cost for Machine 2 will be $50, a 30% chance it will be $100, and a 40% chance it will be $150. Before the purchase decision is made, the department can have a certified technician evaluate the quality of Machine 2. If the technician believes that Machine 2 is satisfactory, there is a 70% chance that its annual maintenance cost will be $50, and a 30% chance it will be $100; if the technician believes that Machine 2 is unsatisfactory, there is a 20% chance that the annual maintenance cost will be $50, a 50% chance it will be $100, and a 30% chance it will be $150. Currently, it is expected that there is a 50% chance that the technician will give a satisfactory report, and the technician will charge $50. What should the HFS department do? What are the EVSI and EVPI? Construct a decision tree to help the HFS department find the correct solution.

11. The human factors department at HFSCorp is trying to hire a regional manager for its newly opened branch in Daytona Beach, Florida. Based on the job duties and description, the hiring team has determined the following four criteria to make the selection: educational background (EB), prior experience (PE), references and recommendations (RR), and long-term career goal (CG). Since these criteria are not equally important to the job, the hiring team has made a pairwise comparison, as shown in the following matrix:

	EB	PE	RR	CG
EB	1.0	1/2	3.0	2.0
PE	2	1.0	5.0	3.0
RR	1/3	1/5	1.0	1.0
CG	1/2	1/3	1.0	1.0

After screening a pool of applicants, the hiring team has narrowed down the selection to three candidates, Joe, Mike, and Anne. Based on the four criteria, the hiring team make a pairwise comparison matrix for each of the hiring criteria among the three candidates. For EB:

EB	Joe	Mike	Anne
Joe	1.0	5	8
Mike	1/5	1.0	3
Anne	1/8	1/3	1.0

For PE:

PE	Joe	Mike	Anne
Joe	1.0	1/4	1/2
Mike	4	1.0	3
Anne	2	1/3	1.0

For RR:

RR	Joe	Mike	Anne
Joe	1.0	1/2	1/4
Mike	2	1.0	1/3
Anne	4	3	1.0

For CG:

RR	Joe	Mike	Anne
Joe	1.0	2	1/3
Mike	1	1.0	1/7
Anne	3	7	1.0

Use AHP to help the hiring team decide which candidate to hire. Check the consistency for each of the matrices.

7

System Optimization Models

Systems engineering is concerned with designing complex systems in an efficient and economic manner. No matter what types of systems are being developed, resources nevertheless have to be consumed; these resources could be human power, capital funds, raw materials, or working hours, to name a few. All resources are limited; as system engineers, we need to obtain the best value from these limited resources, or optimize their utilization. Systems engineering needs to incorporate optimization to achieve the design objectives. As a matter of fact, systems engineers carry out optimization on an almost daily basis; for example, we can think of decision making as a kind of optimization, as we need to pick the best alternatives to achieve the optimum outcome. The optimization methods and models we review in this chapter are more general, and these methods and models originate from operations research.

In this chapter, we will present a high-level introduction to optimization problems and the application of optimization models in systems engineering. Optimization is defined within the systems engineering context, and since this is not an operations research (OR) textbook, we will focus on the two most commonly used optimization models in systems design: unconstrained classic models and constrained optimization based on mathematical programming. More specifically, in this chapter, we will

1. Give a formal definition of optimization and its fundamental characteristics and features. We will give a general mathematical formulation for optimization problems.
2. Review the classic unconstrained optimization model. This type of model is based on calculus operations; some basic concepts will be reviewed.
3. Review constrained optimization models using mathematical programming. More specifically, we will focus on the basic model of mathematical programming, the linear programming (LP) model. We will discuss the formulation process and solutions of LP models based on graphical methods (two decision variables) and the Simplex algorithm, using software (i.e., Excel) to solve them.

This chapter will give an overview of the fundamental concepts of optimization that are commonly used in systems engineering problem solving, giving readers a good starting point and correct mindset for optimization formulations in complex system designs.

7.1 What Is Optimization?

As mentioned earlier, optimization is one of the most important topics in the area of applied mathematics, particularly in operations research. Optimization seeks the value of certain factors or variables to optimize the objective function value (maximization or minimization). In the simplest case, optimization is a mathematical model/algorithm/procedure to determine the value of the input variables with regard to some constraints to achieve optimal system performance, often measured as the value of an objective function(s). Optimization is a critical part of OR; many branches of OR fall under optimization, such as mathematical programming, combinatorial optimization, queuing theory, decision making, game theory, engineering design and control, and so on.

We should mention here that the generalized theory of optimization differs from the decision-making models we discussed in Chapter 6. In optimization theory, the goal is to find the decision variable values from a defined domain to optimize the objective function. Compared to decision-making theory, in which we have a limited number of activities to choose from, the possible values for decision variables in optimization problems are often from sets/domains that have unlimited values (e.g., from a continuous real interval value). One usually cannot find the solution by enumerating all the possible values into the objective functions.

We can use the following expression to define a general optimization problem:

Given a function $f: A \rightarrow \mathbb{R}$, where A is the subset of Euclidean space $\mathbb{R}^n$, and $\mathbb{R}$ is a real number, we seek an element $x_0 \in A$, such that $f(x_0) \leq f(x)$ for any x in A or $\forall x \in A$ for a minimization optimization; or $f(x_0) \geq f(x)$ for any x in A or $\forall x \in A$ for a maximization optimization.

Optimization methods and models were developed together with calculus long ago, but only in recent years, especially after the 1940s, did they start to bloom and undergo rapid development and application, due to increased demand from the advancement of computer technology, large complex system development and design, and the accelerating speed of product turnover, which made optimization a feasible tool for solving many of these problems. The application of optimization also drives the development of the theory, which led to more advancement of the OR field.

7.2 Unconstrained Optimization Models

In this section, we will introduce the basic concepts of unconstrained optimization by differentiation. Function optimization is the fundamental application of differentiation, which is covered by any first-year college calculus textbook. For readers to understand the concepts better, we give a brief overview of the basic differentiation concepts. Please keep in mind that this review is by no means comprehensive. For a more in-depth review of differentiation calculus, please see a textbook such as *Single Variable Calculus* by James Stewart (7th edn).

7.2.1 Derivatives and Rate of Change

In calculus, the derivative at a certain point is the rate of change of the function with respect to its input variable at that point. For example, a function $y=f(x)$ at the point (x_0,y_0)'s rate of change is the slope of the tangent line that passes the point (x_0,y_0), as shown in Figure 7.1. Using the formal definition of a limit, we have the following definition for the slope (provided that this limit exists).

$$\text{slope} = \lim_{x \to x_0} \frac{f(x) - f(x_0)}{x - x_0} \tag{7.1}$$

For example, the slope at $x=2$ for function $y=f(x)=x^2+3x$ is

$$\lim_{x \to 2} \frac{x^2 + 3x - 2^2 - 3(2)}{x - 2}$$

$$= \lim_{x \to 2} \frac{x^2 + 3x - 2^2 - 3(2)}{x - 2}$$

$$= \lim_{x \to 2} \frac{x^2 - 2^2 + 3(x - 2)}{x - 2}$$

$$= \lim_{x \to 2} \left[(x+2) + 3 \right] = 4 + 3 = 7$$

Using the idea of limits, we can define the derivative as follows: The derivative of a function f at the value of x, denoted as $f'(x)$, is defined as

$$f'(x) = \lim_{h \to 0} \frac{f(x+h) - f(x)}{h} \tag{7.2a}$$

provided that the above limit exists.

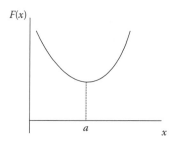

FIGURE 7.1
Local minimum point of a function.

Another equivalent way to define the derivative is

$$f'(x) = \lim_{a \to x} \frac{f(a) - f(x)}{a - x} \tag{7.2b}$$

As another notation for $f'(x)$, it can sometimes also be written in the following format:

$$y = f(x)$$

$$f'(x) = \frac{dy}{dx}$$

As a prerequisite, we assume that readers have a basic understanding of derivatives, so we will not spend too much time on deriving the derivative functions; here we just list the formulas for the most commonly used functions, as the derivatives for other functions can be derived from a combination of these common formulas.

1. Derivative of the constant c

$$y = f(x) = c, \quad f'(x) = \frac{dy}{dx} = 0 \tag{7.3}$$

Furthermore, if c is a constant and $y = cf(x)$, then

$$\frac{dy}{dx} = cf'(x) \tag{7.4}$$

2. Derivative of power functions

$$y = f(x) = x^n, \quad f'(x) = \frac{dy}{dx} = nx^{n-1} \tag{7.5}$$

3. The sum/subtract rule:
   If $f(x)$ and $g(x)$ are both differentiable, and $h(x)=f(x)\pm g(x)$, then

$$h'(x) = f'(x) \pm g'(x) \tag{7.6}$$

4. The product rule:
   If $f(x)$ and $g(x)$ are both differentiable, and $h(x)=f(x)g(x)$, then

$$h'(x) = f'(x)g(x) + f(x)g'(x) \tag{7.7}$$

5. The quotient rule:
   If $f(x)$ and $g(x)$ are both differentiable, and $h(x)=f(x)/g(x)$, then

$$h'(x) = \frac{g(x)f'(x) - f(x)g'(x)}{\left[g(x)\right]^2} \tag{7.8}$$

6. Derivatives of trigonometric functions

$$y = f(x) = \sin(x), \quad f'(x) = \cos(x) \tag{7.9}$$

$$y = f(x) = \cos(x), \quad f'(x) = -\sin(x) \tag{7.10}$$

$$y = f(x) = \tan(x), \quad f'(x) = \sec^2(x) \tag{7.11}$$

7. The chain rule:
   If g is differentiable at x and f is differentiable at $g(x)$, then the derivative for the composite function $y=f[g(x)]$ is given by

$$\frac{dy}{dx} = f'\left[g(x)\right]g'(x) \tag{7.12}$$

8. Some special functions:

$$y = f(x) = e^x, \quad f'(x) = e^x \tag{7.13}$$

$$y = f(x) = \ln(x), \quad f'(x) = \frac{1}{x} \tag{7.14}$$

Using the above basic derivatives functions, we can solve some derivative-related problems. For example, in economics, the concept of marginal cost is defined as the instantaneous rate of change of the total cost function, or the first-order derivative of the total cost function $C(x)$; the marginal cost function is $dC(x)/dx$, or $C'(x)$. Suppose that a company has estimated the cost function of producing x items is as follows:

$$C(x) = 0.05x^2 + 50x + 1000$$

The marginal cost function (or rate of change function) is given by the first-order derivative

$$C'(x) = \frac{dC(x)}{dx} = \frac{d(0.05x^2)}{dx} + \frac{d(50x)}{dx} + \frac{d(1000)}{dx} = 0.1x + 50$$

The marginal cost at $x = 100$ items is

$$C'(100) = 0.1(100) + 50 = 60 \ (\$ \text{ per item})$$

If the function has more than one independent variable, that is, $y = f(x_1, x_2, \ldots, x_n)$, to find the rate of change, we need to use partial differentiation. The process of partial differentiation is that all other variables except for the variable of interest are treated as constant, and the differentiation procedure is exactly the same as the single-variable function. The partial differentiation is denoted as $\partial y/\partial x$. For example,

$$y = f(x_1, x_2) = 4x_1^2 + 5x_1x_2 + 6x_2^3 + 8$$

The partial differentiation of y with respect to x_1 is

$$\frac{\partial y}{\partial x_1} = 8x_1 + 5x_2$$

And the partial differentiation of y with respect to x_2 is

$$\frac{\partial y}{\partial x_2} = 5x_1 + 18x_2^2$$

For a more comprehensive review of partial differentiation and related concepts, readers please refer to any calculus textbook such as *Calculus* by James Stewart (2011).

Understanding the basic concept of derivatives, we now see how derivatives can help us to find the optimal (maximum or minimum) value. Let us first look at the function values around the optimum points. We denote a as the local minimum point if and only if $f(a) \leq f(x)$, for all x around c, as illustrated in Figure 7.1.

Similarly, we define b as the local maximum point if and only if $f(b) \geq f(x)$ for all x around b, as illustrated in Figure 7.2.

From Figures 7.1 and 7.2, one can easily see that the tangent line at the local minimum or maximum is horizontal; or, using derivatives, we have $f'(a) = 0$, and $f'(b) = 0$. This is the necessary condition for the local optimum, as described in Fermat's theorem.

Local Optimum Theorem (or Fermat's Theorem)

If function f has a local minimum or maximum at c, and if the first-order derivative of f exists, then $f'(c) = 0$.

The condition $f'(c) = 0$ is only necessary for an optimum point, but not sufficient, because if a function $y = f(x)$ is differentiable at point $x = c$, and $f'(a) = 0$, the c could be a minimum or maximum, or another possibility is that c is a point of reflection (a horizontal line at c). In order to determine whether it is a minimum or maximum, we must look at the second-order derivative. For the local minimum point $x = c$, for the value less than c, the rate of change is negative, until it reaches zero at $x = c$, and then it starts to increase (becoming positive); the rate of change of the rate of change is positive (Figure 7.1), or $f''(c) > 0$; for the local maximum point, we can easily see from Figure 7.2 that the rate of change of the rate of change at the maximum point is negative, or $f''(c) < 0$. If the second-order derivative is still zero, then a higher-order derivative is sought until a nonzero value is found at the nth-order derivative. Then a similar rule applies to determine the optimum point:

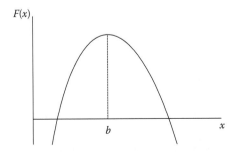

FIGURE 7.2
Local maximum point of a function.

$x = c$ is the local minimum if $f^{(n)}(c) > 0$, n is even

$x = c$ is the local minimum if $f^{(n)}(c) < 0$, n is even

$x = c$ is the point of reflection if $f^{(n)}(c) = 0$, n is odd

For functions $y = f(x_1, x_2, \ldots, x_n)$ with more than one independent variable, the necessary condition that the point $x^* = [x_1, x_2, \ldots, x_n]$ is a stationary point is that the first-order partial derivative at x^* equals zero, or,

$$\left[\frac{\partial y}{\partial x_1}, \frac{\partial y}{\partial x_2}, \ldots, \frac{\partial y}{\partial x_n} \right] = 0 \qquad (7.15)$$

To determine whether x^* is the local minimum or maximum, the second-order partial derivatives are similarly required; namely, the *Hessian matrix*, which is derived as

$$H(f) = \begin{bmatrix} \dfrac{\partial^2 f}{\partial x_1^2} & \dfrac{\partial^2 f}{\partial x_1 \partial x_2} & \cdots & \dfrac{\partial^2 f}{\partial x_1 \partial x_n} \\[2mm] \dfrac{\partial^2 f}{\partial x_2 \partial x_1} & \dfrac{\partial^2 f}{\partial x_2^2} & \cdots & \dfrac{\partial^2 f}{\partial x_2 \partial x_n} \\[2mm] \cdots & \cdots & \cdots & \cdots \\[2mm] \dfrac{\partial^2 f}{\partial x_n \partial x_1} & \dfrac{\partial^2 f}{\partial x_n \partial x_2} & \cdots & \dfrac{\partial^2 f}{\partial x_n^2} \end{bmatrix} \qquad (7.16)$$

If the determinant of $H(f(x^*))$ or $|H(f(x^*))| > 0$, then $x^* = [x_1, x_2, \ldots, x_n]$ is a minimum point and if $|H(f(x^*))| < 0$, then $x^* = [x_1, x_2, \ldots, x_n]$ is a maximum point.

As mentioned earlier, in systems engineering, we often need to put limited resources to the best use; differentiation-based optimization is one of the most commonly used models to find the lowest cost, maximum profit, least use of materials, and so on. Let us see how we can use the above conditions to solve an optimization problem.

Example 7.1

Imagine that we are trying to build a cylindrical container. The container is required to have a volume of V. What is the radius of the base circle (r) that uses the minimum amount of materials to build the cylinder? What is the ratio of the height h to r?

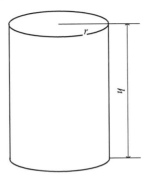

FIGURE 7.3
Cylinder container.

Solution: A cylinder is shown in Figure 7.3. If we ignore the thickness of the materials, the amount of the materials to build the cylinder can be represented by the area of the cylinder. Let us denote the area of the cylinder as S, and the height of the cylinder is h, and assume that the container is sealed, so we have the following formula for the total area:

$$S(r) = 2\pi r^2 + 2\pi rh$$

because

$$V = \pi r^2 h$$

so

$$h = \frac{V}{\pi r^2}$$

Substituting into the formula for the total area, we have

$$S(r) = 2\pi r^2 + 2\frac{V}{r}$$

$S(r)$ is continuous and differentiable, so applying the theorem of optimization, we obtain the following equation

$$S'(r) = 4\pi r - 2\frac{V}{r^2} = 0$$

Solving the above equation, we have

$$r = \sqrt[3]{\frac{V}{2\pi}}$$

To verify that this is truly the minimum value, we use the second-order derivative as follows:

$$S''(r) = 4\pi + 4\frac{V}{r^3} > 0$$

so it is truly a minimum value.

$$\frac{h}{r} = \frac{\dfrac{V}{\pi r^3}}{r} = \frac{V}{\pi r^3} = \frac{V}{\pi \dfrac{V}{2\pi}} = 2$$

The ratio $h/r = 2$ when materials used are minimum.

Let us look at an example that involves multiple independent variables: $f(x,y,z) = x^2 + 2y^2 + 3z^2 + 2x + 4y - 12z$. First, apply the necessary condition to find the stationary point:

$$\begin{cases} \dfrac{\partial f}{\partial x} = 2x + 2 = 0 \\[2em] \dfrac{\partial f}{\partial y} = 4y + 4 = 0 \\[2em] \dfrac{\partial f}{\partial z} = 6z - 12 = 0 \end{cases}$$

So, we can obtain the stationary point as (−1, −1, 2).

Next, we need to derive the Hessian matrix by taking the secondary partial derivative as follows:

$$\frac{\partial^2 f}{\partial x^2} = 2, \quad \frac{\partial^2 f}{\partial x \partial y} = 0, \quad \frac{\partial^2 f}{\partial x \partial z} = 0$$

$$\frac{\partial^2 f}{\partial y \partial x} = 0, \quad \frac{\partial^2 f}{\partial y^2} = 4, \quad \frac{\partial^2 f}{\partial y \partial z} = 0$$

$$\frac{\partial^2 f}{\partial z \partial x} = 0, \quad \frac{\partial^2 f}{\partial z \partial y} = 0, \quad \frac{\partial^2 f}{\partial z^2} = 6$$

So, the Hessian matrix is

$$H = \begin{bmatrix} 2 & 0 & 0 \\ 0 & 4 & 0 \\ 0 & 0 & 6 \end{bmatrix}$$

and it is easy to see that $|H|>0$, so $f(x,y,z)$ reach the minimum point at $(-1,-1,2), f_{min}=-15$.

7.3 Constrained Optimization Model Using Mathematical Programming

With most of the optimization problems with which engineering design is concerned, the resources are often limited. Such resources could be human labor hours, capital funds, materials, or equipment. As system designers, we need to seek the optimal utilization of the limited resources such as maximizing the profit; or, sometimes, when the objectives of the system are set, we need to seek the most optimal planning of production or scheduling of labor to minimize the cost. For example, a production schedule would want to maximize weekly profit by producing as many items as possible, but the schedule is subject to the limited raw materials, the workforce, and the working hours available per week. Similar examples may also be found choosing between different investments that have alternative payment schedules, or work-schedule problems to meet certain days' requirements yet minimizing the number of employees, and many more.

This type of problem is addressed as constrained optimization, and usually solved by mathematical programming. The term *mathematical programming* was first used in the 1940s; the objective of mathematical programming models is optimization. Since the 1940s, with the rapid advancement of computer technology and the demand from systems design problems, mathematical programming has quickly grown as an active, independent discipline. Nowadays, mathematical programming has been applied widely in many fields; its theory and application have been brought into the natural sciences, social sciences, and engineering. Based on the nature of the problems and different methods/algorithms involved, mathematical programming covers a wide range of subdomains, including linear programming, nonlinear programming, multiobjective programming, dynamic programming, integer programming and stochastic programming, and so forth. For a comprehensive review of these subfields, readers can refer to any book such as Winston (2005). In this text, we are primarily going to discuss LP, as it is the most fundamental model in mathematical programming. Understanding LP will help us to grasp the idea of mathematical programming, and, furthermore, LP is one of the most widely used models within it. In the next section, we will first introduce the basic ideas of LP problem formulation, and discuss several different methods for solving an LP problem, including the graphical method, the Simplex algorithm, and the use of software (such as a spreadsheet).

7.3.1 Linear Programming

In mathematical programming, if the objective function and constraints functions are all linear and defined on real numbers, then the optimization becomes an LP problem. LP has been applied extensively in engineering and management science; learning LP will help the designer formulate and solve such problems effectively. Moreover, it will also facilitate solving nonlinear programming problems, as many of these are approximated in linear form in a local region, so that the LP algorithm can be used to solve it approximately. It is essential to learn the basic concepts and models of LP before learning any other mathematical modeling techniques.

In this section, we first give a formal definition of the LP model, and use some examples to illustrate how to formulate a LP problem, followed by a simple graphical way to solve the problem. The graphical method is only applicable to two-variable problems; for problems involving more than two variables, we will introduce the general algorithm called the Simplex method; finally, we will learn how to use software to solve LP problems. Due to limitations of space, this is only a very brief introduction to LP material; further reading is recommended for in-depth understanding of LP models.

What is LP? An LP problem is an optimization problem with the following characteristics:

1. It attempts to maximize (or minimize, depending on the objective) a linear function of *decision variables*, usually denoted by x_i, $i = 1, 2, ..., n$, with n being the number of decision variables. The function that is to be maximized or minimized is called the *objective function*.

2. The values of the decision variables must satisfy a set of constraints. Each constraint must be represented by a linear equation or linear inequality.

3. A sign restriction is associated with each decision variable. That is, for any decision variable x_i, the sign restriction usually specifies that x_i must be nonnegative ($x_i \geq 0$).

4. Since only linear equations are involved, which means that all the properties of linear equations are also assumed in LP, this includes the proportionality assumption (the effect of a decision variable is proportional to a constant quantity), the additivity assumption (the combined effect of different decision variables equals the algebraic sum of the individual variables), and the divisibility assumption

(assumes that fractional values are allowed for decision variables, except where integers are strictly required, which is the integer programming model), and, of course, the certainty assumption requires that each parameter is known with certainty.

7.3.1.1 Formulation

To formulate an LP problem, one first needs to understand the problem, identify the objective of the program, and based on the problem, identify the decision variables, that is, those variables that can be manipulated by us. By knowing the decision variables, we can represent the objectives and constraints of the problem by deriving the objective functions and constraint functions. Generally speaking, an LP model formulation will have the following format:

Objective function:

$$\text{Max}(\text{or min}) \ E = \sum_{j=1}^{n} e_j x_j \tag{7.17}$$

subject to the m constraints

$$\sum_{j=1}^{n} a_{ij} x_j = c_j, \quad i = 1, 2, \ldots, m$$

$$x_j \geq 0, j = 1, 2, \ldots, n \ (\text{sign restriction}) \tag{7.18}$$

Let us use several examples to illustrate how to formulate an LP problem.

Example 7.2

A company manufactures two types of toy furniture: a desk and a chair. A toy desk makes $6 profit and a toy chair makes $5 profit. Each type of toy needs to be assembled/processed and painted before being shipped out to the customer. A toy desk needs 4 h assembly/processing time and 1 h painting time; a toy chair needs 3 h assembly/processing time and 2 h painting time. Each week, the company has 200 h assembly time available and 100 h painting time available dedicated to these two products. The production requirements for the two products are listed in Table 7.1.

TABLE 7.1
Production Requirements for Example 7.2

	Products		Weekly Capacity
	Desk	Chair	
Assembly (h/unit)	4	3	200 h
Painting (h/unit)	1	2	100 h

Assuming all the weekly products will be sold, as a production manager, how do you plan the weekly production to maximize the weekly profit?

Solution: After understanding the problem, we first need to define the decision variables. The decision variables should completely define the decisions to be made. In this example, the decision is to plan the weekly production; more specifically, to determine the numbers of units to be produced weekly of the desk and the chair. With this definition, we define the following decision variables for this problem:

$$x_1 = \text{the weekly production of desks}$$

$$x_2 = \text{the weekly production of chairs}$$

With the decision variables being clearly defined, we can easily write the objective function in terms of the decision variables. Each desk makes a \$6 profit and each chair makes a \$5 profit; thus, the objective for this particular problem is to maximize the weekly profits, that is,

$$\text{maximize } z = 6x_1 + 5x_2$$

If there are no constraints, then the values of x_1 and x_2 would tend to infinity to maximize the profit. Unfortunately, the values of x_1 and x_2 are subject to the following two constraints based on the problem description:

> Constraint 1: There are only 200 h assembly time available for the week (this could be due to the limited workforce and hours for assembly work).
> Constraint 2: There are only 100 h painting time available for each week (This capacity again could be due to the limited workforce/hours available each week).

Based on the information given to us (Table 7.1), we can easily formulate the mathematical equation for the two constraints, written as follows:

$$\text{Constraint 1 } \left(\text{assembly}\right): 4x_1 + 3x_2 \le 200$$

$$\text{Constraint 2 } \left(\text{painting}\right): x_1 + 2x_2 \le 100$$

Finally, we have to consider the sign requirements for the decision variables. Can the decision variables take negative values or do they need to be nonnegative? Based on the nature of the problem, it is clear that both of the decision variables have to be nonnegative, or $x_1 \ge 0$ and $x_2 \ge 0$. Sometimes the decision variables can take a negative value; for example, if the decision variable is a financial account balance value, then it is allowed to be negative. When considering the sign requirements for the decision variables, we have to examine the nature of the problem and the physical meanings of the variables. Combining the objective function, the constraints function, and the sign restriction requirements for the decision variables, we have the following formulation for Example 7.2:

$$\text{Maximize } z = 6x_1 + 5x_2 \text{ (objective function)}$$

Subject to (or s.t.)

$$4x_1 + 3x_2 \leq 200 \text{ (assembly constraint)}$$

$$x_1 + 2x_2 \leq 100 \text{ (painting constraint)}$$

$$x_1, x_2 \geq 0 \text{ (sign restriction for decision variables)}$$

From the above example, we can see that defining the decision variables is the key to formulating the LP problem. Usually the decision variables are very straightforward to define from the problem description. In some special cases, decision variables need to be carefully sought, and defining the decision variables correctly can greatly simplify the LP formulation.

In a real-world application, problems often require us to think critically and seek the best way to define the decision variables so that formulation of the problems can be simplified. The smaller the size of the model, the easier it is to solve. Critical thinking skills do not develop quickly; they require training in mathematics and extensive exercises and practice. We have included some appropriate exercise problems at the end of the chapter to help readers to practice. For a more comprehensive review of different special cases of LP formulation techniques, readers can refer to Winston (2005).

To solve an LP model is essentially to find the optimal value for the linear objective function, under the constraint that a set of linear functions has to be satisfied. In 1939, Soviet mathematician Leonid Knatorovich developed the earliest methods to solve LP problems; his work has been further advanced by George B. Dantzig, who published the Simplex method in 1947, which was a major milestone for solving LP problems. The Simplex method unified LP solutions; with the implementation of a software algorithm, the Simplex method has become a fast and effective solution for LP models.

7.3.1.2 Solving LP Models Using Graphical Method

In this chapter, we will describe the graphical method, the Simplex method and the use of Excel to solve LP models.

If there are only two or three decision variables (in fact, two variables are preferred, because a three-dimensional plot is hard to visualize on a two-dimensional surface), the graphical method can be applied to solve the LP model. The graphical method is simple and intuitive, and thus can be easily understood; it describes the procedure for solving LP models from a geometric perspective. Learning the graphical method can also facilitate understanding of the nature and fundamental theory for solving any LP model, as the principles are the same.

Let us use the previous example to illustrate the detailed procedure for the graphical method. From Example 7.2, we have the following LP model:

$$\text{Maximize } z = 6x_1 + 5x_2 \quad \left(\text{objective function}\right)$$

Subject to (or s.t.)

$$4x_1 + 3x_2 \leq 200 \quad \left(\text{assembly constraint}\right)$$

$$x_1 + 2x_2 \leq 100 \quad \left(\text{painting constraint}\right)$$

$$x_1, x_2 \geq 0 \quad \left(\text{sign restriction for decision variables}\right)$$

We will use the Cartesian coordinates system (or rectangular coordinate system) to make the graphs. The decision variable x_1 will be the x-axis and the decision variable x_2 the y-axis. Since both x_1 and x_2 are nonnegative, we will only use the first quadrant (the upper right part) of the coordinate grid, as shown in Figure 7.4.

7.3.1.2.1 Feasible Region of the LP Problem

Step 1: Plot the feasible regions, as represented by the constraints functions.

The feasible region of the LP problem is the set of all the points (the values of the control variables) that satisfy all the constraints functions of the LP model. To find the feasible region, we need to plot all the individual regions that each constraint function represents; the intersection of all the individual regions (if the intersection exists) is the feasible region.

To find the feasible region, we first change the sign (<, ≤, >, or ≥) of the inequality of the constraint function to the equals sign (=)and plot the line of the equation. For example, for the Constraint (1), $4x_1 + 3x_2 \leq 200$, we first plot the line of $4x_1 + 3x_2 = 200$, as shown in Figure 7.5.

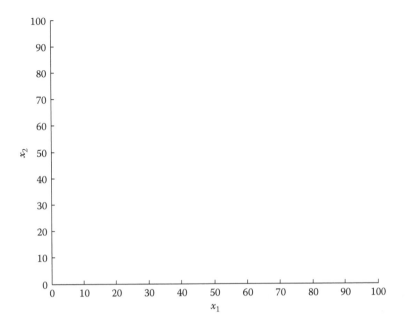

FIGURE 7.4
First quadrant of the coordinate system.

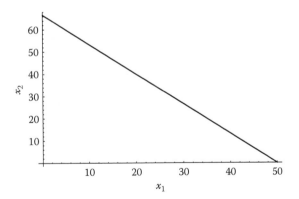

FIGURE 7.5
Line plot for $4x_1 + 3x_2 \leq 200$.

A small hint for plotting this line: a simple way to do so is to find two points, by first letting $x_1 = 0$, from which we get $x_2 = 200/3$, and then letting $x_2 = 0$, from which we get $x_1 = 50$. Connecting $(0, 200/3)$ and $(50, 0)$, a function line can be plotted, as seen from Figure 7.5. The line is solid because the original inequality constraint function is $4x_1 + 3x_2 \leq 200$. If the

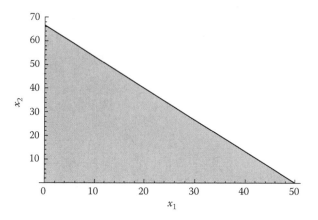

FIGURE 7.6
Feasible region for constraints $4x_1 + 3x_2 \leq 200$.

constraint function is a strict inequality (i.e., < or >), then a dotted line would be used.

Once the linear function is plotted, the region of the constraint function is determined, as shown in the shaded area in Figure 7.6.

Using a similar approach, we can plot the other constraint functions, as seen in Figure 7.7. The intersection of both the constraints is represented by the shaded area in Figure 7.7. All the points in the shaded area satisfy both of the constraints.

Step 2: Plot the objective function to find the optimum value z.

Once we have the feasible region identified, the next step is to plot the objective function $z = 6x_1 + 5x_2$, moving the function within the feasible region to

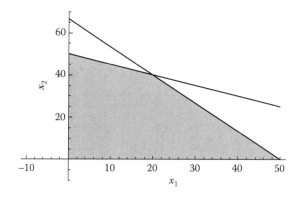

FIGURE 7.7
Feasible region as illustrated in shaded area (the intersection between $4x_1 + 3x_2 \leq 200$ and $x_1 + 2x_2 \leq 100$).

find the optimum value. With different points within the feasible region, the objective function will have different values of z. The objective function value z can be thought of as a constant; to plot the objective function, we can apply an arbitrary value to z (the easiest way to do this is to let $z=0$ initially), then we have

$$x_2 = -\frac{6}{5}x_1 \qquad\qquad (7.19)$$

Equation 7.19 is fairly easy to plot; for example, we can take two points to plot this equation that satisfy it easily, $(0,0)$ and $(5,-6)$, as shown in Figure 7.8 (the thick line at lower left, passing through $(0,0)$).

When we move the starting z-line to the right, as indicated by the arrow in Figure 7.8, the value of the objective function increases; we can think of a series of parallel lines with the same slope as $x_2=-(6/5)x_1$. As long as there are shaded areas untouched by the line, we can still move the line toward the upper right to make the z-value larger, until we arrive at the last point(s) of the feasible region; that is to say, if we move beyond this point, we would leave the feasible region. This means that we have arrived at the maximum value of z, as shown in Figure 7.8. From Figure 7.8 it is easy to see that the optimum point is the intersection point of the two constraint equation lines, $x_1=20$, $x_2=40$. If we substitute this into the objective function, we can obtain the maximum weekly profit, $z=6(20)+5(40)=320$ ($). We have to be careful here in identifying the graphical solution, since, sometimes, when we plot the equations, this might not be precise enough for us to obtain the exact answers visually, especially when we are using paper to plot them. We need to vigorously obtain the solution by solving the intersection points of the two constraint equations analytically, that is, solving

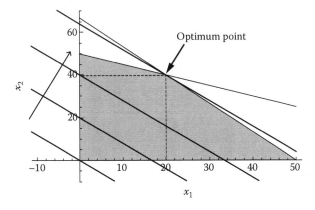

FIGURE 7.8
Illustration of the plot of objective functions.

$$\begin{cases} 4x_1 + 3x_2 = 200 \\ x_1 + 2x_2 = 100 \end{cases}$$

and obtaining the solution $x_1 = 20$, $x_2 = 40$.

From the above example, we can see that usually the optimum points occur at the corner points of the feasible region, assuming that the feasible region is limited and bounded, as in Figure 7.7. However, there are some other possible cases of feasible regions of which readers need to be aware.

7.3.1.2.2 Unlimited Number of Solutions

For the above example, if the objective function becomes $z = 8x_1 + 6x_2$, then this objective function is parallel to one of the constraints functions, so ultimately the objective function will not just touch one corner point, but a line segment, that is, the segment P1P2 on Figure 7.9. So every point between P1 and P2 are optimal solutions for the LP model. Since there are an unlimited number of points on the segment, so there are an unlimited number of solutions for this problem.

7.3.1.2.3 Unbounded Feasible Region

If the feasible region is unbounded, as illustrated in Figure 7.10, no matter how far the z-line moves upward, it is possible that the z-line will meet with a corner point. In this case, the optimal solution for this problem is unbounded and does not exist.

The above example illustrates the maximization of an objective function. For the case of minimizing an objective function, we have a similar approach, as follows:

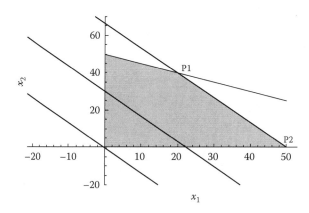

FIGURE 7.9
Unlimited number of optimal solution, as indicated by the segment P1 to P2.

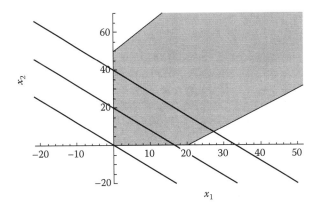

FIGURE 7.10
Illustration of unbounded feasible region, represented by the shaded area.

1. Plot all the equations for all the constraints. Similarly to the above example, we first plot the equation for each of the constraints, and then identify the regions that the inequality represents. If we are not sure which side of the equation line the region should be, a simple way to verify this is to take two points, one from each side of the equation line, substitute them into the equation, and see which point satisfies the inequality; then, the region to which that point belongs should be the correct region for that particular constraint.

2. Find the feasible region by identifying the intersection of all the regions of the constraints.

3. Give the objective function a *z* an initial value, plot the objective function line, then move the objective function parallel to the start line, and toward the direction in which the value of *z* is reduced (usually to the left and below the axis; first make sure it is touching the feasible region) until the parallel line touches the corner point of the feasible region; this will yield the optimum point for the minimization of the objective function.

We have prepared some exercise problems at the end of the chapter to both maximize and minimize the objective functions; readers can practice these procedures to see the difference in solving LP maximization and minimization problems. Nevertheless, the graphical method is very simple and intuitive, but the problem is that it only can solve LP models with no more than two (or three, for a 3D plot) control variables. For LP models that involve more than three variables, a general analytical algorithm is needed. The most common algorithm is the Simplex method, which we will discuss in the next section.

7.3.1.3 Simplex Algorithm

The Simplex algorithm requires that the LP problem first needs to be converted to the standard form. From the previous examples, it is easily seen that most of the constraints have the inequality form (i.e., $\leq$ or $\geq$). The Simplex algorithm requires that all the constraints are equations, without changing the model; this form of LP is called the standard form of LP. A standard form of LP can described as follows:

$$\max z = c_1x_1 + c_2x_2 + \cdots + c_nx_n = \sum_{j=1}^{n} c_j x_j$$

Subject to (s.t.)

$$\begin{cases} a_{11}x_1 + a_{12}x_2 + \cdots + a_{1n}x_n = b_1 \\ a_{21}x_1 + a_{22}x_2 + \cdots + a_{2n}x_n = b_2 \\ \quad\cdot \\ \quad\cdot \\ \quad\cdot \\ a_{m1}x_1 + a_{m2}x_2 + \cdots + a_{mn}x_n = b_m \end{cases} \quad x_1, x_2, \ldots, x_n \geq 0$$

(7.20)

Sometimes, we can write the above set of equations as $\mathbb{A}x = \mathbb{b}$, with

$$\mathbb{A} = \begin{bmatrix} a_{11} & a_{12} & \cdots & a_{1n} \\ a_{21} & a_{22} & \cdots & a_{2n} \\ \cdot & \cdot & & \cdot \\ \cdot & \cdot & & \cdot \\ a_{m1} & a_{m2} & \cdots & a_{mn} \end{bmatrix}$$

and

$$x = \begin{bmatrix} x_1 \\ x_2 \\ \cdot \\ \cdot \\ \cdot \\ x_n \end{bmatrix}, \quad b = \begin{bmatrix} b_1 \\ b_2 \\ \cdot \\ \cdot \\ \cdot \\ b_m \end{bmatrix}$$

There are four characteristics for the standard form of LP:

1. The objective function is a maximization function, $\max z$.
2. All the constraints functions are equations (=).
3. All the decision variables are nonnegative, that is, $x_j \geq 0, j = 1, 2, \ldots, n$.
4. All the right-hand side constants of the constraint functions are non-negative, that is, $b_i \geq 0, i = 1, 2, \ldots, m$.

Based on the definition of the standard form of the LP model, we can easily convert any LP model to a standard form; this is the standardization process of the LP model.

1. Since it is required to maximize the objective function, if the objective happens to be a minimization problem, or $\min z = \sum_{j=1}^{n} c_j x_j$, then we can multiply both sides of the equation by -1 and turn it into a maximization problem; that is to say, the original minimization problem is equivalent to

$$\max z' = \min(-z) = -\sum_{j=1}^{n} c_j x_j \tag{7.21}$$

2. We need to convert inequalities to equations. Here, we need to use a new concept of slack variables ($s_i \geq 0, i = 1, 2, \ldots, m$), with one slack variable for each constraint.

   If the constraint has the following form,

$$\sum_{j=1}^{n} a_{ij} x_j \leq b_i$$

then we use the slack variable $s_i \geq 0$, so that the constraint becomes

$$\sum_{j=1}^{n} a_{ij} x_j + s_i = b_i$$

On the other hand, if the constraint has the $\geq$ form, as follows,

$$\sum_{j=1}^{n} a_{ij} x_j \geq b_i$$

then we use the slack variable $s_i \leq 0$, to convert the constraint to

$$\sum_{j=1}^{n} a_{ij}x_j - s_i = b_i$$

By using the slack variables, we can convert inequality constraints to equation forms without changing the constraints themselves.

3. If one of the right-hand side values of a constraint equation is negative, we just need to multiply both sides of the equation by -1 and change its value to nonnegative.

Using the above procedure, we can convert Example 7.2 to the following standard form by adding two slack variables, s_1 and s_2.

$$\text{Maximize } z = 6x_1 + 5x_2$$

Subject to (or s.t.)

$$4x_1 + 3x_2 + s_1 = 200 \quad \left(\text{assembly constraint}\right)$$

$$x_1 + 2x_2 + s_2 = 100 \quad \left(\text{painting constraint}\right)$$

$$x_1, x_2, s_1, s_2 \geq 0$$

7.3.1.3.1 Concept of LP Solutions

For the LP problem illustrated by Equation 7.20, if $X = (x_1, x_2, \ldots, x_n)^{\mathrm{T}}$ satisfies all m constraints, then we call $X = (x_1, x_2, \ldots, x_n)^{\mathrm{T}}$ a feasible solution for the LP problem; the set of all feasible solutions is called the feasible set/region of the LP problem.

7.3.1.3.2 Basic and Nonbasic Variables

Consider that $\mathbb{A}x = \mathbb{b}$ has n variables and m linear equations and assume $n \geq m$. We can define a basic solution to $\mathbb{A}x = \mathbb{b}$ by setting $n - m$ variables equal to zero and solving the remaining m equations. (Assuming the remaining m variables are linear independent, then according to Cramer's rule, a unique solution of the m variables exists.)

We call the $n - m$ variables (those equal to zero) nonbasic variables (NBVs) and the solution of the remaining m variables are called basic variables (BVs).

From the above definition, it is easy for us to see that any basic solution in which all variables are nonnegative is a basic feasible solution (BFS). By using the concept of the basic solution, we can easily derive a set of basic solutions for the LP problem initially. We can arbitrarily set $n - m$ variables equal to zero, or if we have m slack variables, we can set all the nonslack variables $x_j = 0$, and make the slack variables s_i as BVs. The procedure of the Simplex algorithm is an iterative process that substitutes the NBVs with BVs in the direction of increasing the objective function value, yet maintaining the constraints not being violated. It is proven that if the optimal solution for LP exists, then there must exist a basic feasible solution that is optimal. The theorems of optimal basic feasible solutions, and direction and unboundedness, can be found in depth in Winston (1995). In the next section, we will use an example to illustrate how the Simplex algorithm is used to find the optimal solution of LP problems.

7.3.1.3.3 Simplex Algorithm Procedures

The Simplex algorithm consists of five steps to find the optimal solution, described as follows (Winston 1995):

1. Covert the LP to standard form.
2. Obtain an initial basic feasible solution (BFS) from the standard form.
3. Determine if the current BFS is optimal. If yes, then stop. If no, proceed to Step 4.
4. If the current BFS is not optimal, determine which of the NBVs should become a BV, and meanwhile determine which one of the BVs should leave the set of BVs and become an NBV.
5. Use linear equation operations to obtain the value for the new BV and new improved objective value.

Let us use Example 7.2 to illustrate the procedure.

1. Convert the LP to its standard form.

    As illustrated in the previous section, we have converted the LP model to its standard form by using slack variables, as follows:

$$\text{Maximize } z = 6x_1 + 5x_2$$

Subject to (or s.t.)

$$4x_1 + 3x_2 + s_1 = 200 \quad \left(\text{assembly constraint}\right)$$

$$x_1 + 2x_2 + s_2 = 100 \left(\text{painting constraint}\right)$$

$$x_1, x_2, s_1, s_2 \geq 0$$

2. Obtain a BFS.

For this example, the simplest way to obtain a BFS is to let $x_1 = 0$, $x_2 = 0$, and the slack variable can be easily solved as $s_1 = 200$, $s_2 = 100$.

To perform the iteration of Steps 2, 3, and 4, it is more convenient to use a tabular format to help with the iteration; the table is called a Simplex table, as shown in Figure 7.11.

In Figure 7.11, Columns 2 and 3 (inside the dotted oval area) are the basic feasible solutions (BFSs) and their values (solutions), and those BFSs correspond to the unit matrix in the table, as indicated by the dotted rectangle in the table. The top row is the parameter list of the variables in the objective function, and the left-hand row lists the parameters of the BFSs in the objective functions. When performing the Simplex iteration, only the portion enclosed by the rounded rectangle will perform the linear transformation required; the first two columns change as the BFSs change.

To illustrate the idea, let us use our example to see how to construct the Simplex table, as shown in Table 7.2.

$c_j \rightarrow$			c_1	c_2	...	c_m	...	c_n
c_B	BFS	b	x_1	x_2	...	x_m	...	x_n
c_1	x_1	b_1	1	0	...	0	...	a_{1n}
c_2	x_2	b_2	0	1	...	0	...	a_{2n}
...	...	...	...	...	...	...	...	...
c_m	x_m	b_m	0	0	...	1	...	a_{mn}
z_j			$\sum\limits_{i=1}^{m} a_{i1} c_i$	$\sum\limits_{i=1}^{m} a_{i2} c_i$	...	$\sum\limits_{i=1}^{m} a_{im} c_i$	...	$\sum\limits_{i=1}^{m} a_{in} c_i$
			$= c_1$	$= c_2$		$= c_m$		
$c_j - z_j$			0	0	...	0	...	c_n $-\sum\limits_{i=1}^{m} a_{in} c_i$

FIGURE 7.11
Simplex table format.

TABLE 7.2

Initial Setup for Simplex Algorithm for
Example 7.2

$c_j \rightarrow$			6	5	0	0
C_B	BFS	b	x_1	x_2	s_1	s_2
0	s_1	200	4	3	1	0
0	s_2	100	1	2	0	1
	z_j		$0(4)+0(1)=0$	0	0	0
	$c_j - z_j$		6	5	0	0

3. Test for optimality.

   If all the $c_j - z_j \leq 0$, $i = 1, 2, \ldots, n$, then we have found the optimal BFSs; stop the iteration and report the results. If there is at least one $c_j - z_j > 0$, then a better BFS is possible. If this is the case, move on to step 4. In Table 7.2, at least two $c_j - z_j$ are positive, 6 and 5. So, we need to find a better BFS.

4. Find a better BFS to increase the objective function value, and update the Simplex table.

   There are three substeps involved in this step; first, we need to find which variable in the BFS set needs to leave the set, and then we need to determine which variable in the non-BFS set needs to enter the BFS set.

   a. Determine the variable to enter the BFS set. One thing we need to mention first is that all the values of $c_j - z_j$ for all the BFSs are zero (which can be easily seen from the table, since all the BFS correspond to a unit basic matrix, so for each variable in the set of BFS, $z_j = c_j$). As mentioned earlier, as long there is one $c_j - z_j > 0$, the corresponding x_j is eligible to enter the set of BFSs. If there is more than one variable such that $c_j - z_j > 0$, we choose the one with the largest $c_j - z_j$ value; namely, we find k such that

   $$c_k - z_k = \max_j (c_j - z_j)$$

   From Table 7.2, it is easy to see that x_1 should be the variable to enter the set of BFSs, since it has the largest value of $c_j - z_j = 6$, so we mark it as Column k, as seen in Figure 7.12.

   b. Determine the variable to leave the set of BFSs. Based on the result from (a), calculate the value of θ for Column k as

   $$\theta_i = \frac{b_i}{a_{ik}}, \quad i = 1, 2, \ldots, m$$

   And the row corresponding to the *least positive* value of θ_i should be the variable to leave the set of BFSs; namely

Systems Engineering: Design Principles and Models

c_B	BFS	b	$c_j \rightarrow$ 6 x_1	5 x_2	0 s_1	0 s_2	θ
0	s_1	200	4	3	1	0	$\dfrac{200}{4} = 50$
0	s_2	100	1	2	0	1	$\dfrac{100}{1} = 100$
	z_j		0	0	0	0	
	$c_j - z_j$		6	5	0	0	

L

K

FIGURE 7.12
First step result of the Simplex algorithm.

$$\theta_l = \min\left\{ \frac{b_i}{a_{ik}} \,\Big|\, a_{ik} > 0 \right\} = \frac{b_l}{a_{lk}}$$

From Figure 7.12, we can easily see that s_1 should be the variable to leave the set of BFSs.

  c. Update the Simplex table to obtain a new BFS. First, we need to update Column BFS by replacing s_1 with x_1, and update the c_i column by replacing the corresponding parameter of s_1 with that of x_1 (changing 0 to 6). A basic row linear transformation is performed in the Simplex table, to change Row k to be part of the basic unit matrix, or to make $a'_{lk} = 1$ (in this case divide the row by 4). To do that, we need to

  i. Divide row l by a_{lk}; we have for all the elements in Row l,

$$b'_l = \frac{b_l}{a_{lk}}$$

$$a'_{ij} = \frac{a_{ij}}{a_{lk}}$$

(7.22)

So, from Figure 7.12, we can obtain Table 7.3.

  ii. For Column k, transform the element $a_{ik} = 0$, for all $a_{ik} \neq a_{lk}$. Applying the basic linear equation transformation, we have, for all other rows, $i \neq l$,

TABLE 7.3

Step 1 Iteration Calculation

c_B	BFS	b	x_1	x_2	s_1	s_2
	$c_j \rightarrow$		6	5	0	0
6	x_1	$200/4 = 50$	1	3/4	1/4	0
0	s_2	100	1	2	0	1
	z_j					
	$c_j - z_j$					

TABLE 7.4

Step 1 Iteration Calculation Results

c_B	BFS	b	x_1	x_2	s_1	s_2
	$c_j \rightarrow$		6	5	0	0
6	x_1	50	1	0.75	0.25	0
0	s_2	50	0	1.25	−0.25	1
	z_j		6	4.5	1.5	0
	$c_j - z_j$		0	0.5*	−1.5	0

c_B	BFS	b	x_1	x_2	s_1	s_2	θ
	$c_j \rightarrow$		6	5	0	0	
6	x_1	50	1	.75	.25	0	200/3
0	s_2	50	0	1.25	−.25	1	40
	z_j		6	4.5	1.5	0	
	$c_j - z_j$		0	1.5*	−1.5	0	

FIGURE 7.13
Second step result of the Simplex algorithm.

$$b'_l = b_i - \frac{b_i}{a_{lk}} a_{ik}$$

$$a'_{ij} = a_{ij} - \frac{a_{ij}}{a_{lk}} a_{ik} \tag{7.23}$$

So, from Table 7.3 we obtain Table 7.4.

Using the procedures in Step 3, we can see that there is one $c_j - z_j > 0$, so optimality is not reached yet. Repeat Step 4 to find the variables to enter the set of BFS and leave it; we reach the situation shown in Figure 7.13.

TABLE 7.5

Step 2 Iteration Calculation Results

	$c_j \rightarrow$		6	5	0	0
c_B	BFS	b	x_1	x_2	s_1	s_2
6	x_1	20	1	0	0.4	−0.6
5	x_2	40	0	1	−0.2	0.8
	z_j		6	5	1.4	0.4
	$c_j - z_j$		0	0	−1.4	−0.4

So, we know x_2 is the variable entering and s_2 is the variable leaving. Applying the basic linear transformation operation from Step 4, we obtain the Simplex shown in Table 7.5.

From Table 7.5, all the $c_j - z_j$ in the last row have nonpositive values, so the BFS from Table 7.4 with $x_1 = 20$ and $x_2 = 40$ are optimal solutions, and substituting into the objective function we obtain $z_{max} = 6(20) + 5(40) = \$320$ (Table 7.5).

A comprehensive review of the Simplex algorithm and its extended concepts (including shadow price, duality and sensitivity analysis, and integer programming) are beyond the scope of this text. For a more in-depth review of subject, please refer to any OR text book such as Winston (1995).

7.3.2 Solving LP Using a Spreadsheet

The Simplex algorithm is very straightforward and easy to understand, but it may be tedious, especially when the model size is large; thus, it is suitable to be implemented using computer software. As a matter of fact, there are many scientific computing software packages that include the Simplex algorithm, which makes it easier for us to solve LP models. In this section, we will show how we can use a spreadsheet such as Excel to solve an LP problem.

7.3.2.1 Step 1. Activation of Solver Function in Excel

In Excel, there is an add-in function called Solver that is developed to solve mathematical program models. To use the Solver add-in function, once Excel is open, go to the "File" menu and select "Options"; in the "Excel Options" window, select "Add-Ins", then in the lower left corner of the "Excel Options" window, choose "Excel Add-Ins" and click the "Go…" button, as shown in Figure 7.14.

FIGURE 7.14
Excel add-ins selection.

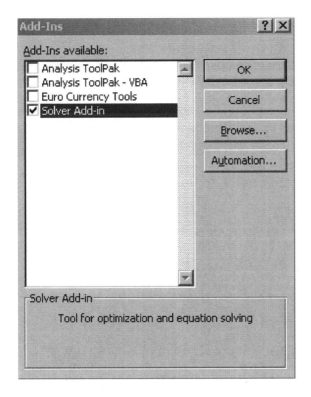

FIGURE 7.15
Add-ins dialog window.

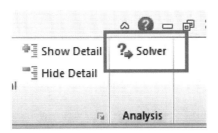

FIGURE 7.16
Illustration of the new Solver function button.

Once the "Go..." button is clicked, the dialog window shown in Figure 7.12 will pop up; check the option of "Solver Add-in" and then click on "OK", as shown in Figure 7.15.

Once the "Solver Add-in" option is selected, go back to the Excel main menu, and select the "Data" tab; you will notice that a new analysis button has been added at the upper right of the menu bar, as shown in Figure 7.16.

We will be using this function to solve the LP problem. But, before that, we need to set up (formulate) the LP model properly in Excel.

7.3.2.2 Step 2. Set-Up of LP Model in Excel

Let us use Example 7.2 to illustrate how to set up the LP model. Open a new worksheet, select an area for control variables and use these variables to express the objective function, as shown in Figure 7.17.

In another area (usually below the control variables), set up the constraints, as shown in Figure 7.18.

In cells E6 and E7, we need to insert the formula on the left-hand side of the constraint inequality, as shown in Figure 7.19.

7.3.2.3 Step 3. Solving the LP Model

Once the model has been properly set up, click on "Solver" on the "Data" tab, click the ▣ button on the right and select the right cells, as indicated in Figure 7.20.

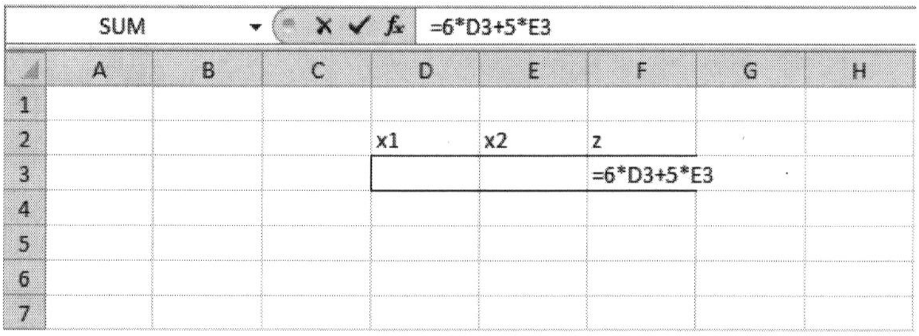

	SUM		▾	✕ ✓ ƒx	=6*D3+5*E3			
◢	A	B	C	D	E	F	G	H
1								
2				x1	x2	z		
3						=6*D3+5*E3		
4								
5								
6								
7								

FIGURE 7.17
Setting up the objective function.

	G10		▾	ƒx				
◢	A	B	C	D	E	F	G	H
1								
2				x1	x2	z		
3							0	
4								
5			x1		x2	Constraints		RHS
6		Assembly	4		3	0 <=		200
7		Painting	1		2	0 <=		100
8								
9								
10								

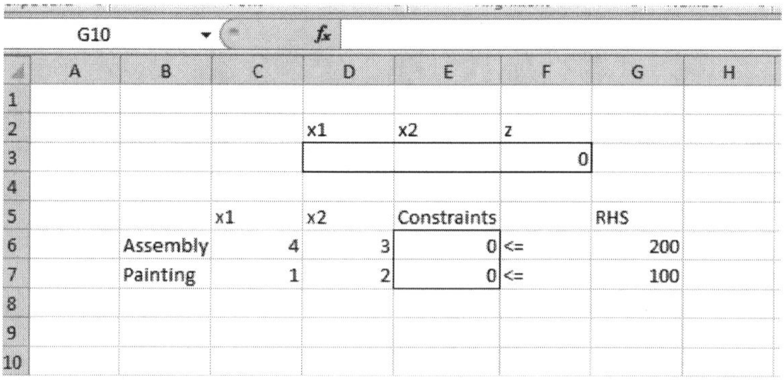

FIGURE 7.18
Setting up the constraints.

	x1	x2	z	
				0

	x1	x2	Constraints		RHS
Assembly	4	3	=$C$6*$D$3+$D$6*$E$3		
Painting	1	2	0	<=	100

	x1	x2	z	
				0

	x1	x2	Constraints		RHS
Assembly	4	3	0	<=	200
Painting	1	2	=$C$7*$D$3+$D$7*E3		100

FIGURE 7.19
Formula for constraints.

Next, click on the "Add" button, adding the constraints, again by using the ▣ button and select the appropriate cells from the worksheet, choose the "select a Solving Method" as "Simplex LP"; after all these are complete, click the "Solve" button (Figure 7.21). The results of the control variables and objective values will be updated in the worksheet, as shown in Figure 7.22.

From the results, $x_1 = 20$ and $x_2 = 40$, the maximum objective value $z = \$320$.

7.4 Summary

Systems engineering makes efficient use of resources to develop complex system functions, since most resources are limited and scarce. Optimizing the utilization of these limited resources has always been a challenge for system designers. Optimization can be defined as a procedure/algorithm to determine the value of independent (or control) variables to maximize (or minimize) the value of objective functions. In this chapter, we reviewed some of the most common used models in systems engineering practice. First, we introduced the concepts of unconstrained optimization models. These models are based on basic differentiation theory. If a function $f(x)$ is differentiable, then the local optimum can be obtained by Fermat's theorem; that is to say, the first-order derivative of f equals zero. Fermat's theorem only gives a necessary condition for finding the local optimum; it only shows if a point is a reflection point. To confirm that the reflection is truly a local maximum (or minimum) point, a higher order of differentiation is needed. If the second-order

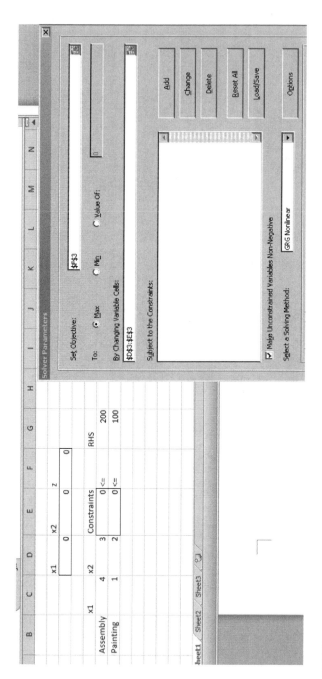

FIGURE 7.20
Setting up the variables and objective values in Solver.

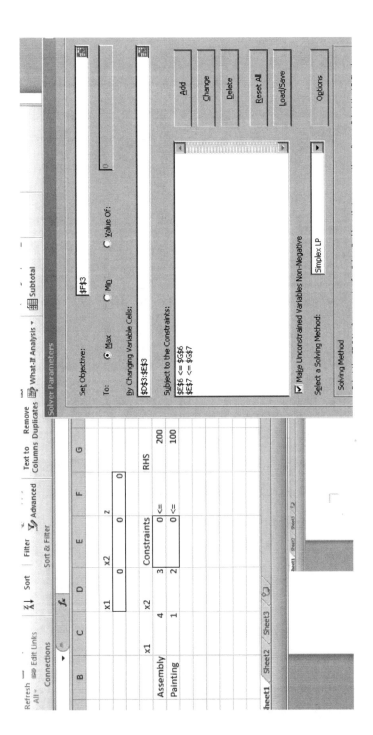

FIGURE 7.21
Input the formula for constraints.

	A	B	C	D	E	F	G
1							
2				x1	x2	z	
3				20	40	320	
4							
5			x1	x2	Constraints		RHS
6		Assembly	4	3	200 <=		200
7		Painting	1	2	100 <=		100
8							
9							

FIGURE 7.22
Results of the LP problem.

derivative is negative, then the point is a local maximum; if the second-order derivative is positive, then the point is a local minimum.

In the second part of the chapter, we described constrained optimization models using mathematical programming. The fundamental form of mathematical programming is LP. With LP models, both the objective functions and constraint functions are linear. LP models are formally defined and we used some examples to illustrate how to formulate a LP model. If an LP model has only two or three decision variables, we can solve it by using graphical methods. If more than three variables are involved, we need to apply a more general algorithm called the Simplex method. A tabular format of the Simplex algorithm was described in detail, and at the end of the chapter, we briefly discussed how to use spreadsheet software to implement the Simplex method.

Optimization plays an important role in the system engineering design process. This chapter provides a basic review of the most commonly used optimization models for systems engineers to understand the nature of optimization problems. For a more in-depth review of these materials, it is recommended to look at an OR text as mentioned in the chapter.

PROBLEMS

1. Define optimization. Why is optimization important in systems engineering?

2. If $f(x) = \ln\sqrt{1+x^2}$, what is $f'(0)$?

3. If $f(x) = x^n + 2e$, find $f^{(n)}(x)$.

4. If $x \in \mathbb{R}$, what is the reflection point for function $y = f(x) = 3x^3 - 27x$? Is this point a minimum point or maximum point? Why?

5. Using unconstrained conditions, find the minimum point for the function $f(x_1, x_2)$

$$f(x_1, x_2) = \left(x_1^2 - 1\right)^2 + x_1^2 + 2x_2^2 - 2x_1)$$

6. The cost of a product A is $C(x) = 16x + 250$ (thousand dollars), with x being the number of items produced; the total revenue after selling all the x products is $R(x) = 24x - 0.008x^2$ (thousand dollars). What should x be to maximize the total profit (revenue-cost), and what is the maximum profit?

7. Formulate the following using an LP model:

A company makes two types of toy product, Product A and Product B. Product A makes $50 profit/unit, and Product B makes $40 profit/unit. There are four work stations that are needed to produce A and B; Product A uses Stations 1, 2 and 3 and Product B uses Stations 2, 3, and 4. The hours needed for the two products are listed in the following table as well as the daily production capacity for each of the stations (in hours). Make a daily production plan for this company to maximize the profit (Table 7.6).

8. A small furniture company makes four different types of furniture, using wood and glass as the materials. The production requirements and daily capacity for labor hours and materials are listed in the following table. Formulate it using LP (Table 7.7).

TABLE 7.6

Station	Product A Requirement (h)	Product B Requirement (h)	Daily Capacity (h)
1	2	0	300
2	0	3	540
3	2	2	480
4	1	2	300

TABLE 7.7

Furniture Type	Labor Hours Needed (h/unit)	Wood (per unit)	Glass (per unit)	Profit ($/unit)	Maximum Allowable Production (units/day)
1	2	4	6	60	100
2	1	2	2	20	200
3	3	1	1	40	50
4	2	2	2	30	100
Daily capacity	400	600	100		

9. Solve the following LP problems graphically.

   a. max $z = 2x_1 + 3x_2$

$$\text{s.t.} \begin{cases} x_1 + x_2 \leq 4 \\ 4x_1 + 3x_2 \leq 16 \\ 4x_2 \leq 12 \\ x_1, x_2 \geq 0 \end{cases}$$

   b. max $z = x_1 + 3x_2$

$$\text{s.t.} \begin{cases} 6x_1 + 7x_2 \leq 50 \\ x_1 \geq 3 \\ x_2 \geq 2 \\ x_1, x_2 \geq 0 \end{cases}$$

   c. min $z = 6x_1 + 4x_2$

$$\text{s.t.} \begin{cases} 2x_1 + x_2 \geq 1 \\ 3x_1 + 4x_2 \geq 3 \\ \\ x_1, x_2 \geq 0 \end{cases}$$

10. Solve the above LP problems using a spreadsheet.
11. Solve Problem 6 (the one associated with Table 7.6) using Excel. Is it economically feasible for the company to pay $10 for an employee to work one extra hour every day?

8

Process Modeling Using Queuing Theory and Simulation

In our everyday life, there are many so-called "waiting line systems"; for example, customers waiting in the checkout line in a grocery store, passengers waiting in line to go through the security checkpoint at an airport, and customers calling customer service and waiting in a queue to be answered in the order received. If the demand for processing exceeds the capacity of serving units, we often encounter a "wait line" system. If you look at these systems, it is not difficult to find the common features involved in these systems; (1) there is a demand for service and the demand is stochastic, that is to say, the time when a customer arrives is somewhat random and unpredictable; (2) there are limited resources (or servers) to fulfill the demand; when there are more customers than the available servers, waiting occurs and the order of being served follows a predetermined rule, usually the first-in-first-out (FIFO) order. In systems operations and processes, it is very common to observe the characteristics of waiting line systems, such as in manufacturing systems; raw materials waiting in the queue to be processed, machines in queues to be serviced, and finished products waiting in line to be shipped out; in service systems, entities (such as customers) forming a line, waiting for their turn to be served. As system designers, we are tasked to control the length of the waiting line; on the one hand, entities that request processing and service need to have shorter waiting times, but, on the other hand, we need to control the number of process/service channels to make sure they provide enough capacity with the most cost efficiency. In waiting line system design, the objective is to optimize the number of service channels to achieve the desired systems performance. The mathematical models addressing the waiting line system behaviors are called queuing theory and models.

In this chapter, we will review the fundamental concepts of queuing theory and its application in systems engineering; more specifically, we will

1. Define queuing systems and the related decision factors that are involved in a queuing system
2. Introduce the basic queuing theory model, describe the birth-death process, and study the $M/M/1$ and $M/M/c$ systems, deriving the basic parameters for these models
3. Introduce the application of queuing models in systems design

4. Introduce simulation-based queuing systems analysis; namely, we will use discrete event simulation (DES) to investigate queuing models with more complexity

8.1 Basic Queuing Theory

Queuing theory was originally developed by Agner Krarup Erlang in 1909. Erlang was a Danish engineer who worked in the Copenhagen telephone exchange. When studying the telephone traffic problem, he used a Poisson process as the arrival mechanism and, for the first time, modeled the telephone queues as an $M/D/1$ queuing system. Ever since the first queuing model, queuing theory has been well developed and extended to many complex situations, even with complicated queuing networks. These models, together with the advancement of computer technology, have been used widely now in many fields and have shown significant benefits in optimizing behavior within these systems. In this chapter, we will introduce the most fundamental queuing theory and models to make readers familiarized with the concepts of queuing theory and its application in system design. For a more detailed review of the subject of queuing theory, readers can refer to any other book on stochastic processes or advanced queuing theory.

8.1.1 Queuing System

A typical queuing system (waiting line system) is illustrated in Figure 8.1. The dotted line indicates the boundary of a waiting line system, hereafter referred to as a queuing system. In a queuing system, generated from a source, or a *population*, individual *customers* arrive randomly at the queuing system and request service from one of the *service channels* (or servers). If the service channels are all busy, then the arriving customers wait in a line for

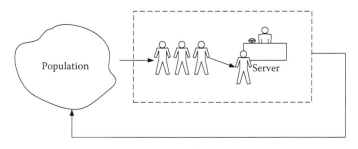

FIGURE 8.1
Typical queuing system.

their turn to be served; the order to be served is based on the predetermined waiting rule (e.g., first in, first out, or FIFO).

In general, the entities that request services are called customers, and the processes that provide services and fulfill customers' needs are called service channels. It is obvious that the capacity is the key factor in influencing system behaviors. If the service channels have less capacity that cannot satisfy customer demand, then the waiting line will form, and, systems will become more and more crowded; thus, the quality of service will be degraded and many customers may choose to leave the system before getting served. From the standpoint of customers, the more service channel capacity the better; this implies less time to wait to be served and a high service quality is perceived. On the other hand, if the service channels have more capacity than needed, then from the service provider's perspective, more service channels mean more investment, capital expenditure, and human labor involved, which increases the operations cost of the service or the manufacturing process. So, one of the most important purposes of studying queuing theory is to find a balance between these two costs, the *waiting cost* and the *service cost*. If the customer waits for too long, he/she may not be happy with the service, thus might not return in the future, causing loss of potential profit; or, parts may be waiting too long, increasing production cycle time, again losing potential sales and profits. These costs are considered to be waiting costs. Service costs are those that increase service capacity such as salary paid to the servers. Queuing theory application balances these two costs by determining the right level of service so that the total cost of the operations (waiting cost + service cost) can be optimized.

From Figure 8.1, it is easily seen that a queuing system consists of three processes that interact and regulate each other's behavior.

8.1.1.1 Arrival Process

Customers arrive from a population. To simplify the model at the beginning, we first need to make the following two assumptions:

1. Population size is infinite. There is no doubt that any population is finite in size. For example, the number of customers at a local market is subject to the population size of the area, which obviously is finite. However, if the departure rate is small relative to the population size, then the change of the population size due to the service (i.e., returning the customer back to the population) is negligible. Under this assumption, we can have a constant arrival rate in the queuing model, which is much easier to formulate and study. In some cases, the service portion is too large relative to the population size to be ignored; we cannot treat the population size as infinite as it will not reflect the true behavior of the queuing system. For example, if we have a population of 30 trucks and if

one truck needs maintenance, it will be removed from the population and wait to be serviced. In this case, the impact of service on the population size cannot be ignored; thus, the arrival rate of the trucks in the service center will be affected by how many trucks are actually being serviced and how many are in operation. In this case, a finite queuing model can be used to study the behavior of such a system.

2. One customer arrives at a time. In a queuing system, we usually assume that at any given time, there is only one customer arriving; bulk arrival is not considered. We might argue that in real life there are cases of bulk arrival; to make the model simple, we can always increase the precise unit of time, so that no two arrivals occur at the exact same time. Bulk arrival will be addressed in Section 8.3 using simulation. For this reason, a queuing system is also called a discrete event system, as customers arrive at discrete time points, not continuously like the flow of water.

3. Customer arrivals are independent of each other. In other words, the arrival of one individual customer is entirely independent, and does not rely on other customers' behavior.

4. The arrival process is stationary. That is, we assume that the parameters of the arrival mechanism, such as the probability distribution function (p.d.f.), the mean arrival rate, usually denoted as λ, and the variance remain unchanged during operations; they do not vary with time, and are not independent of the time of the operations. If these parameters change with time, then the arrival process is considered nonstationary; for example, a restaurant may have an hour of peak customer arrival at lunchtime and dinnertime; the arrival parameters are not the same in the peak hour as in the other time periods. In the mathematical models of queuing theory, we assume a stationary arrival process, and will introduce the concept of the nonstationary arrival in Section 8.3 using simulation.

8.1.1.2 Queue

When the arriving customer finds all the servers (or service channels) are busy, this customer has to enter an area/place to form a line to wait, waiting for his/her turn; this waiting line is called a queue. There are many types of rules involving the queue, depending on the real-world situation the queuing system is modeling. The most commonly used queue rule is that customers wait in a first-in-first-out (FIFO) queue. We can observe this rule almost everywhere in our daily life; it seems to be a fair process for all customers/units. However, there are some other types of priority rules that can be utilized; for example, sometimes a last-in-first-out (LIFO) rule is used for a parking lot system, airplane boarding, or, sometimes, the stacking order of computer data structure. In some other cases, when each customer does not have the same priority, then the ranking of the queue is determined by some type of priority attribute. For example, in an airport boarding process, the VIP passengers always get placed in the front of the boarding queue when they arrive. What priority rule to use to model the queue is solely dependent on the nature of the system being studied.

The above rule assumes that there are unlimited spaces in the buffer area for every unit arriving to wait (of course, for a stable queuing system, the queue length will not tend to infinity; we will talk about this condition in Section 8.1.2), so every customer chooses to wait when all servers are busy. There are occasions when customers see that the queue too long and they do not want to wait; instead, they leave the system immediately. This customer behavior is called reneging and the queue has a lost mechanism. There are some queuing systems that have a mixed model of waiting and lost mechanisms, depending on the customers' tolerance levels. For example, if waiting for period of time x, a customer will choose to leave; or, if the queue length exceeds y, then newly arriving customers will depart immediately. This mostly occurs in queuing systems with limited waiting areas, as seen at many drive-through service windows.

Besides the different priority rules used in the queuing system, there are some other variations of waiting mechanisms in a queue. The most common mechanism is a single queue; there are also multiple queues and recurring queues. Again, this depends on the nature of the systems being modeled.

8.1.1.3 Service Process

The service process is the process in which activities are performed by the server with the customer, to fulfill the needs of the customer; once the needs of the customer are fulfilled, the customer leaves the system. For this reason, the service process is also called the output process of the queuing system. Similarly to customer arrival, the service process also presents random behavior (think about the experience one has at a store checkout; some customers have only one or two items, but some have a full shopping cart). Thus, the service time for each customer is random, usually following a particular probability distribution. The service quality is measured by the time taken to complete the service. The service provided by the queuing system is measured by its capacity, or the number of service channels available. In queuing models, the arrangement of the service channels is usually either in a *series* structure or a *parallel* structure. The series queuing structure is addressed by the queuing network model and the parallel structure is handled by multiple service channel models (assuming all the channels are identical). As mentioned earlier, the service process is where the decision factors are; decision makers control the number of service channels (i.e., the number of clerks to hire for the checkout process) to control the output rate for customer needs, increasing/decreasing the service cost to decrease/increase the cost incurred in customer waiting.

To facilitate the introduction of the different queuing models, we shall use a unified system of notation, as follows:

t: Time period of the system

x: Number of units of customers in the system

λ: Arrival rate; measures the mean number of units arrives per time period

μ: Service rate; measures the mean number of units completed per time period per service channel

With this notation, we can easily derive some other parameters of interest. For example, $1/\lambda$ is the mean interarrival time between two customer arrivals and $1/\mu$ is the mean service time. For a queuing system to be stable (i.e., the queue length does not grow infinitely), we usually require that $\lambda < \mu$, or that the service rate is faster than the arrival process. We will see why this is a necessary condition in Section 8.1.2 when we talk about the $M/M/1$ and $M/M/c$ queues. A general convention for a queuing system uses a format of six symbols in total: $X/Y/Z/A/B/C$; specifically,

X: indicates the time distribution of the arrival process. Some of the most commonly used arrival processes include M, indicating the Poisson arrival process (we will explain this process shortly); D means the arrival process is deterministic; E_k is an Erlang(k) distribution of arrival process; G means a general distribution for the arrival process (other than Erlang, deterministic, or Poisson).

Y: indicates the time distribution of the service process. For example, M implies an exponential distribution of the service time, while G means a general distribution.

Z: indicates the number of service channels.

A: indicates the queuing system capacity. A default value for the capacity is infinity, ∞.

B: indicates the population size. A default value for the population size is infinity, ∞.

C: indicates the queuing rules, such as FIFO.

So an $M/M/1/\infty/\infty/\text{FIFO}$ queue refers to a Poisson arrival process, exponential service time distribution, one server, infinite queue capacity, and infinite population size, and the waiting line rule is FIFO. In the next section, we will investigate this system, as it represents the most fundamental queuing model.

8.1.2 *M/M/*1 Queuing System

8.1.2.1 *Exponential Time Arrival and Poisson Process*

As mentioned earlier, $M/M/1/\infty/\infty$/FIFO is considered to be the basic queuing structure and since infinity and FIFO are the default values for the capacity and queuing rules, we just use $M/M/1$ to simplify the representation. In an $M/M/1$ queuing system, the arrival process is a *Poisson* process, as mentioned earlier. If you have studied probability theory, you will recall that the Poisson process is a counting process of random arrivals, with the interarrival time following an *exponential* distribution.

Let us look at the exponential distribution. The reason an exponential distribution is used to model the queuing arrival process is due to its *memoryless* (or *no-memory*) property. Stated in plain language, this implies that any individual customer's arrival time does not depend on any other customer's arrival time. To state this property mathematically, if x is the exponentially distributed random variable for the interarrival time between two consecutive customer arrivals, we have the following relationship:

$$P(x > s+t \mid x > s) = P(x > t) \tag{8.1}$$

The above conditional probability expresses the no-memory property of the exponential distribution, and it can be proven that no distribution except the exponential distribution has this property (Winston 1995).

To demonstrate this property, let us take a look at the exponential distribution first. An exponential distribution with parameter λ has the following density function:

$$f(x) = \begin{cases} \lambda e^{-\lambda x}, & x \geq 0 \\ 0, & \text{otherwise} \end{cases} \tag{8.2}$$

and $f(x)$ is illustrated in Figure 8.2.

Since we know the cumulative probability,

$$P(x > t) = \int_{t}^{\infty} \lambda e^{-\lambda x} dx = 1 - \int_{0}^{t} \lambda e^{-\lambda x} dx \tag{8.3}$$

This is easy to understand, since

$$1 = \int_{0}^{\infty} \lambda e^{-\lambda x} dx = \int_{0}^{t} \lambda e^{-\lambda x} dx + \int_{t}^{\infty} \lambda e^{-\lambda x} dx$$

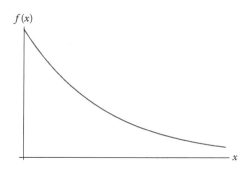

FIGURE 8.2
Illustration of exponential function.

It is easy to derive the integral

$$\int_0^t \lambda e^{-\lambda x} dx = 1 - e^{-\lambda t}$$

(Readers can refer to any calculus book to review the basic integral.)
So, for Equation 8.3, we have the following results:

$$P(x > t) = 1 - \int_0^t \lambda e^{-\lambda x} dx = 1 - (1 - e^{-\lambda t}) = e^{-\lambda t} \qquad (8.4)$$

That is to say, the probability that the interarrival time is greater than t is $e^{-\lambda t}$ if the interarrival time is exponentially distributed. It is not difficult to derive the mean and variance for $f(x)$:

$$E(x) = \frac{1}{\lambda} \qquad (8.5)$$

and

$$Var(x) = \frac{1}{\lambda^2} \qquad (8.6)$$

The mean value of $E(x)$ is the mean value for the interarrival time, and is the reciprocal of λ, so λ is the mean value of the arrival rate. You might find this concept of rate a little confusing; if you think about the units of the interarrival time and arrival rate, it will be easier to understand. λ, as the mean arrival rate, describes the numbers of arrival per time period, and, of course, the reciprocal

$1/\lambda$ is the average time per arrival, which is the mean interarrival time. We have to be clear about the difference between time and rate; you will also see that understanding these concepts is very critical in modeling and simulations.

Now we can easily prove the memoryless property of the exponential distribution:

Proof:

$$P\left(x > s+t \mid x > s\right) = \frac{P(x > s+t, x > s)}{P(x > s)} = \frac{P(x > s+t)}{P(x > s)}$$

According to Equation 8.4, we know that

$$P\left(x > s+t\right) = e^{-\lambda(s+t)}$$

and

$$P\left(x > s\right) = e^{-\lambda s}$$

So, we have

$$P\left(x > s+t \mid x > s\right) = \frac{e^{-\lambda(s+t)}}{e^{-\lambda s}} = e^{-\lambda t} = P(x > t)$$

∎

It needs to be noted here that

1. The exponential distribution is the only distribution that has this no-memory property; no other distribution functions possess this property.

2. The no-memory (or memoryless) property is very important for modeling queuing systems. It implies that the distribution of each individual arrival is totally independent of any other, and to predict the arrival time of the next arrival, we do not need to rely on when the last arrival occurred.

When the interarrival time follows an exponential distribution, the arrivals follow a Poisson process, and their relationships are defined as follows (Winston 1994):

If the inter-arrival time follows an exponential distribution with parameter λ, if and only if the random variable of numbers of arrivals within time period of t follows a Poisson distribution with parameters of λt.

To express this more particularly in mathematical format, a discrete random variable N follows a Poisson distribution with parameter λ, if for $N=0$, 1, 2, ...,

$$P(N = n) = \frac{e^{-\lambda}\lambda^n}{n!} \tag{8.7}$$

It can easily be shown that the mean of N, $E(N)=\lambda$. So, for an arrival time following an exponential distribution with a mean of λ, the number of arrivals N_t within time t follows a Poisson distribution with the parameter of λt, expressed as follows:

$$P(N_t = n) = \frac{e^{-\lambda t}(\lambda t)^n}{n!}, \quad n = 0, 1, 2, \dots \tag{8.8}$$

and obviously $E(N_t)=\lambda t$.

Regarding the service time in an $M/M/1$ system, we also assume that each individual customer's service time is totally independent of any other and has no memory of any other. Thus, we use the exponential distribution to model the random variable for the service time. That is, if the random variable y for the service time is exponentially distributed, the probability density function is given by $f(y)=\mu e^{-\mu y}$, the mean service time equals $1/\mu$, and μ is called the service rate (i.e., the number of service tasks completed per time period). In the next section, we will show how $M/M/1$ is modeled using the Poisson process and exponential distribution, by using a birth-death process.

8.1.2.2 Birth-Death Process

The *birth-death process* is an important tool to model the steady-state behavior of the queuing system. As commonly seen in most queuing systems, if a steady state of the system exists, its steady-state behavior will be stabilized. Just imagine a large grocery store; in the morning when it first opens, there are no customers in the store. Customers start to arrive (customer arrival) and the store starts to serve the customers (customer departure). If the arrival rate λ and the service rate (or departure rate) μ do not change over time, and $\lambda<\mu$ (i.e., service rate is faster than arrival rate—why? We will explain this later), then soon the store will reach a state where the distribution of the numbers of customers in the store $N(t)$ will remain unchanged; we call this a steady-state behavior. The steady-state behavior represents the true characteristics of the queuing system (i.e., the average time a customer spends in the system, and the average length of the queue); thus, when studying a queuing system, we need to know the behavior of its steady state. In other words, under the steady state, the probability that there are j customers in the queuing system at the time t, given the initial number of the customers in the system i, denoted as $P_{ij}(t)$, will not depend on the initial state and current time; $P_{ij}(t)$ will converge to an equilibrium probability of P_j. So, if we can derive the value of P_j, $j=1, 2, 3, \ldots n, \ldots$, then we can completely describe the behavior of $M/M/1$ queuing system.

In deriving the probability of P_j, we can use the birth-death process to describe the *transition* relationships between different system states; here, we define the system state as the number of customers in the system at time t, $N(t)=i$, $i=0, 1, 2, \ldots$. There are basically two events involved in the queuing system: arrival (the birth process) and departure (the death process). If we assume that only one customer can arrive at any time, then for a small time period $t + \Delta t$ from t, only one of the following three events could happen (one can make the value of Δt small enough to ensure one of these will happen and exclude other events):

1. There is one customer arrival and no customer departure; system moves from state "i" to "$i+1$".
2. There is no customer arrival and there is one customer departure; system moves from state "i" to "$i-1$".
3. There is one arrival and one departure; system remains at state i.

These three events are illustrated graphically in Figure 8.3.

A detailed mathematical illustration of these transition probabilities can be found in Winston (1995); if readers are interested in the proof of these state transition diagrams, we suggest they read chapter 20 of Winston (1994). Based on the same principle, we can derive the complete birth-death process for an $M/M/1$ queue as shown in Figure 8.4.

If the system is in the steady state, the probability of being in any system state n, $n=0, 1, 2, \ldots$ remains unchanged; this implies from the birth-death process diagram above that the chances of an incoming event to state n equals the chances of an outgoing event from state n. For state n, $n \geq 1$,

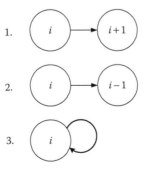

FIGURE 8.3
Three possible events for queuing system status.

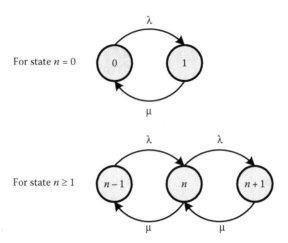

For state $n = 0$

For state $n \geq 1$

FIGURE 8.4
Birth-death process for $M/M/1$ queue. (Redrawn from Winston, W. L., *Operations Research: Application and Algorithms*, Belmont, CA: Duxbury Press, 1994.)

incoming events are from state "$n-1$" to state n and from state "$n+1$" to state n; the probability of entering n is $\lambda P_{n-1} + \mu P_{n+1}$. The outgoing events are from state n to state "$n-1$" and state "$n+1$" to state n; the outgoing probability is $\lambda P_n + \mu P_n$. So, under the steady state, the outgoing probability equals the incoming probability, that is,

$$\lambda P_{n-1} + \mu P_{n+1} = (\lambda + \mu)P_n, \quad n = 1, 2, \ldots \quad (8.9)$$

For $n = 0$, since this value only has one state to connect to the birth-death process, the steady-state transition function is

$$\lambda P_0 = \mu P_1 \quad (8.10)$$

So, we obtain the complete set of transition functions for the steady states of the $M/M/1$ queue:

$$\lambda P_0 = \mu P_1$$

$$\lambda P_0 + \mu P_2 = (\lambda + \mu)P_1$$

$$\lambda P_1 + \mu P_3 = (\lambda + \mu)P_2$$

$$\cdots\cdots$$

$$\lambda P_{n-1} + \mu P_{n+1} = (\lambda + \mu)P_n$$

$$\cdots\cdots$$

With these transition functions, we can easily derive the value for P_n:
From Equation 8.10, we have

$$P_1 = \frac{\lambda}{\mu} P_0$$

If we let

$$\rho = \frac{\lambda}{\mu}$$

then we can rewrite the above equation as

$$P_1 = \rho P_0 \tag{8.11}$$

From the birth–death equations, for $n = 1$, we have

$$\lambda P_0 + \mu P_2 = (\lambda + \mu) P_1$$

Substituting $P_1 = \rho P_0$ into the above equation, we can obtain the relation between P_0 and P_2, as follows:

$$\lambda P_0 + \mu P_2 = (\lambda + \mu) P_1 = (\lambda + \mu) \rho P_0$$

Dividing by μ on both sides of the equation, we obtain

$$\frac{\lambda}{\mu} P_0 + P_2 = \left(\frac{\lambda}{\mu} + 1\right) P_1 = \left(\frac{\lambda}{\mu} + 1\right) \rho P_0$$

or

$$\rho P_0 + P_2 = \rho^2 P_0 + \rho P_0$$

Canceling ρP_0, we obtain the following equation

$$P_2 = \rho^2 P_0 \tag{8.12}$$

We repeat the same procedure for the next equation of the steady-state equations; we will derive the following:

$$P_1 = \rho^1 P_0$$

$$P_2 = \rho^2 P_0$$

$$P_3 = \rho^3 P_0$$

$$P_4 = \rho^4 P_0$$

......

$$P_n = \rho^n P_0$$

......

Since all the probabilities are mutually exclusive,

$$\sum_{i=0}^{\infty} P_i = 1 \qquad (8.13)$$

so we can easily obtain P_0 as

$$P_0\left(1+\rho+\rho^2+\rho^3+\cdots+\rho^n+\cdots\right)=1$$

So, if we know the sum of $1+\rho+\rho^2+\rho^3+\cdots+\rho^n+\cdots$, then we can obtain the value of P_0.

Let us denote $G=1+\rho+\rho^2+\rho^3+\cdots+\rho^n+\cdots$, then $P_0=1/G$. Solving for G is quite straightforward; if you have had basic algebra, it should be easy for you to understand. Let us see how to use some simple procedures to work out G.

$$G = \sum_{i=0}^{\infty} \rho_i = 1+\rho+\rho^2+\rho^3+\cdots+\rho^n+\cdots \qquad (8.14)$$

Multiplying by ρ on both sides of Equation 8.14, we obtain

$$\rho G = \rho \sum_{i=0}^{\infty} \rho_i = \rho \left(1 + \rho + \rho^2 + \rho^3 + \cdots + \rho^n + \cdots\right)$$

$$= \rho + \rho^2 + \rho^3 + \cdots + \rho^n + \cdots \tag{8.15}$$

Subtracting Equation 8.15 from Equation 8.14, we have

$$(1-\rho)G = \left(1 + \rho + \rho^2 + \rho^3 + \cdots + \rho^n + \cdots\right) - \left(\rho + \rho^2 + \rho^3 + \cdots + \rho^n + \cdots\right) \tag{8.16}$$

We assume $\rho = (\lambda/\mu) < 1$, then $\lim_{n\to\infty}(\rho)^n = 0$, so on the right-hand side of Equation 8.16, everything is canceled except 1, that is,

$$(1-\rho)G = 1$$

or

$$G = \frac{1}{1-\rho} = \frac{1}{1 - \dfrac{\lambda}{\mu}} = \frac{\mu}{\mu - \lambda} \tag{8.17}$$

So, we obtain P_0 as

$$P_0 = \frac{1}{G} = \frac{\mu - \lambda}{\mu} \tag{8.18}$$

Readers should now have some idea why we mentioned that a necessary condition for a stabilized queuing system is that $\mu > \lambda$. From Equation 8.18, this condition is necessary for P_0 to be solved, so that a steady-state probability P_i can be found.

Using Equation 8.18, we can obtain all the probabilities for the steady state of an $M/M/1$ queue:

$$P_n = \rho^n P_0 = \left(\frac{\lambda}{\mu}\right)^n \frac{\mu - \lambda}{\mu}, \quad n = 1, 2, \ldots \tag{8.19}$$

8.1.2.3 Measures of M/M/1 Queue

Based on the results from Equation 8.19, we can easily derive the main measures of the $M/M/1$ queue. The main measures for $M/M/1$ include

L_S: Mean number of customers in system

L_Q: Mean number of customers in queue

W_S: Mean time that customers spend in system (including waiting time and service time)

W_Q: Mean time that customers spend in the queue

L_S is the mean (expected) number of customers in system, so according to the definition of the mean value,

$$L_S = \sum_{n=0}^{\infty} nP_n = \sum_{n=0}^{\infty} n(1-\rho)\rho^n \tag{8.20}$$

This can be rewritten as

$$\sum_{n=0}^{\infty} n(1-\rho)\rho^n = \sum_{n=0}^{\infty} n\rho^n - \sum_{n=0}^{\infty} n\rho^{n+1} = \left(\rho^1 + 2\rho^2 + 3\rho^3 + 4\rho^4 + \cdots\right)$$

$$-\left(\rho^2 + 2\rho^3 + 3\rho^4 + 4\rho^5 + \cdots\right) = \rho + \rho^2 + \rho^3 + \rho^4 + \cdots = \left(1 + \rho + \rho^2 + \rho^3 + \rho^4 + \cdots\right) - 1$$

From Equation 8.14 we already know that

$$1 + \rho + \rho^2 + \rho^3 + \rho^4 + \cdots = \frac{1}{1-\rho}$$

So

$$L_S = \frac{1}{1-\rho} - 1 = \frac{\rho}{1-\rho} = \frac{\lambda}{\mu - \lambda} \tag{8.21}$$

Since, for the $M/M/1$ queuing system, the queue length is one fewer than the number of customers in the system, we can obtain L_Q as

$$L_Q = \sum_{n=0}^{\infty} (n-1)P_n = \sum_{n=0}^{\infty} nP_n - \sum_{n=0}^{\infty} P_n = L_S - \sum_{n=0}^{\infty} P_n \tag{8.22}$$

We know that

$$L_S = \frac{\lambda}{\mu - \lambda}$$

and

$$\sum_{n=0}^{\infty} P_n = 1 - P_0 = \rho = \frac{\lambda}{\mu}$$

So

$$L_Q = \frac{\lambda}{\mu - \lambda} - \frac{\lambda}{\mu} = \frac{\lambda^2}{\mu(\mu - \lambda)}$$

Often, we are interested in knowing the time a customer spends in the system. To derive the mean time spent in the system W_S and the mean time spent in waiting in the queue W_Q, we need to derive the distribution of the time in the system, which can be proven to follow an exponential distribution with a parameter of $-(\mu - \lambda)$; another way to solve for the time in the system and the queue is to use L_S and L_Q by applying the powerful *Little's law*:

> *For any queuing system in which the steady state exists, the following relationship is true*

$$L = \lambda W \qquad (8.23)$$

where:
- L is the length (or number of customers) in the queuing system (or subsystem)
- λ is the arrival rate of the queuing system
- W is the time spent in the corresponding system (or subsystem).

So, intuitively, we can derive the following two equations:

$$L_S = \lambda W_s$$

$$L_Q = \lambda W_Q$$

Little's law tells us that it is true regardless of the number of servers, as long as the steady state can be achieved. By using Little's law, we can derive the formulas for W_S and W_Q as follows:

$$W_S = \frac{L_S}{\lambda} = \frac{\frac{\lambda}{\mu - \lambda}}{\lambda} = \frac{1}{\mu - \lambda} \qquad (8.24)$$

$$W_Q = \frac{L_Q}{\lambda} = \frac{\frac{\lambda^2}{\mu(\mu - \lambda)}}{\lambda} = \frac{\lambda}{\mu(\mu - \lambda)} \qquad (8.25)$$

From the above formula, we can observe that as $\rho \to 1$, that is, $\lambda \to \mu$, the waiting time W_Q will approach infinity, and if $\rho \to 0$, $W_S \to 1/\mu$, which implies that the time the customer spends in the system is only the service time (approximately).

We will give an example to show how to use these formulas to measure the performance of the queuing system.

Example 8.1: A product service center has only one technician to provide a repair service. It is known that the arrival of customers follows a Poisson process, with two arrivals per hour on average; the service time follows an exponential distribution with a mean of 12 min. Assuming that there are unlimited spaces for customers to wait:

  a. What is the probability that the technician is idle (i.e., not busy)?
  b. What is the probability that there are three customers in the store?
  c. What is the probability that there is at least one customer in the store?
  d. What is the average number of customers in the store?
  e. On average, how long does one customer spend in the store?
  f. What is the average number of customers waiting in the queue?
  g. What is the average time that a customer waits in the queue?

Solution: This is an M/M/1 queuing system; from the information given, we know that $\lambda = 2/h$ and $\mu = 1/12$ per min $= 5/h$ (please note here that we need to convert everything into the same units [in this case, hours]). So, $\rho = \lambda/\mu = 0.4$.

  a. If the server is idle, this means that there is no customer in the store, so the probability of the server being idle is

$$P_0 = 1 - \rho = 1 - 0.4 = 0.6$$

  b. The probability of three customers in the store is

$$P_3 = \rho^3 (1 - \rho) = (0.4)^3 (1 - 0.4) = 0.038$$

  c. The probability that there is at least one customer in the store is

$$1 - P_0 = 1 - 0.6 = 0.4$$

  d. The average number of customers in the store, L_S, according to Equation 8.21, can be obtained as

$$L_S = \frac{\lambda}{\mu - \lambda} = \frac{2}{5 - 2} = 0.67 \,(\text{person})$$

e. The average time a customer spends in the system, W_s, can be found using Equation 8.23:

$$W_s = \frac{1}{\mu - \lambda} = \frac{1}{5-2} = \frac{1}{3}(h) = 20(\text{min})$$

We can also derive this by applying Little's law:

$$W_s = \frac{L_s}{\lambda} = \frac{0.67}{4}(h) = 20(\text{min})$$

f. The average number of customers waiting in the queue is, according to Equation 8.21,

$$L_Q = \frac{\lambda^2}{\mu(\mu - \lambda)} = \frac{(2)^2}{5(5-2)} = \frac{4}{15} = 0.267(\text{persons})$$

g. The average waiting time of a customer in the queue is

$$W_Q = \frac{\lambda}{\mu(\mu - \lambda)} = \frac{2}{5(5-2)} = \frac{2}{15}(h) = 8(\text{min})$$

8.1.3 Multiserver Queuing Systems

Now let us look at a more complicated queuing system with multiple servers. In multiserver systems, we assume customers still arrive independently and individually; the arrival process for each customer is still a Poisson process with a parameter of λ, the same as in the single-server system. When a customer arrives and there is at least one server idle (i.e., not busy), then this customer receives service immediately; if all servers are busy, customers wait in one single queue and follow the rules of FIFO to be served. The service time for each individual customer is the same as the single-server system, that is, exponential with a rate of μ. This system is often seen in service systems such as banks and post offices; we denote such a system as $M/M/s/\infty$, with s being the number of service channels in the system.

In solving the multiserver system, we can also use the birth-death process, and derive the probability distribution for P_n.

Let

$$\rho = \frac{\lambda}{s\mu} \tag{8.26}$$

It is easy to see that the necessary condition for the system to achieve a steady state is $\rho < 1$; that is to say, the arrival rate has be less than the overall service rate, otherwise, the system will increase to infinity. If $\rho < 1$, we can derive the steady-state probability as follows:

$$P_0 = \frac{1}{\sum_{i=0}^{s-1} \frac{(s\rho)^i}{i!} + \frac{(s\rho)^s}{s!(1-\rho)}} \tag{8.27}$$

and

$$P_n = \begin{cases} \frac{(s\rho)^n P_0}{n!}, & n \le s \\ \frac{(s\rho)^n P_0}{s!s^{n-s}}, & n > s \end{cases} \tag{8.28}$$

From Equation 8.28, it is easy to show that the probability that all servers are busy is

$$P(j \ge s) = \frac{(s\rho)^s P_0}{s!(1-\rho)}$$

One thing to keep in mind is that Little's law still applies here. We can use Little's law to derive the average number in the system, L_S, the average waiting time in the system, W_S, the average number in the queue, L_Q, and the average waiting time in the queue, W_Q. The Little's law relations still hold for L_S/W_S and L_Q/W_Q as in Equation 8.23; that is to say,

$$L_S = \lambda W_S$$

$$L_Q = \lambda W_Q$$

It can be shown that (Winston 1994)

$$L_Q = \frac{P(j \ge s)\rho}{1-\rho} \tag{8.29}$$

So, using Little's law,

$$W_Q = \frac{L_Q}{\lambda} = \frac{P(j \geq s)}{s\mu - \lambda} \tag{8.30}$$

Since the average service time is $1/\mu$, then we know that the average time a customer spends in the system is given by

$$W_S = W_Q + \frac{1}{\mu} = \frac{P(j \geq s)}{s\mu - \lambda} + \frac{1}{\mu} \tag{8.31}$$

and, by Little's law, we can obtain the average number in the system, L_S, by

$$L_S = \lambda W_S = \frac{P(j \geq s)\rho}{1 - \rho} + \frac{\lambda}{\mu} \tag{8.32}$$

Let us use an example to illustrate how to use Equations 8.26 through 8.32.

Example 8.2: A bank has three service windows. Assume that customers arriving to the windows follow a Poisson process, with mean interarrival time of 1 min. The service time follows an exponential distribution, with mean service time of 2.5 min. Customers wait in one queue if all the servers are busy, the queue rule is FIFO.

1. What is the probability that the bank is idle?
   From the question we know that $\lambda = 1$ min, $\mu = 0.4$ min, $s = 3$ so

$$\rho = \frac{\lambda}{s\mu} = \frac{1}{1.2} = 0.833 < 1$$

and

$$s\rho = 3(0.833) = 2.5$$

So, the probability that the bank is idle (i.e., there is no customer in the bank) is, according to Equation 8.27,

$$P_0 = \frac{1}{\sum_{i=0}^{s-1} \frac{(s\rho)^i}{i!} + \frac{(s\rho)^s}{s!(1-\rho)}}$$

$$= \left[\frac{(2.5)^0}{0!} + \frac{(2.5)^1}{1!} + \frac{(2.5)^2}{2!} + \frac{(2.5)^3}{3!(1-0.833)} \right]^{-1} = 0.045$$

2. What is the average number of customers waiting in the queue and the average number of customers in the bank?

$$P(j \geq 3) = \frac{(s\rho)^s P_0}{s!(1-\rho)} = \frac{(2.5)^3 (0.045)}{3!(1-0.0833)} = 0.131$$

then, according to Equation 8.29,

$$L_Q = \frac{P(j \geq s)\rho}{1-\rho} = \frac{0.131(0.833)}{1-0.833} = 0.653$$

The average number of customers in the bank system is given by

$$L_S = \frac{P(j \geq s)\rho}{1-\rho} + \frac{\lambda}{\mu} = L_Q + \frac{\lambda}{\mu} = 0.653 + 2.5 = 3.153$$

3. What is the average time that a customer will wait in the queue? How long does each customer spend in the bank? (All times in minutes.)

Using Little's law, we can obtain the average waiting time as follows:

$$W_Q = \frac{L_Q}{\lambda} = \frac{0.653}{1} = 0.653$$

and

$$W_S = W_Q + \frac{1}{\mu} = 0.653 + \frac{1}{0.4} = 3.153$$

A multiserver system with *one* single waiting line is more efficient than one with individual waiting lines for each of the servers. As you can imagine, a single queue for multiple servers will balance the workload for all of them; as the customers follow a FIFO rule, they will always choose the next available server. This selection process is "fairer" than individual waiting line systems, not just for the servers, but for the customers as well, as it truly follows the FIFO rule; while with an individual waiting line, it follows FIFO for that line, but considering all the queues together, the FIFO rule might not be followed. If you choose a "wrong" queue with a customer who takes a long time ahead of you, you may get served after people arriving after you, if they choose a faster queue. The difference between the two different queuing systems can be verified by comparing the average time a customer waits in the queue and the average queue length. Using the previous example, if customers form a separate queue for each server, then the arrival rate for each server becomes $\lambda' = \lambda/3$ and the service rate is still the same, μ. We leave this as homework in the Problem section for readers to practice and compare with the results from Example 8.2.

8.1.4 Queuing Systems with Finite Population

In the previous section, we stated that one of the assumptions that we made is to assume that we have unlimited population. This assumption is necessary for a constant arrival rate λ to simplify the problems. As we discussed earlier, although a queuing system with infinite population may never be found, however, if the departure rate is relatively small enough compared to the population, the small changes in the arrival rate can be ignored so the assumption is valid. That is the assumption we used to solve the M/M queuing systems problems above.

In some real-world queuing systems, however, the population size is so small that it has very significant effects on the individual arrival rate; assuming an infinite population size will no longer make the model valid. For example, a company has 15 trucks; each truck may fail at any time, and if the trucks are identical with identical failure rates, then the failure occurrence rate (if you think of failure as a "customer," a failure occurs as a "customer" arrives for repair service) with all 15 trucks running would be triple the arrival rate if only five trucks were operational and running. In cases like this, we can no longer treat population size as unlimited, but rather need to consider the population size in the formulation.

Let us denote the following:

N: Population size

λ: Individual arrival rate

λ_n: Actual arrival rate to the service when there are n customers in the service, $n = 0, 1, 2, ..., N$

M: Number of service channels

μ: Individual service rate per service channel

μ_n: Actual service rate when there are n customers in the service, $n = 0, 1, 2, ..., N$

P_n: Probability that there are n customers in the service, $n = 0, 1, 2, ..., N$

It is easy to see that

$$\lambda_n = \lambda(N - n), \quad n = 0, 1, 2, ..., N$$

and

$$\mu_n = \begin{cases} n\mu, & n \le M \\ M\mu, & n > M \end{cases}$$

using the steady-state birth-death process, similar to Figure 8.4, we can establish the steady-state relationships; that is, in any of the steady states, the probability of incoming to that state (i.e., n) from other states (i.e., from state "$n-1$" and state "$n+1$") equals the probability of leaving that state (i.e., from state n to state "$n-1$" or "$n+1$"). So we have

$$\lambda P_0 = \mu P_1$$

For $n \le M-1$, we have

$$(N-n+1)\lambda P_{n-1} + (n+1)\mu P_{n+1} = \left[(N-n)\lambda + n\mu\right]P_n$$

For $M \le n \le N-1$, we have

$$(N-n+1)\lambda P_{n-1} + M\mu P_{n+1} = \left[(N-n)\lambda + M\mu\right]P_n$$

and for $n=N$, we have the special relationship

$$\lambda P_{N-1} = M\mu P_N$$

plus the mutually exclusive relations for all the probabilities

$$\sum_{i=0}^{N} P_i = 1$$

We can solve for the distribution of P_n as follows (Blanchard and Fabrycky 2006):
Let

$$\rho = \frac{\lambda}{\mu}$$

Denote

$$C_n = \begin{cases} \dfrac{N!}{(N-n)!n!}\rho^n & n = 1, 2, \ldots, M \\ \dfrac{N!}{(N-n)!M!M^{n-M}}\rho^n, & n = M+1, \ldots, N \end{cases} \tag{8.33}$$

So, we obtain the distribution of P_n as follows:

$$P_n = \begin{cases} \dfrac{N!}{(N-n)!n!}\rho^n P_0 & n = 1, 2, \ldots, M \\[3mm] \dfrac{N!}{(N-n)!M!M^{n-M}}\rho^n P_0, & n = M+1, \ldots, N \end{cases} \tag{8.34}$$

with

$$P_0 = \left[\sum_{i=0}^{M-1} \frac{N!}{(N-i)!i!}\rho^i + \sum_{j=M}^{N} \frac{N!}{(N-j)!M!M^{j-M}}\rho^j \right]^{-1} \tag{8.35}$$

The mean performance parameters are given as follows:
Average number waiting in the service queue:

$$L_Q = \sum_{i=M}^{N} (i-M)P_i \tag{8.36}$$

Average number in the system (waiting + being served):

$$L_S = L_Q + \sum_{i=0}^{M-1} iP_i + M\left(1 - \sum_{j=0}^{M-1} P_j \right) \tag{8.37}$$

For $M = 1$ (single service channel), it is easy to show that

$$L_S = L_Q + (1-P_0) = N - \frac{\mu}{\lambda}(1-P_0)$$

With an average of L_S customers in the system, there are still $N-L_S$ customers outside of the system, so the effective arrival rate of the customers, or λ_e, is

$$\lambda_e = \lambda(N-L_S) \tag{8.38}$$

So, using Little's law, we can obtain the average time waiting and average time spent in the system as

$$W_S = \frac{L_S}{\lambda_e} \tag{8.39}$$

and

$$W_Q = \frac{L_Q}{\lambda_e} \tag{8.40}$$

Let us use an example to see how these formulas can be used to solve the queuing system problem with a finite population.

> **Example 8.3:** Assume there is one repair man to service five machines. Every machine's operation time follows an exponential distribution with a mean of 12 min. When the machine fails, the repair man fixes the problem, taking a time following an exponentially distributed time of 10 min.
>
>   1. What is the probability that the repair man is idle?
>   2. What is the mean number of machines that are not operational?
>   3. What is the average number of machines that are waiting to be serviced?
>   4. When a failure occurs, what is the average time that a machine has to wait until it becomes operational again?
>   5. What is the average time that each failed machine has to wait before being serviced?
>
> *Solutions*: From the problem description, we know that $\lambda = 1/12$ (per min), $\mu = 1/10$ (per min), $\rho = \lambda/\mu = 0.833$, $N = 5$, and $M = 1$.
>
>   1. If and only if there is no machine failure, then the repair man is idle. So the probability that the repair man is idle is, according to Equation 8.34,

$$P_0 = \left[\sum_{i=0}^{M-1} \frac{N!}{(N-i)!i!} \rho^i + \sum_{j=M}^{N} \frac{N!}{(N-j)!M!M^{j-M}} \rho^j \right]^{-1}$$

$$= \left[\frac{5!}{5!}(0.833)^0 + \frac{5!}{4!}(0.833)^1 + \frac{5!}{3!}(0.833)^2 + \frac{5!}{2!}(0.833)^3 + \frac{5!}{1!}(0.833)^4 + \frac{5!}{0!}(0.833)^5 \right]^{-1}$$

$$= 0.0073$$

>   2. The mean number of machines that are not operational equals the mean number of machines that are in the service system L_S. According to Equation 8.36, with $M = 1$,

$$L_S = L_Q + (1 - P_0) = N - \frac{\mu}{\lambda}(1 - P_0) = 5 - \frac{1}{0.833}(1 - 0.0063) = 3.81$$

3. The average number in the waiting line is

$$L_Q = L_S - (1 - P_0) = 3.81 - (1 - 0.0063) = 2.82$$

4. The average time that a machine waits until it becomes operational again is actually the average time a failed machine spends in the service system, so according to Little's law,

$$W_S = \frac{L_S}{\lambda_e} = \frac{L_S}{\lambda(N - L_S)} = \frac{3.81}{\frac{1}{12}(5 - 3.76)} = 38.42 \,(\text{min})$$

5. The average time that each failed machine has to wait before getting serviced can also be obtained by Little's law, using Equation 8.39:

$$W_Q = \frac{L_Q}{\lambda_e} = \frac{2.82}{\frac{1}{12}(5 - 3.81)} = 28.44 \,(\text{min})$$

There are many other types of queuing systems, such as systems with limited space s (customers will leave if there are already s customers in the system; for example, at the drive-through window of a fast-food restaurant); queuing systems with varying arrival rates (i.e., some customers might choose to leave the system when they find that the system already has many customers waiting); queuing systems with non-Poisson arrival or nonexponential service time distributions (usually indicated as $G/G/c$ queues, as G indicates a non-exponential distribution). Also, there are different types of queuing networks (such as a series of different queuing systems connected in a series or parallel structure). For more information about queuing theory, readers may refer to a specialized queuing theory textbook for a more in-depth review.

8.2 Application of Queuing Theory

Queuing theory may be used in systems design and process optimization. By formulating the system (subsystem) processes as queuing systems, one can analyze and control the process parameters to optimize the output of the system; often, the objective of the queuing output is its total cost per time period. This is also called the static optimization of queuing systems.

The goal of process design is to increase the efficiency of the operation at minimum cost. In queuing systems, as we mentioned above, the cost incurred primarily comes from two sources: the waiting cost and the service cost. There is a trade-off between these two costs; the waiting cost occurs when customers spend time in the queuing system. They are either waiting to be served or actually being served. The loss of time of the customers in the queuing system is considered as waiting cost; the longer time they spend in the system, the higher the waiting cost. The only way to reduce waiting cost is to increase more service capacity, either by adding more service channels or increasing the service rate per channel (i.e., hiring more experienced employees to provide faster service); this greater capacity leads to a higher service cost. When the service cost is increased, then customers wait less time in the system and the waiting cost is reduced, and vice versa. Through the trade-offs between service cost and waiting cost, the total cost of the queuing system can be minimized. Let us use an $M/M/1$ queue as an example to illustrate how the total cost is optimized.

Let

z = total cost of queuing system per time period

c_W = waiting cost per customer per time period

c_s = service cost per serving customer per time period

Then we have

z = service cost + waiting cost = $c_s \times$ (number of customers served per time period) + c_W (average number of customers in the system).

So

$$z(\mu) = c_s\mu + c_W L_s = c_s\mu + c_W \frac{\lambda}{\mu - \lambda}$$

$z(\mu)$ is a function of μ, since the service rate is our only decision variable. To minimize z, using the necessary conditions for an unconstrained optimization problem, we have

$$\frac{dz}{d\mu} = c_s - c_W\lambda \frac{1}{(\mu - \lambda)^2} = 0$$

Solving this, we can obtain the optimal service rate as

$$\mu^* = \lambda + \sqrt{\frac{c_W}{c_s}}\lambda \qquad (8.41)$$

We can verify this is truly a minimum by deriving the second-order derivative, that is

$$\frac{d^2z}{d\mu^2} = 2c_W\lambda\frac{1}{(\mu-\lambda)^3} > 0$$

Example 8.4: Machines come to a service center for a regular checkup service. The arrival process is a Poisson process with an arrival rate of 5/h. Each machine can make $100 profit/h. The service charge is $75/h/machine. What should the service rate be to minimize the total cost?

Solution: $\lambda=5/h$, $c_W=\$100/h$ and $c_s=\$75/h/machine$. According to Equation 8.40, the optimal service rate μ^* is

$$\mu^* = \lambda + \sqrt{\frac{c_W}{c_s}}\lambda = 5 + \sqrt{\frac{100}{75}}5 = 10.77 \cong 11$$

So, the average service rate is roughly 11 machines/h. The total cost per hour is minimized, which can be obtained as follows:

$$c_s\mu + c_W\frac{\lambda}{\mu-\lambda} = 75(5) + 100\frac{5}{11-5} \cong 458.33$$

The minimum cost per hour is $458.33.

For a simple system like the one above, the analytical method is sufficient to optimize the system performance. However, if the queuing system is complex and dynamic, involving many variables and a large degree of uncertainty, it is very difficult, sometimes even impossible, to solve it by using the analytical model. In this case, a simulation-model-based solution is more appropriate. In the following section, we will introduce the basic concept of simulation and illustrate how a simulation model can be used to solve a complex queuing system empirically.

8.3 Queuing System Analysis Using Discrete Event Simulation (DES)

As shown above, for queuing systems such as the $M/M/c$ queue, the analytical models are well developed and we can predict their steady-state performance without too much difficulty. We made many assumptions (i.e., Poisson arrival process) about the analytical queuing models in the previous sections, to simplify the problem so that a mathematical model could be formulated. However, in the real world, the situation becomes more dynamic and complicated than those mathematical models can handle; even though queuing network models are available, many situations are simply beyond the capabilities of the analytical mathematical model. To give some examples, employees in a manufacturing facility work on different shifts and they have coffee breaks every 2 h and a 1 h lunch break for every 8 h shift; a restaurant opens from 8 a.m. to 8 p.m., and customers' arrival rate is not stationary throughout the day (i.e., peak hours at lunchtime and dinnertime). When modeling these systems, information such as the shift patterns, lunch breaks, machine break-downs, arrival rate, and so forth cannot be ignored, as they will have significant impacts on system performance; also, some systems might never arrive at a steady state and do not operate on a 24/7 basis. So, it is nearly impossible to study these types of systems using queuing theory and models; a theoretical solution for those queuing systems would be difficult to obtain. An alternative to the mathematical model is to use the simulation model instead.

8.3.1 Introduction to Simulation

Simulation, generally speaking, is the replication or imitation of the operation of a real-world process or system over time (Bank et al. 2001). It can be regarded as a process of modeling that represents complex systems from the real world. By investigating systems empirically, modeling and simulation can provide insight into systems behavior, as well as allowing testing of alternative solutions through system experimentation.

With the advance of computer technology, simulation is a widely used and increasingly popular method for studying complex systems. Some advantages to simulation that may account for its growing popularity include (Law and Kelton 2000):

- Mathematical models cannot accurately describe complex and dynamic stochastic elements in a system; as mentioned above, these elements are difficult to quantify.
- Simulation allows the estimation of performance of an existing system under different operating conditions, through a valid representation model. Through a validated simulation model, the various operating conditions can be investigated even if such a condition does not exist in the real-world situation.
- Support of alternative comparison and experimental designs. With a validated simulation model, different design/redesign alternatives can be compared in a very efficient way.
- Simulation offers better control over experimental conditions compared to experimenting with the system itself, especially as direct manipulation is costly or unfeasible, or the system may not even yet exist.
- Simulation allows the study of a system over a long time frame in compressed time. Since the simulation model computes the events, and does not actually "replicate" events in the real time frame.

However, there are also disadvantages to modeling and simulation:

- Stochastic simulations only produce estimates of how the real system will perform; most of the time the results are not generalizable, since simulation models are constructed based on individual systems.
- Complex simulations can be time consuming and expensive to develop, especially when there are large numbers of processes and variables involved.
- The large volume of data that is produced by the simulation can sometimes place greater confidence in the study's results than is justified. Especially when the model is not valid, the data produced can be misleading.
- Simulation modeling can be used to compare alternatives, but not to find the optimal solution to a problem.

8.3.2 Discrete Event Simulation (DES)

Despite these issues, simulation is still one of the most popular methods used to solve real-world systems problems. Particularly for simulating queuing systems, *discrete event simulation* (DES) has been used widely as the primary simulation method. DES, as the name indicates, uses a sequence of discrete events occurring to simulate the events (i.e., arrival and departure) in the queuing system in the real world; it is a technique for the study of systems whose state changes at discrete points in time. The fundamental concepts in DES are that a system is composed of objects, known as entities (such as customers in the queuing system), and each entity has unique properties called attributes. The system state is a collection of attributes that represent the entities of the system; by investigating the systems variables and analysis using statistics, the performance of the queuing system can be studied and optimized.

Typical procedures (Figure 8.5) include:

1. Formulate the problem and plan the simulation study. Here, the real-world problem is that the issues that need to be addressed by the model are defined so that the level of detail adequately reaches the model's objectives.
2. Collect data on the system of interest and determine the model's logic. The data is analyzed and represented in a proper format; this format has be validated to represent the true system parameters from the real world. Validating input data is very important to ensure it is complete, correct, and consistent. Data analysis and validation should occur before any coding begins to avoid reprogramming the model. We will review some of the data analysis techniques in Section 8.3.3.
3. Develop the computer simulation model and verify the model. Check and debug the model to make sure it is error free. As mentioned earlier in the book, a verified model is not the same as a

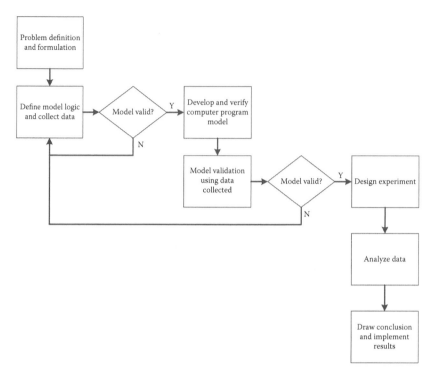

FIGURE 8.5
Steps of a simulation study. (Redrawn from Law, A.M. and Kelton, W.D. *Simulation Modeling and Analysis*, 3rd edn. New York: McGraw-Hill, 2000.)

validated model; a verified model means that the model is error free and is doing what the developer wants it to do, or *doing things right*, but sometimes a verified model does not necessarily perform in the same way as the real-world system. In that case, we need to validate the model to make sure it is *doing the right things*.

4. Validate the DES model. Statistical analysis that compares the model's output to actual system data will ensure that the model is valid and performs similarly to the actual system.

5. Experiment with the valid model. This is where alternative designs are developed. Comparative results should be documented, presented, and implemented.

8.3.3 Data Analysis Using Goodness of Fit

Simulation is considered empirical; in other words, the model depends on the actual data collected from the real world, and a distribution function is developed to represent the nature of the data collected. In the real world, data present a large degree of variability, and we cannot assume that the data follows a certain distribution without analyzing it first. As a matter of fact, an accurate estimation of the distribution to represent the data is critical, as this distribution will be used by the simulation model to generate simulation events, as it is well known that "garbage in, garbage out"; without an accurate data model, one cannot produce valid simulation results.

With collected data, the first thing to do is plot the histogram of the data to observe the pattern of the distribution; based on the observed pattern, a hypothesis is made. For example, if one observes a symmetrical bell shape, then a normal distribution is reasonably hypothesized; if no particular shape can be concluded from the histogram, one always can use an empirical fit for the data. We will talk about this later in this section.

With the hypothesized distribution function, a maximum likelihood estimation (MLE) method is used to estimate the parameters of the distribution. An MLE is a method to estimate the parameters of a statistical model. The idea behind the MLE method is that for the collected data set and the hypothesized underlying model for the data, the distribution parameters are sought to maximize the likelihood function, which can be defined as follows:

Suppose there is a sample of n independent and identically distributed (IID) observations, $x_1, x_2 \ldots, x_n$; then, the likely function is the joint density function, which can be defined as

$$\mathcal{L}(\theta; x_1, x_2, \ldots, x_n) = f(x_1, x_2, \ldots, x_n \mid \theta) = f(x_1 \mid \theta) \times f(x_2 \mid \theta) \times \cdots f(x_n \mid \theta)$$

with θ being the parameter for the distribution functions. The MLE method determines the parameter θ such that $\mathcal{L}(\theta; x_1, x_2, \ldots, x_n)$ or sometimes the log-likelihood function $\ln(\mathcal{L}(\theta; x_1, x_2, \ldots, x_n))$ is maximized. Due to the nature of this book, we are not going to delve into the details of MLE methods; for more information, readers can refer to any advanced statistics book such as Pfanzagl (1994).

Once the parameters have been estimated, the next step is to test if the fitted distribution model from the estimated parameters truly represents the data. A commonly used test is the goodness-of-fit test. Specifically, goodness-of-fit tests allow for the testing of the null hypothesis of the data fitting a particular distribution. The null hypothesis is as follows:

H_0: *The x_i's are random variables with distribution function* $\hat{F}$

The goodness-of-fit test verifies the null hypothesis, that is, whether or not that the hypothesized theoretical distribution is a good fit for the actual data (or, in other words, that there is no difference between the actual and proposed distributions). A popularly used statistics test for the hypothesis is to use chi-square testing, illustrated in Equation 8.42:

$$\chi^2 = \sum \frac{(O - E)^2}{\sigma^2} \tag{8.42}$$

where:

σ^2 is the variance of the observed data

O is the observed data frequency count

E is the fitted data frequency count from the theoretical model if the null hypothesis is true

Chi-square testing is also called "nonparametric" statistics as it does not depend on the assumption of normality, but rather compares the frequencies of the occurrence.

A larger corresponding p-value for the goodness-of-fit tests indicates a good fit. A rule of thumb is that the p-value should be 0.15 or higher for a theoretical distribution to be used. If the p-value is less than 0.15, this implies that the hypothesized model is not really a good fit for the data, so the null hypothesis above cannot be accepted. In this case, an empirical distribution should be used.

Distributions in simulation models fall into two main types: theoretical and empirical. Theoretical distributions use mathematical formulas to generate samples (such as the normal distribution or the exponential distribution), while empirical distributions simply divide the actual data into groups and calculate the proportion of values in each group. Law and Kelton (2000) explain that if a theoretical distribution can be found to fit the actual data reasonably well, this is preferable to using an empirical distribution for the following reasons:

- Theoretical distributions "smooth out" irregularities found in empirical distributions.
- It is not possible to generate values outside of the range of the observed data in the simulation using an empirical distribution.
- The theoretical distribution is a more compact way of representing a set of data values.
- A theoretical distribution is easier to change.

Equation 8.42 gives a continuous, piecewise-linear distribution function F by first sorting the $X_{(i)}$'s into increasing order, letting $X_{(i)}$ denote the ith smallest member of the data set so that $X_{(1)} \leq X_{(2)} \leq \cdots \leq X_{(n)}$. Then, the empirical distribution F is given by:

$$F(x) = \begin{cases} 0, & \text{if } x < X_{(1)} \\ \dfrac{i-1}{n-1} + \dfrac{x - X_{(i)}}{(n-1)\left(X_{(i+1)} - X_{(i)}\right)}, & \text{if } X_{(i)} \leq x < X_{(i+1)}, i = 1, 2, \ldots, n \\ 1, & \text{if } x \geq X_{(n)} \end{cases}$$

As we can see, $F(x)$ is actually a cumulative distribution function to generate a random value from the distribution. For example, if we found that the service time x of an operator has the following cumulative empirical distribution

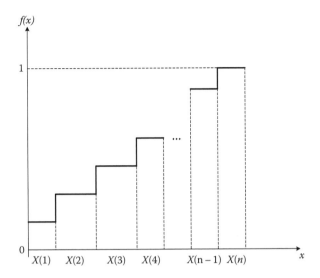

FIGURE 8.6
Illustration of cumulative distribution function.

distribution (see Figure 8.6). Then we can generate a uniform random number in [0, 1], mapping to the x axis to generate a random number in x_i.

These distribution functions, either theoretical or empirical, will be used as input to the simulation model to generate random numbers to create the event for the model. In the next section, we use an example using Arena software to illustrate how a DES model can be constructed.

8.3.4 DES Model Using Arena

Here we use a DES software package, Rockwell Automation's Arena v.14 computer simulation software, to give an example of how a DES simulation is built. Arena is a graphical user interface (GUI)-based DES software package, which allows us to easily and logically model real-world queuing systems in an intuitive way, by using the appropriate process blocks (Kelton et al. 2010). With a short learning curve, almost any user can easily learn to build a model, without going through long learning lessons; meanwhile, Arena also offers options for advanced users who prefer to use basic code to build the system from the nuts-and-bolts level. Moreover, Arena provides the ability to experiment on a system, either by creating or manipulating certain processes or entities involved in the system's operations and analyzing the results (Rockwell Automation 2011). For example, new operational techniques, new machines, altered schedules, or "what if" scenarios can be implemented into the system without disturbing the current operational flow, using unnecessary resources, or putting the system and its components in danger.

Before illustrating the Arena simulation model, let us first review the elements/concepts that are used in Arena.

8.3.4.1 Entity

Entities are the "players" of simulation systems; they correspond to the customers in queuing systems. In DES models, entities are created at the beginning of the simulation; they pass through the model, trigger the occurrence of events (i.e., entity arrives, waits in a queue, departs the system, etc.), and when these events occur, systems update their status through variables and attributes, so that the performance of the queuing systems can be recorded and measured.

8.3.4.2 Attribute

Attributes are the properties of the entities; they characterize the entity to track/record its behaviors. For example, the entity's name, type, arrival time, and even pictures of it are all instances of entity attributes.

8.3.4.3 Resources

Resources are the servers in the queuing systems, as mentioned in the previous sections, since all the resources (service channels) have limited *capacity* (number of service channels), so entities compete for the resources. Resources have four states; busy, idle, inactive, and failed. When a resource is not available (either busy, inactive, or failed), customers wait in a queue until the resource is available again, so that they can be served.

8.3.4.4 Queue

Queues are the places for the entities to wait for service from the resources. A queue has a rule, the default rule is FIFO; there are also other rules that can be applied to the queue, such as LIFO or by priority (e.g., as defined in some attribute of the entity; it could be the lowest value of that attribute first, or the highest value first).

8.3.4.5 Variable

Variables record the behaviors of the queuing systems; comparing attributes, variables are global while attributes are "private" to each entity. For example, the number of entities within the system is a variable; it is not tied to any entity. Although entities' behavior could may change the value of a variable, the copy of the variable exists at system level, while the copies of attributes only coexist with the entity. When there is no entity in the system, there will be no attributes, but the variable will still exist.

Let us use a simple example to illustrate how to build a model using Arena.

> **Example 8.5:** Parts arrive at a manufacturing facility to be processed (drilled). Data shows that parts arrive according to a Poisson process, with a mean interarrival time of 6 min, and one part arrives at a time. When parts arrive, they will wait in a queue to be processed in a first-come-first-served basis. Currently, the facility has *one* drilling machine available. The processing time per part follows a triangular distribution, with a minimum possible time of 4 min, maximum possible time of 6 min, and most likely processing time of 5 min. When the drilling is completed for a particular part, it is shipped out if it is of good quality. However, the data shows that about 15% of the parts will have defects that prevent them from being shipped out. If parts are defective, they will be sent to a station for rework. Currently, there is *one* employee available to carry out the manual rework, which takes an exponential distributed time with a mean of 10 min; after the rework, all the parts will be salvaged and shipped out. We want to know the average cycle time for completed parts and salvaged parts (the time in the system), the average time parts wait to be processed, the average time parts wait to be reworked, and the utilization (i.e., the percentage of time that the drilling machine and the rework station are busy).

As you can see from the example description above, it is quite challenging to solve the problem analytically, as it involves a nonexponential time distribution and other complexities such as the rework rate. This problem is very suitable for the use of a simulation model to solve it empirically.

Let us use Arena as the simulation software to illustrate how to build a simulation model to solve this problem.

1. A good start would be to sketch the problem using a flow chart; in this way, the logic is easily visualized, as the simulation model will look like the flow chart. The flow chart for the above example is illustrated in Figure 8.7.

2. Build the Arena model. Building the model is very straightforward; for simple problems like Example 8.5, we can just use the basic process module, as shown in Figure 8.8.

Modules are pre-packaged basic modeling units, which will satisfy the most fundamental needs for model development. The Arena software has a typical window style interface; all the modules in the different levels of the template will fulfill the modeling needs of the simulation models. Arena has also provided the basic "block" and "element" for advanced users, enabling them to build the model from a near-scratch level, providing maximum flexibility and capabilities for different kinds of modeling needs. For a more detailed description of Arena blocks and elements, readers can refer to Kelton et al. (2009).

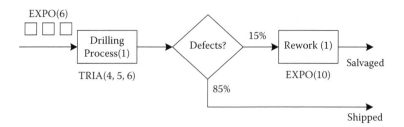

FIGURE 8.7
Flow chart sketch for Example 8.4.

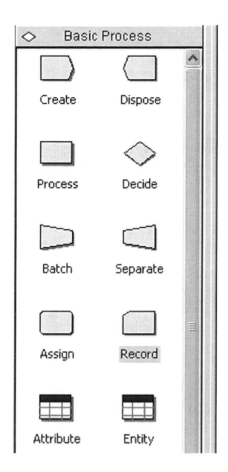

FIGURE 8.8
The Arena basic process module. (Courtesy of Rockwell Automation, Inc. Milwaukee, WI)

Let us use modules from the "basic process" template to build the simulation model for Example 8.5.

8.3.4.6 Step 1: Generate Parts Arrival and Assign Parts Attributes

To generate the arrival of the parts according to the Poisson process, with the interarrival time following the exponential distribution with a mean of 6 min, we use the "create" module. Drag and drop a "create" module on the working area (the big white area in the middle), and double-click on the "create" to change the parameters, as seen in Figure 8.9.

One thing that readers have to keep in mind is that Arena does not allow two items to have the exact same name, so, when entering names for each module, or for anything defined within the model, it is a good idea to make the name unique and meaningful. This is extremely useful when debugging the model, and, later, an easy-to-understand name will make the reading of the model much easier.

Once the parts enter the system, before carrying out any other action, it is a good idea to assign some "attributes" to facilitate the recording of entity behavior. In this example, we want to collect the average time that each entity (customer) stays in the system. To collect this information, we need to record, for each of the entities, the times that they arrive and depart; the time interval between the departure and arrival is the time spent in the system. Because each customer arrives and departs at different times, it is not feasible for the model to setup an observer to keep track of all these different times. It is more efficient for each of the customers to carry their own arrival time value themselves and, just as they depart, we can look at the current time, subtract the arrival time (attribute) they carry with them, and thus obtain the time in the system for each customer. Attributes are attached to each individual entity; each entity carries a copy of the attribute. When this entity departs the system, its copy of the attribute will be deleted with the entity's departure. You will find using the attributes very handy and convenient when collecting tally statistics for the entities. Arena has predefined

FIGURE 8.9
Illustration of "Create" module. (Courtesy of Rockwell Automation, Inc.)

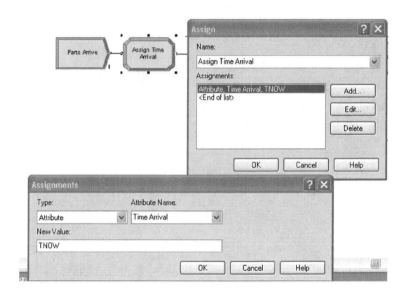

FIGURE 8.10
Illustration of "Assign" module. (Courtesy of Rockwell Automation, Inc.)

some of the attributes, including the entity number, entity picture, and so on. To add a new attribute, drag and drop an "Assign" module connect after the "Create," and double-click on it, change the name, and click on the "Add" button; then select "Attribute" in the type drop-down list, and type the new attribute name "Time Arrival," and in the new value field, type the value "TNOW." (This is the system variable to record the current simulation clock. Beginners who do not know this variable can point their mouse to the new value field and right-click, and then choose "Build expression," and a pop-up window will appear, allowing them to find the right variable.) Figure 8.10 illustrates the "Assign" module parameters.

8.3.4.7 Step 2: Drilling Process

After the entities arrive and are assigned attributes, they then move on to the drilling process. Drag and drop a "process" module and connect it to the "Assign" module. The parameters in the "Process" module are illustrated in Figure 8.11. In the logic drop-down list, choose "Seize delay release"; since there is a limited resource (one person), only one entity can be processed at a time. So each entity needs to "Seize" the resource (thus making it unavailable for other entities), "Delay" a processing time (according to the triangular distribution [4, 5, 6] minutes) and then "Release" the resource for the section, click on the "Add" button and a new resource window will pop up. In the new window, type the name of the resource "Machine 1" with a quantity of 1. (Please pay attention here that this

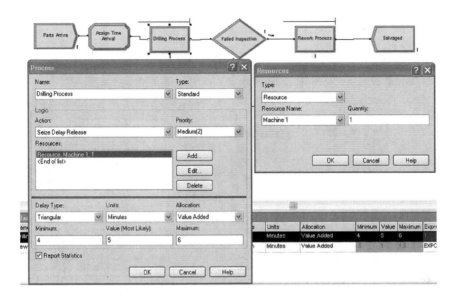

FIGURE 8.11
Illustration of "Process" module. (Courtesy of Rockwell Automation, Inc.)

quantity is not the capacity of the resource, but the number of resources needed to complete service for one entity. If you need to change the capacity of the resource, you need to change it in the "Resource" data element in the Basic Process Template.)

8.3.4.8 Step 3: Decide for Rework

A "Decide" module is needed to branch off the parts that need rework from the rest. Figure 8.12 shows the screen shot of the "Decide" module. Note that "2-way by Chance" is chosen as the branch type, and the value of the percentage of the portion being reworked, "15," is entered. Since this is a percentage value, we put "15" instead of "15%" in the field.

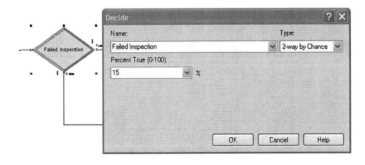

FIGURE 8.12
Illustration of "Decide" module. (Courtesy of Rockwell Automation, Inc.)

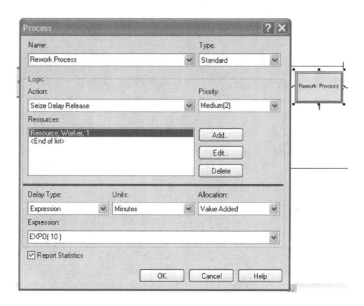

FIGURE 8.13
"Rework process" module details. (Courtesy of Rockwell Automation, Inc.)

The "Process" module for rework is illustrated in Figure 8.13. It is similar to the "Drilling" process, except the names and parameters are different.

8.3.4.9 Step 4: Collect Time in System and System Exit

When entities complete the service, before they exit the system, we need to collect the time they have spent in the system. For this purpose, a "Record" module is needed. The detailed information for the "Record" module is shown in Figure 8.14. We choose "Time Interval" as the type, and in the attribute name drop-down list, we choose the attribute "Time Arrival," which is defined earlier in the "Assign" module. Arena maintains the list of all the defined

FIGURE 8.14
"Record" module details. (Courtesy of Rockwell Automation, Inc.)

attributes; we just need to choose it from the drop-down list, instead of the typing the name ourselves. In this way, some typo errors can be avoided.

On leaving the "Record" module, we use "Dispose" to have the entity depart the system. The final model structure is illustrated in Figure 8.15.

8.3.4.10 Step 5: Execute the Model and Obtain Results

From the Arena menu, go to "Run," then choose "Setup" from the menu, and choose the "Replication parameters" tab in the menu window, as shown in Figure 8.16.

As seen in Figure 18.16, we want to change the replication length to 800 h and also change the hours per day from 24 to 8 h, and change the base time units from hours to minutes, as all the time data we used in the model are in units of minutes.

Once the run parameters are set, run the model if there is no error within it. The results of the model can be found in the "Report" section, as seen in Figure 8.17.

Clicking through the different sections of the "Report," we will find the results we need, as seen in Figure 8.18 and Tables 8.1 and 8.2.

This example demonstrates how to use simulation to solve a complex queuing system problem; the basic steps are presented. However, for real-world problems, the simulation model needs to be validated before the results can be used to draw any conclusions. For more details about DES model validation, readers can refer to Kelton et al. (2009).

8.4 Summary

In this chapter, we reviewed queuing theory and its application in systems engineering. In systems design, since the resources are always limited, in many situations we have to deal with a waiting line system. Queuing theory studies the behavior of the waiting line system to optimize its behavior. We started with the $M/M/1$ queue; by using the birth-death process, we derived the distribution of the $M/M/1$ queue, and obtained its fundamental statistics, including the average number of customers in the system, the average number waiting in the queue, the average time spent in the system, and the average time spent in the queue. Little's law ($L = \lambda W$) is very useful to describe the relationships between average numbers (L) and average times (W) spent within the system and queues, and this law is universal for all types of queuing systems, regardless of the distributions of the arrival time and process time. We also briefly introduced the multiserver $M/M/c$ queue and gave some examples to illustrate how to use Little's law to solve for the basic statistics.

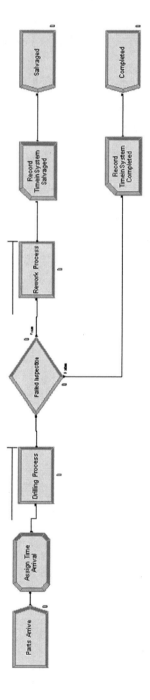

FIGURE 8.15

Example 8.5 final model structure. (Courtesy of Rockwell Automation, Inc.)

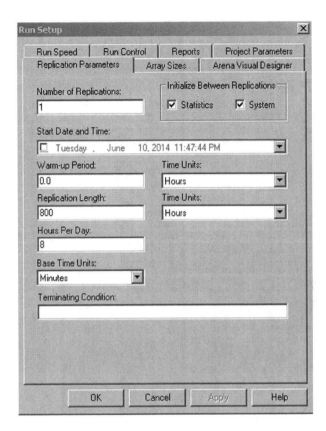

FIGURE 8.16
Setup window for model parameters. (Courtesy of Rockwell Automation, Inc.)

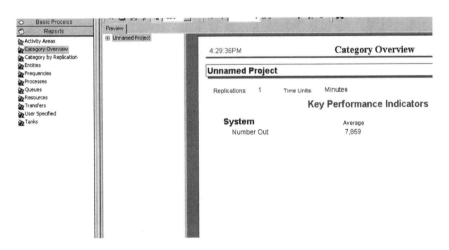

FIGURE 8.17
Illustration of model reports. (Courtesy of Rockwell Automation, Inc.)

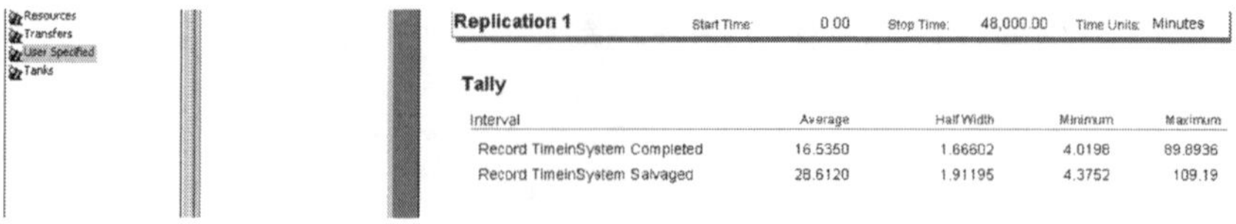

Tally				
Interval	Average	Half Width	Minimum	Maximum
Record TimeinSystem Completed	16.5350	1.66602	4.0198	89.8936
Record TimeinSystem Salvaged	28.6120	1.91195	4.3752	109.19

FIGURE 8.18

Average cycle time for completed parts and salvaged parts (time spent in the system); for completed parts, the average time is 16.5350 min, and for salvaged parts, this time is 28.6120 min. (Courtesy of Rockwell Automation, Inc.)

TABLE 8.1

Queue Detail Summary

	Waiting Time
Drilling process queue	11.46
Rework process queue	2.47

Note: This time can be found in the "Queue" category of the report.

TABLE 8.2

Resource Detail Summary

	Usage				
	Inst Util	Num Busy	Num Sched	Num Seized	Sched Util
Machine 1	0.82	0.82	1.00	7,860.00	0.82
Worker	0.25	0.25	1.00	1,197.00	0.25

Note: The utilization (i.e., the percentage of time that drilling machine and rework station are busy) for Machine 1 is 82% and worker (rework resource) is 25%.

In the last section of the chapter, we described how to use simulation to solve more complex queuing system problems. Some basic simulation concepts were introduced followed by a simulation demonstration using Arena software. The simulation method is very useful if the queuing system is dynamic and involves many factors that make an analytical approach impossible.

PROBLEMS

1. Compare the performance of the separate waiting line systems in Example 8.2, comparing the average queue length, the average number of customers in the system, the average waiting time spent waiting in the queue, the average time spent in the system, and the probability that a customer has to wait. What is your conclusion based your results?

2. Customers arrive at a service shop following a Poisson process with a mean arrival rate of 6/h; the service time per customer is exponentially distributed with a mean service time of 8 min. Assuming there is enough space for customers to wait,

   a. What is the probability that the shop has no customer in it?

   b. What is the probability that there are four customers in the shop?

   c. What is the probability that there is at least one customer in the shop?

   d. What is the mean number of customers in the shop?

   e. What is the mean number of customers waiting in the queue?

   f. What is the mean waiting time in the queue?

3. A service center has customers arriving in a Poisson process, and the service time is exponentially distributed with a mean of 15 min. If customers spend more than 1.25 h in the center on average, the manager will consider hiring additional service resources. What is the maximum arrival rate of customers to make the hiring decision?

4. In a manufacturing facility, machines break down at an exponential rate of 4/h. There is one service person who repairs these machines at an exponential rate of 5/h. The cost of losing production for these failed machines is $30/h/machine. What is the average cost rate ($/h) for the factory due to these machine failures?

5. Machines come to a service center for a regular repair service; the arrival process is a Poisson process with an arrival rate of 10/h. Each machine can make $500 profit/h and the service charge is $200/h/machine. What should the service rate should be to minimize the total cost? What is the minimum cost?

6. At a post office, a newly hired person works at a rate of 10 customers/h exponentially; an experienced employee works at a rate of 20 customers/h. It has been estimated that customer waiting time is worth $5/h, and customers arrive at a Poisson process at a rate of 6 customers/h. If the new employee is paid $10/h, what is the pay rate for the experienced employee so that the average cost would break even between novice and experienced employees?

7. Simulate Problem 2 using Arena and obtain the results for (d), (e), and (f). Compare the results obtained from Problem 2. Discuss the differences between the two sets of results.

8. Students at a university come to the bursar's office entrance following a Poisson process, with a mean interarrival time of 5 min (the first student arrives at Time 0). After entering the door, there is a hallway that takes students uniformly between 1.5 and 2 min to walk to the service window. Upon arrival at the service window, students wait in a single queue to be served by one of the three clerks; the service

time per student is distributed in a triangular distribution with a minimum of 3 min, maximum of 5 min, and most likely 5 min. Upon completion of the service, students leave the window and walk back to the door and exit. Using ARENA, simulate for 16 h, obtain the average number of students completing the service, the average time students spend in the office, and the average number of students waiting in the waiting line.

9. Parts arrive at a cleaning facility for processing. The arrival of parts follows a Poisson distribution, with a mean of 1 min (first arrival occurs at Time 0). Upon arrival, parts are processed in the cleaning station with one machine; the cleaning time is triangularly distributed (minimum 30 s, maximum 90 s, and most likely 1 min). After cleaning, 80% pass inspection and are shipped out directly; 20% are considered defective and sent to a rework station (with one machine) to be fixed. Assume 100% of the rework is successful and the rework time is exponentially distributed with a mean of 5 min. Use ARENA to simulate the process for 1000 min and find the average time good parts and salvaged parts spend in the system and the queue length for the two workstations and utilizations.

9

Engineering Economy in Systems Engineering

In systems engineering, we are concerned with the efficient utilization of limited resources. There are two types of efficiency that systems designers need to take into account: technical efficiency and economic efficiency. In previous chapters, we have discussed the various types of systems technical performance measures, such as systems functionality, reliability, maintainability, and so on. On the other hand, we need to achieve the design technical objectives economically. Systems design requires capital investment, and every activity and effort occurring in the design requires resources, such as human labor and materials, and eventually these efforts are associated with cost. One of the most important goals in systems design is to minimize the cost involved and maximize the utility that is generated from the cost investment. Thus, designing for economic efficiency is very critical for systems engineering.

In systems engineering, we use the concepts of engineering economy to measure and control the cost involved in systems design and project management. Engineering economy is an applied economic science that addresses the economic issues within the engineering field. It is based on the principle of fundamental economics, and applying it in engineering design and related activities. The basis for the engineering economy is the time value of the money and opportunity cost. In this chapter, we will review some of the basic concepts of engineering economy and become familiar with the application of engineering economy in making decisions in systems engineering. More specifically, we will

1. Understand the concept of time value of the money and interest rate
2. Understand the concept of the present value (P), future value (F), annual equivalent value (A), and know how to convert from one to another
3. Describe how to use the values to compare different alternatives and make decisions
4. Introduce break-even analysis, based on the different values of money, and the decision-making process using break-even analysis

9.1 Interests and Time Value of Money

Time increases the value of money. We all know that if we deposit an amount of money in the bank, after a certain time, the amount deposited will increase. The bank pays interest on the money we deposit; this interest is paid for the time value of the deposited money. The reason that time increases the value of money is because money, as the general format of resources, can be invested to generate new values; for example, paying human labor for services provided, or investing in an engineering project to make more profit from it. We know that resources are limited; with money in hand, we can either consume now, or give the opportunity to someone else (i.e., the bank), so they can loan the money to someone who needs it to start a new project. Because we give up the opportunity of using the money now, and give the opportunity to others to use it first, in return, the opportunity we give needs to be compensated, in the form of interest.

What is interest? Interest is a fee paid by the borrower of the money (or asset) as compensation to its owner for the opportunity to use it. Interest can be considered as the price of borrowing the money. The amount of interest paid is determined by the interest rate and elapsed time for the borrowing of the money. We use i to denote the interest rate. The interest rate has to be specified according to the time it covers; for example, we need to specify whether it is the annual interest rate, monthly rate, or daily rate, as these rates of interest will be different.

There are two types of interest rate; namely, the simple interest rate and the compound interest rate. The simple interest rate, as the name implies, is simply the interest on the principal, and the interest earned will not generate further interest. For example, if the annual interest rate is a simple interest rate of 10%, for a deposit amount of $200, at the end of Year 1, the total amount in the bank account will be

$$200 + 200\,(10\%) = \$220$$

At the end of Year 2, since only the principal amount ($200) generates interest, we have

$$220 + 200\,(10\%) = \$240$$

and so on. With compound interest, on the other hand, the interest earned previously will generate future interest as well. So if 10% is the compound interest rate, at the end of Year 2, the total amount becomes

$$220 + 220\,(10\%) = \$242$$

If we denote the initial amount as x_0 and the compound interest rate as i, then it is easy to derive the total amount x_n after n years as

$$x_n = x_0 (1+i)^n \qquad (9.1)$$

The compound interest rate is that most commonly used for calculating the time value of money in the current economy; this is the interest rate we will be using in this chapter. Unless specified, the interest rate we refer to hereafter is the compound interest rate. With the concept of the interest rate, we can now measure the value of money between different points in time, or in other words, the equivalent worth of the money.

9.2 Economic Equivalence

In this section, we will be deriving the formulas for the equivalent money value for different points in time. Before we start, let us list all the terminologies and notions that will be used throughout the chapter.

N: Total time period: this can be years, months, weeks, days, and so on.

n: Time period: this can be years, months, weeks, days, and so on.

i: Interest rate per time period.

P: Present time equivalent value: This is the worth of the money at beginning of the time period, that is, $n=0$; this value is also referred to as present worth, present value, or present equivalence.

F: Future time equivalent value: This is the worth of the money at end of a future period, that is, $n \geq 1$; this value is also referred as future worth, future value, or future equivalence.

A: Annual equivalent value: This is the consecutive equal amount of the money's worth at the end of each period. This value is also referred to as the annual worth or annual uniform worth/value of the money. One needs to pay attention here; although it is called the annual worth, it is actually the equal amount of the value at the end of the time period, so it could be the monthly equal amount if the time periods are measured in months.

With these notations being defined, we now can describe the formulas that can convert between the values of P, F, and A. In solving engineering economy problems, a useful method is to use graphical tools to illustrate the money flow on a timeline axis. We call this tool the money flow diagram, as illustrated in Figure 9.1.

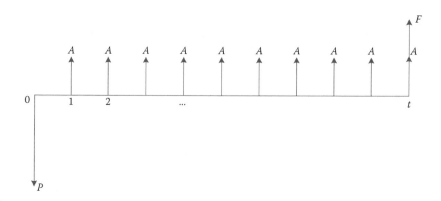

FIGURE 9.1
Illustration of money flow diagram.

As shown in Figure 9.1, each of the time marks t represents the end of that period t, as we assume that money values are computed only at the end of that time period. If this money flow is a positive one (i.e., income or revenue), we draw the amount of the money using an upward arrow, to represent its positive value; if it is a negative one (i.e., investment, cost or loss), we draw the amount downward. By using the money flow diagram, we can visualize the profile of the money flow in a very intuitive way; thus, it is easier for us to carry out the analysis.

9.2.1 Present Value (*P*) and Future Value (*F*)

Let us first derive the relationship between the present value (P) and future value (F). A single amount of P occurs at the beginning of the time period (time $= 0$); what is the equivalent future value of F at the end of the time period n given the compound interest rate of i? The money flow diagram converting P to F is illustrated in Figure 9.2.

From Equation 9.1, it is easy to obtain the equation for converting P to F as

$$F = P(1+i)^n \qquad (9.2)$$

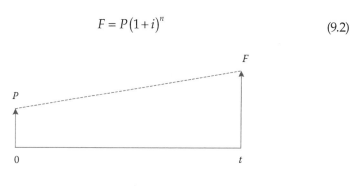

FIGURE 9.2
Money flow diagram converting P to F.

We can use a factor to represent the relationship between P and F, written as

$$F = P\left(\frac{F}{P}, i, n\right) \quad (9.3)$$

The multiplier factor $(F/P,i,n) = (1+i)^n$ implies a conversion from P to F, given the values of n and i. From Equation 9.2, it is easy to find the reciprocal equation from F to P, written as in Equation 9.4,

$$P = F\frac{1}{(1+i)^n} \quad (9.4)$$

or

$$P = F\left(\frac{P}{F}, i, n\right) \quad (9.5)$$

with

$$\left(\frac{P}{F}, i, n\right) = \frac{1}{(1+i)^n}$$

Example 9.1

A deposit of $100 is made now, how much will it be worth after 10 years with an annual interest rate of 5%?

Solution: According to Equation 9.2, the total future amount of the deposit after 10 years is

$$F = P(1+i)^n = 100(1+0.05)^{10} = 162.89\,(\text{USD})$$

In Appendix III of the book, the conversion factors have been precalculated for some of the most commonly used values of i and n, so readers can use the value directly from the tables for those i and n.

9.2.2 Annual Value (*A*) and Future Value (*F*)

The annual value is the equal amount of money that occurs at the end of each time period, starting at $n=1$; its money flow diagram can be illustrated in Figure 9.3.

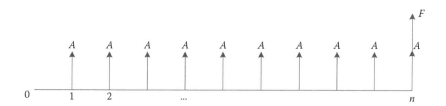

FIGURE 9.3
Money flow diagram converting A to F.

   With the equal amount of A at the end of each time period, what is the equivalent future value F? From the money flow diagram, we can convert every A to the future value at n, starting from the last A, which occurs at n, all the way back to the first A, which occurs at $n=1$. There are, in total, $n-1$ time periods from the first occurrence to the last occurrence of A. So the future value of all the As can be obtained as follows:

$$F = A + A(1+i) + A(1+i)^2 + A(1+i)^3 + \cdots + A(1+i)^{n-1}$$
$$= A\left[1 + (1+i) + A(1+i)^2 + \cdots + A(1+i)^{n-1}\right]$$

Let

$$S = 1 + (1+i) + (1+i)^2 + \cdots + (1+i)^{n-1} \tag{9.6}$$

So we have

$$F = AS$$

Multiplying by $(1+i)$ on both sides of Equation 9.6, we obtain

$$(1+i)S = (1+i) + (1+i)^2 + \cdots + (1+i)^n \tag{9.7}$$

Subtracting Equation 9.6 from Equation 9.7, we have

$$iS = (1+i)^n - 1$$

We can obtain S by

$$S = \frac{(1+i)^n - 1}{i}$$

So, substituting S back, we obtain Equation 9.8:

$$F = A\frac{(1+i)^n - 1}{i} \tag{9.8}$$

or, we can say that the factor from A to F is

$$\left(\frac{F}{A}, i, n\right) = \frac{(1+i)^n - 1}{i} \tag{9.9}$$

Similarly, the reciprocal of Equation 9.8 gives the formula of F to A, described in Equation 9.10:

$$A = F\frac{i}{(1+i)^n - 1} \tag{9.10}$$

or

$$\left(\frac{A}{F}, i, n\right) = \frac{i}{(1+i)^n - 1} \tag{9.11}$$

Example 9.2

If I deposit an equal amount of \$100 every year for a time period of 20 years, with a compound interest rate of 5%, how much money will I have in total after 20 years?

Solution: From the problem description, we know that $A=\$100$, $i=5\%$ and $n=20$ years. So, according to Equation 9.8, we can obtain the total amount future value by

$$F = A\frac{(1+i)^n - 1}{i} = 100\frac{(1+0.05)^{20} - 1}{0.05} = \$3306.60$$

The factors of $(A/F,i,n)$ and $(F/A,i,n)$ are also precalculated in Appendix III for some of the most commonly used rates.

9.2.3 Annual Value (A) and Present Value (P)

This question related to A and P is very common as well; an example is when we purchase a house at a price of P, if we have a mortgage over n years at

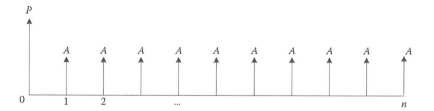

FIGURE 9.4
Money flow diagram converting between A and P.

an interest rate of i, what is the annual equal payment of A? The relation between A and P is illustrated in Figure 9.4.

The formula for converting between A and P can be easily obtained by using the equations between A to F and P to F. From Equation 9.8 we have

$$F = A\frac{(1+i)^n - 1}{i}$$

And from Equation 9.2, we know that

$$F = P(1+i)^n$$

Substituting Equation 9.2 in Equation 9.8 we have

$$F = P(1+i)^n = A\frac{(1+i)^n - 1}{i}$$

$$A = P\frac{i(1+i)^n}{(1+i)^n - 1} \tag{9.12}$$

or, we can say that the factor from P to A is

$$\left(\frac{A}{P}, i, n\right) = \frac{i(1+i)^n}{(1+i)^n - 1} \tag{9.13}$$

The reciprocal of Equation 9.12 gives us the equation from A to P, as illustrated in Equation 9.14:

$$P = A \frac{(1+i)^n - 1}{i(1+i)^n} \tag{9.14}$$

So, the factor from A to P is

$$\left(\frac{P}{A}, i, n \right) = \frac{(1+i)^n - 1}{i(1+i)^n} \tag{9.15}$$

Example 9.3

A house is sold for a cash price of \$150,000; if the buyer has a loan for 30 years with zero down payment, with an interest rate of 4%, what is the annual mortgage payment?

Solution:

$$P = \$150,000$$

$$i = 4\%$$

and

$$n = 30$$

So, according to Equation 9.12,

$$A = P \frac{i(1+i)^n}{(1+i)^n - 1} = 150,000 \frac{4\%(1+4\%)^{30}}{(1+4\%)^{30} - 1} = \$8,674.51$$

The buyer needs to pay \$8,674.51 every year for 30 years to pay off the mortgage.

The factors of $(A/P,i,n)$ and $(P/A,i,n)$ are also precalculated in Appendix III for some of the most commonly used rates.

Table 9.1 summarizes all the formulas described above. Readers may use it as a quick reference.

TABLE 9.1

Summary of Interest Formulas

Conversion between P, F, and A	Formula
$P \rightarrow F$	$\left(\dfrac{F}{P},i,n\right) = (1+i)^n$
$F \rightarrow P$	$\left(\dfrac{P}{F},i,n\right) = \dfrac{1}{(1+i)^n}$
$A \rightarrow F$	$\left(\dfrac{F}{A},i,n\right) = \dfrac{(1+i)^n - 1}{i}$
$F \rightarrow A$	$\left(\dfrac{A}{F},i,n\right) = \dfrac{i}{(1+i)^n - 1}$
$A \rightarrow P$	$\left(\dfrac{P}{A},i,n\right) = \dfrac{(1+i)^n - 1}{i(1+i)^n}$
$P \rightarrow A$	$\left(\dfrac{A}{P},i,n\right) = \dfrac{i(1+i)^n}{(1+i)^n - 1}$

9.3 Decision Making Using Interest Formula

The interest formulas are very useful for decision making and comparing alternatives. When making decisions involving the time value of money, we have to keep in mind that

1. We need to convert the money occurring at different times into a common basis—in this case, converting to the same time point.
2. The decision should be the same regardless of the time point to which we choose to convert the time value of money. In other words, if we use the present value to find an alternative to be more favorable, then using the future value or the annual value should give us the same results. The "same results" means the same decision regarding which alternative is favorable. Whether we use P, F, or A as our criterion is solely dependent on the nature of the problem and complexity of the calculation for each value. We need to be flexible and choose one value that makes the calculation the easiest.

The decision-making process using the time value of money involves the following steps:

1. Determine the decision-making objective. Based on the problem, determine the objective of the decision making; usually the objective is to maximize income and minimize expenditure. If both income and expenditure are involved, then the objective is to maximize the overall net income (income minus cost).

2. Plot the money flow diagram for the problem. As mentioned earlier, the money flow diagram is a very useful tool for us to visualize the profile of the project to review the overall process of the project in terms of when the income and costs will occur and how much they will be, then facilitating the selection of the right value (present, future, or annual) as the common basis for comparison.

3. Based on the money flow diagram, select an equivalence value (present value, future value, or annual equal value) as the comparison basis. As we mentioned in the previous sections, no matter what value is being chosen, the decision should be the same (i.e., if the present value shows that one alternative is better, then the other two values should have the same comparison results, if no mistake is made in the conversion process). The criterion used for selecting the value should be based on the level of complexity of the computation and conversion of the time value. For example, if the majority of expenses occur in the present and at the end of the project period, then present value or future value equivalence should be used; if the projects have lots of annual equal money flow, then using the annual equal value may be a good idea. Although using a different criterion would give the same results, it could cause the calculation complexity to vary greatly.

4. Once the value of the criterion has been chosen, convert all the activities to the selected time value, by using the equation in Table 9.1 or from the interest factor table in Appendix III if the interest rate is listed there. Keep in mind that if both income and expenditure (costs) are involved, the convention is to use a positive sign for the income and a negative sign for the costs.

5. Based on the results for the converted value in step 4, select the alternative with the highest net profit, and make recommendations.

In the following section, we will use an example to illustrate how to use the formulas to compare alternatives and make decisions.

9.3.1 Present Value, Future Value, and Annual Value Comparison

Example 9.4

Company XYZ needs to install a cleaning system. There are currently two candidate systems that meet the basic requirements:

> System 1: With an initial purchasing cost of $15,000, System 1 has a net save (value generated minus operation cost) of $2,000 annually. At the end of Year 4, System 1 will need a major overhaul, which will cost $1,500. The life span of System 1 is 8 years; after 8 years, System 1 will have a salvage value of $5,000.
>
> System 2: With an initial purchasing cost of $25,000, system 2 has a net save of $3,300 annually. At the end of Year 4, System 2 will need a major overhaul, which will cost $2,500. The life span of System 2 is also 8 years; after 8 years System 2 will have a salvage value of $8,500.

With an interest rate of 5%, which system should be selected?

We will use present value, future value, and annual value to compare these two systems. Since the system has both positive incomes (i.e., the net save every year and salvage value) and negative costs (i.e., purchasing cost and overhaul cost), the goal of the decision making is to maximize the overall net income (income − costs). The money flow diagrams for System 1 and System 2 are illustrated in Figure 9.5.

9.3.1.1 Present Value Equivalence (PE) Criterion

Using the present value equivalence (PE) criterion, we need to convert all the activities to present value. By looking at Figure 9.5, it is obvious that, for example, for System 1, it includes a present cost of $15,000, a 4th year future cost of $1,500, an 8th year salvage value (income) of $5,000, and finally an annual equal save (income) of $2,000. System 2 has a similar cost and income structure. Knowing this structure, we can easily convert everything to present cost, shown as follows.

For System 1,

$$PE(1) = -15,000 - 1,500 \left(\frac{P}{F},5\%,4 \right) + 2,000 \left(\frac{P}{A},5\%,8 \right) + 5,000 \left(\frac{P}{F},5\%,8 \right)$$

$$= -15,000 - 1,500(0.8227) + 2,000(6.463) + 5,000(0.6768) = \$75.95$$

For system 2,

$$PE(2) = -25,000 - 2,500 \left(\frac{P}{F},5\%,4 \right) + 3,300 \left(\frac{P}{A},5\%,8 \right) + 8,500 \left(\frac{P}{F},5\%,8 \right)$$

$$= -25,000 - 2,500(0.8227) + 3,300(6.463) + 8,500(0.6768) = \$23.95$$

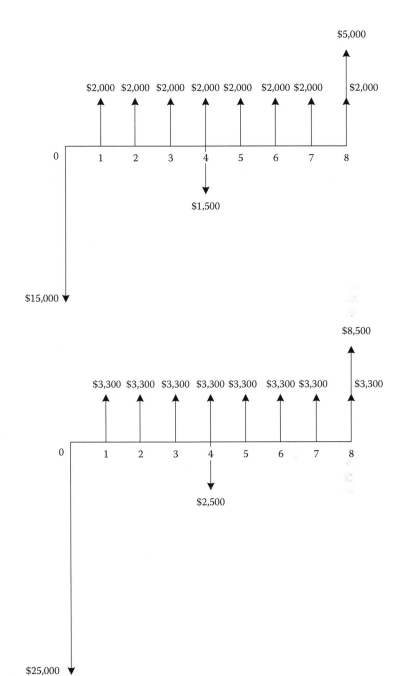

FIGURE 9.5
Money flow diagram for Example 9.4.

Since PE(1) > PE(2), system 1 should be selected when the interest rate is 5%.

9.3.1.2 Future Value Equivalence (FE) Criterion

Using the future value as the comparison basis, we need to convert everything into the future time value, as follows.

For System 1,

$$\text{FE}(1) = -15,000\left(\frac{F}{P},5\%,8\right) - 1,500\left(\frac{F}{P},5\%,4\right) + 2,000\left(\frac{F}{A},5\%,8\right) + 5,000$$

$$= -15,000(1.477) - 1,500(1.216) + 2,000(9.549) + 5,000$$

$$= \$119.00$$

For System 2,

$$\text{FE}(2) = -25,000\left(\frac{F}{P},5\%,8\right) - 2,500\left(\frac{F}{P},5\%,4\right) + 3,300\left(\frac{F}{A},5\%,8\right) + 8,500$$

$$= -25,000(1.477) - 2,500(1.216) + 3,300(9.549) + 8,500$$

$$= \$46.70$$

And again, FE(1) > FE(2); the decision-making result is the same as that using the PE criterion.

9.3.1.3 Annual Value Equivalence (AE) Criterion

Using the annual value equivalence (AE) criterion, everything needs to be converted to an annual equal amount, shown as follows.

For System 1,

$$\text{AE}(1) = -15,000\left(\frac{A}{P},5\%,8\right) - 1,500\left(\frac{P}{F},5\%,4\right)\left(\frac{A}{P},5\%,8\right)$$

$$+ 2,000 + 5,000\left(\frac{A}{F},5\%,8\right)$$

$$= -15,000(0.1547) - 1,500(0.8227)(0.1547) + 2,000 + 5,000(0.1047)$$

$$= \$12.09$$

For System 2,

$$AE(2) = -25{,}000\left(\frac{A}{P},5\%,8\right) - 2{,}500\left(\frac{P}{F},5\%,4\right)\left(\frac{A}{P},5\%,8\right)$$

$$+ 3{,}300 + 8{,}500\left(\frac{A}{F},5\%,8\right)$$

$$= -25{,}000\,(0.1547) - 2{,}500\,(0.8227)(0.1547) + 3{,}300 + 8{,}500\,(0.1047)$$

$$= \$4.27$$

For this problem, it is easy to see that using AE is a little more complex than PE and FE. This is mainly because that the overhaul cost that occurs in the middle of the life span needs to be converted to either present or future value first before it can be converted to the annual value.

9.3.2 Rate of Return

The rate of return is another important measure of value besides the time value of money measures. In some situations, the rate of return is believed to be the best evaluation criterion for comparing alternatives, as it does not rely on the external interest rate, but rather represents the cost-benefit value of the money flow. Rate of return is defined as the interest rate i^* that causes the two alternatives to be the same, or

$$PE_1(i^*) = PE_2(i^*)$$

or

$$FE_1(i^*) = FE_2(i^*), \quad AE_1(i^*) = AE_2(i^*)$$

Rate of return is also considered as a universal measure as it gives us the interest rate that makes the alternatives equal, and we can use rate of return as a threshold to compare with the current interest rate to make decisions. As the market interest rate changes, the rate of return does not change; thus, we can adjust the decision made according to the new market rate, without going through the calculation again; while using other criteria, every time the market rate changes, a new calculation is needed as the result will change. The rate of return is also useful for projects that have different life cycles; it serves as the only criterion by which to compare them.

Let us use Example 9.4 to show how to obtain the rate of return. As mentioned above, the rate of return i^* is the interest rate that satisfies

$$PE_1(i^*) = PE_2(i^*)$$

or

$$PE_1(i^*) - PE_2(i^*) = 0$$

This is obviously a high-order nonlinear equation, and it would be difficult to obtain the analytical solution unless applying some numerical analysis. Here we use an approximation technique called linear interpolation. Using this technique, we find two points that make the results change sign. It is assumed that if the two points are close enough, then a straight line can be used to approximate the curve connecting these two points. Since the two rates make $PE_1(i^*) - PE_2(i^*)$ change sign (either from positive to negative or vice versa), then the line will intersect $PE_1(i^*) - PE_2(i^*) = 0$ and the intersection point will be the rate of return.

Example 9.5

From the previous calculation, we know that with $i = 5\%$, $PE(1) - PE(2) = 75.95 - 23.95 = +52$.

We then use $i = 2\%$ to recalculate the new present value again.
At a 2% interest rate,

$$PE(1) = -15,000 - 1,500\left(\frac{P}{F}, 2\%, 4\right) + 2,000\left(\frac{P}{A}, 2\%, 8\right) + 5,000\left(\frac{P}{F}, 2\%, 8\right)$$

$$= -15,000 - 1,500(0.9238) + 2,000(7.325) + 5,000(0.8535) = 2,431.80$$

$$PE(2) = -25,000 - 2,500\left(\frac{P}{F}, 2\%, 4\right) + 3,300\left(\frac{P}{A}, 2\%, 8\right) + 8,500\left(\frac{P}{F}, 2\%, 8\right)$$

$$= -25,000 - 2,500(0.9238) + 3,300(7.325) + 8,500(0.8535) = 4,117.75$$

$$PE(1) - PE(2) = -1,585.95$$

So, clearly, the rate of return is between 2% and 5% since $PE(1) - PE(2)$ has changed sign. We apply linear interpolation, illustrated in Figure 9.6. Since it is linear, the following equation holds:

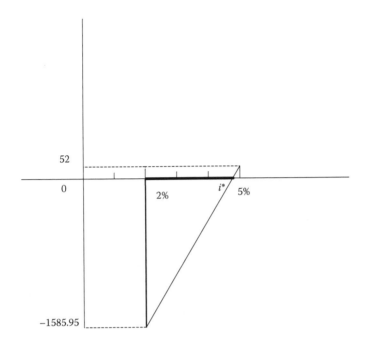

FIGURE 9.6
Illustration of rate of return for Example 9.6.

$$\frac{i^* - 2\%}{5\% - 2\%} = \frac{0 - (-1585.95)}{52 - (-1585.95)}$$

so

$$i^* = 2\% + 3\% \frac{1585.95}{1637.95} = 4.9\%$$

So, at an interest rate of 4.9%, the two systems will yield the same result. When the interest rate is larger than the rate of return level, the system with less initial cost will be more desirable, since with a higher interest rate, the higher the initial cost, the greater impact it will have on the overall cost throughout the life cycle. We can also say that with a greater interest rate, the initial cost will generate more interest; if we need to borrow the money, it would be more expensive, and thus leaves the higher initial cost as the less desirable option. When the interest rate is less than the rate of return, the system with the higher initial cost will be the better choice.

From the results, we can see in the original question that the market interest rate is 5% > i^*; this means that System 1 (with lower initial cost) will outperform System 2 (with higher initial cost).

9.4 Break-Even Analysis

Decisions made in systems engineering are primarily based on predictions of future events; engineers use current understanding of the systems design project and its anticipated future environment to estimate the economic benefits for the system to be designed. However, due to various levels of uncertainty that are involved in the decision factors in economic analysis, the estimation of the cost and benefits have various levels of uncertainty and risk. Examples of these factors include global and local political and economic situations, technological advancement, and the limitations of economic analysis methods and models. We have addressed decision-making uncertainty and risks in previous chapters; some of the models can certainly be used in engineering economy analysis to address these. In dealing with uncertainty in systems design, there is no guarantee that a project estimation will be accurate; what can be done is to provide a bigger picture to incorporate the risk and uncertainty factors into consideration, providing different scenarios or baselines for decision makers to deal with unexpected risks in the future. Generally speaking, there are several ways that a systems designer can prepare for future risks/uncertainty:

1. Provide break-even analysis to illustrate the relationship between costs, profits, and quantities, finding the break-even points for these factors, thus enabling decision makers to develop an overall picture of the profitability of the project.

2. Develop a sensitivity analysis model; by using "what-if" scenarios in the model, major uncertainties and risks can be anticipated. In sensitivity analysis, the range from the best-case to the worst-case scenario can be covered, enabling decision makers to develop strategies for possible upcoming scenarios. When one of these scenarios occurs, a corresponding strategy can be applied with confidence.

Break-even analysis has been used widely in business operation management; in this section, we shall use some examples to illustrate how to conduct a break-even analysis for business decisions.

9.4.1 Cost Volume Break-Even Analysis

In product design and capacity planning, we need to focus on the relationship between costs (fixed and variable), revenue, and production volume. In product design, if there is a fixed cost involved, then the product volume must achieve a certain level to break even.

Example 9.5

Company XYZ is adding a new assembly line. Investing in the new line requires a monthly payment of $10,000, the cost of the parts that the new assembly line produces is $10, and they are sold for a price of $35 per piece. Assuming all parts can be sold, how many parts must be assembled per month to break even?

Solution: In this problem the fixed cost (FC) per month is $10,000, the variable cost (VC) is $10 per piece and the revenue (R) is $35 per piece, so the break-even volume (Q) is determined by the point at which the total monthly profit breaks even with the fixed cost, shown as follows:

$$FC = Q(R - VC) \tag{9.16}$$

or

$$Q = \frac{FC}{R - VC} \tag{9.17}$$

So, the break-even volume is

$$Q = \frac{10,000}{35 - 10} = 400$$

Company XYZ must sell at least 400 pieces per month to break even.

9.4.2 Rent-or-Buy Break-Even Analysis

Sometimes, system designers face the decision of renting a resource or buying it; this is similar to the cost volume analysis above, if we treat the purchasing cost as the fixed cost for the resource.

Example 9.6

An automobile can be purchased at $15,000 and lasts 10 years. After 10 years, the automobile can be sold at $2,000. The annual operation and maintenance cost is $400.

As an alternative, the same automobile can be leased at the rate of $50 per day. For how many days each year may the automobile be used for these two options to break even?

Solution: If we denote the break-even number of days as N, then the average annual cost of owning a vehicle equals the annual renting cost, that is,

$$15,000\left(\frac{A}{P}, 10\%, 10\right) - 2,000\left(\frac{A}{F}, 10\%, 10\right) + 400 = 50N$$

$$15,000(0.1627) - 2,000(0.0627) + 400 = 50N$$

$$2715.1 = 50N$$

$$N \cong 55(\text{days})$$

So, the number of days to break even with the annual cost is 55 days.

The above question can also be asked in a different way: What is the minimum number of days of usage of the vehicle to make purchasing it more desirable? We have given readers some exercises at the end of the chapter to practice.

9.5 Summary

Systems design requires systems to be developed in an efficient manner, both technically and economically. In this chapter, basic concepts and measures for systems engineering economy were reviewed. The key concept in engineering economy is the time value of money. Time will increase the value of a money amount. Since monetary activities occur throughout the whole system life cycle, this requires us to analyze and compare the alternatives from a common time basis. In this chapter, we reviewed three fundamental time values of money: present value equivalence (PE), future value equivalence (FE), and annual value equivalence (AE); the formulas to convert between these values were derived and summarized. We used some examples to illustrate how to use these three values as criteria for the basis to compare multiple alternatives. As a universal measure, the rate of return criterion was also presented. At the end of the chapter, we introduced the break-even model in economic analysis; two examples, volume of production and rent-or-buy, were used to illustrate this analysis method. Break-even analysis provides a reference point for decision makers to relate costs and profits, providing a baseline for determining the profitability of the project. We need to keep in mind that the material covered in this chapter is very basic in nature; engineering economy is a big subject that itself needs a textbook to cover. For more in-depth review, readers can refer to any other engineering economy books that provide a comprehensive coverage of the subject.

PROBLEMS

1. Explain why economic analysis is important in systems engineering.
2. What is interest? Compare the simple interest rate and the compound interest rate.

3. When taking a loan from the bank, there are two options. Option 1: the annual interest rate is 8%, compounded monthly. Option 2: the annual interest rate is 15%, compounded quarterly. Which option should the borrower choose?

4. How many years does it take for a deposited sum to double itself, given the interest rate of 10%, compound annually?

5. What annual compound interest rate is necessary for a deposit to triple itself in 10 years?

6. Assuming that the interest rate is 5%, compounded annually,
   a. What is the present equivalent value of an overhaul cost of $2,000 that will occur 5 years from now?
   b. The initial investment is $20,000. What is the annual equal amount of saving over the 5-year period to justify this investment?

7. The same amount of money occurs at different time points; the amount occurring first has a higher equivalent value than the one occurring next. Is this statement true or false? Justify your answer.

8. A company has a loan of $150,000; the annual compound interest rate is 5%. The company begins to pay back the loan at the end of the first year, at an equal amount, and hopes to pay it off after 10 years. What is the annual equal amount of the payment?

9. In Problem 8, if there were no payment for the first year, and the company started to pay the loan back at the end of the second year, and paid it off after 10 years, what is the annual equal amount of the payment?

10. The HFS department is considering two computer servers for their lab. Server A costs $7,000 initially and has a salvage value of $1,500 after 10 years; Server B costs $5,000 initially and has a salvage value of $1,000 after 10 years. Assuming that they both satisfy the needs of the department, with an annual compound interest rate of 6%, which server should be chosen?

11. For a 5-year investment project, the initial investment at time 0 is $100,000, and the investment at beginning of the second year is $80,000. The annual revenue after the second investment is $50,000 (from Year 2 to Year 5); the salvage value after Year 5 is $70,000. Assuming an annual compound interest rate of 5%, should this project be invested in?

12. Calculate the rate of return for the project in Problem 11. (Hint: the rate of return makes the net value of the project equal 0.)

13. The following table shows the money flow for a project, with $i = 10\%$, compounded annually.

Year	0	1	2	3	4	5
Money flow	−1000	200	250	250	250	250

Is this project feasible?

14. A company is planning to build a 10,000 sq. ft. warehouse in 3 years. Option 1 is to buy a piece of land for $80,000 and build a temporary warehouse on it, the cost of which is $70/sq. ft.; after 3 years, it is estimated that the company can sell the land for $90,000, and sell the warehouse for $120,000. Option 2 is to rent a warehouse. The annual rent is $150/sq. ft., to be paid annually. The annual compound interest rate is 10%. Which option should the company take? Draw the money flow diagram and solve it using the PE, FE, and AE criteria.

15. A person borrows $20,000 at the beginning of Year 1. Every year he is required to pay $4,000 so that the loan is paid off after 10 years. What is the annual compound interest rate? (Hint: use linear interpolation.)

16. Company XYZ is considering buying new power supply equipment. Currently there are two types of equipment are under consideration. Type A: the initial cost is $14,000, the salvage value is $2,000 after 4 years, the hourly cost is $0.84/h, and the annual maintenance fee is $1,200. Type B: the initial cost is $5,500, there is no salvage value after 4 years, and the hourly cost is $0.67/h plus a human labor fee of $0.80/h. Assuming that the interest rate is 10%, compounded annually, how many hours per year does the equipment need to be used to make the two types break even?

17. A company has a production capacity of 50,000 tons annually; the annual fixed cost is $8,000,000, the sale price for each ton is $1,500, tax and other costs are 10% of the sale price, and the variable cost is $1,150/ton. What is the annual percentage of the production capacity required to make it break even?

Section III

Systems Management and Control Methods

10

Systems Management and Control

As mentioned many times in previous chapters, systems engineering is a well-defined, requirement-driven process that brings systems to life. With increasing levels of complexity, the system design life cycle could last for years, even decades, with hundreds or even thousands of people involved globally. There are many ways in which things could go wrong, as unexpected events occur almost all the time. Dealing with these uncertainties in a dynamic environment puts good planning and management efforts in demand. The systems engineering management plan (SEMP) is such a document that has been developed and widely adopted in complex systems design, to plan, schedule, and control system engineering design efforts and activities throughout the system life cycle, mainly focusing on the technical aspects. In this chapter, we will use SEMP as the basis and template to illustrate how system engineering design is planned and controlled; more specifically, we will

1. Describe the system design teams and responsibilities
2. Describe SEMP and its content, including major sections and information included in each of the sections
3. Describe system control concepts; introduce two commonly used project control models, the critical path method (CPM) and the program evaluation and review technique (PERT), providing examples of both models

The purpose of this chapter is to introduce the fundamental concepts of systems management and models that can be used to schedule and control the progress of systems engineering technical activities.

10.1 Systems Management Planning

10.1.1 Engineering Design Project Management and Design Teams

Systems engineering is a team effort, especially for large, complex systems; it is not uncommon to see a design project involving thousands of people from different countries. Managing these people and their daily activities is certainly a challenging task, not to mention the uncertainties and risks that occur all the time, which makes management even more difficult. It is no myth that a successful system development largely depends on a well-thought-out plan of the project, careful selection of the design teams, and efficient coordination of all the different players/stakeholders in the design life cycle. That is to say, systems engineering requires good project planning and management.

Project management plays a key role in the systems design life cycle; depending on the nature of the systems design project, different skills may be required for the different phases of the life cycle, but, generally speaking, the basic functions are similar; the ultimate goal is to manage the design personnel and efforts to be both technically and economically efficient. In a team-based approach for managing systems design projects, the key measures for success are time, cost, and technical performance metrics. The major administrative issues for project management include selecting the project team members and defining their responsibilities, choosing team leads, and managing/allocating resources.

As we mentioned earlier in Chapter 2, systems engineering is multidisciplinary in nature, which implies that the composition of the teams is also multidisciplinary; it is very common that a design team contains a variety of skills and expertise. It may include representatives from management (e.g., chief technical officer, human resources person, manager/directors, etc.), finance and accounting personnel, systems engineers (even in this field, there are different types of skills; e.g., industrial engineers, systems designers, modeling and simulation engineers, engineering psychologists, human factors engineers, etc.; it all depends on the nature of the systems being developed), engineers from related fields (e.g., aerospace engineers, civil engineers, electrical engineers, software engineers, mechanical engineers, etc.) and personnel with specialized skills such as graphical designers, architects, or computer programmers, plus facility and support personnel at various levels. As a project manager/team leader, one's great responsibility is to control the team personnel's efforts on a daily basis, to make sure the design efforts are on the right track in a timely and cost-efficient manner. The manager bears important responsibility for the success of the system design and implementation; he/she serves as the link in the chain of command, controlling work efforts and progress, managing human labor within the team, maintaining communication, promoting collaboration, safeguarding the work atmosphere and environment, monitoring work quality, and, most importantly, keep track of the time and costs spent on the design. A good manager is not only a good system engineer who understands the complexity of systems design and applies a systems engineering approach, but also possesses strong leadership and personal skills to bring all the team members together as a cohesive group. These skills cannot be learned solely in the classroom, but rather from personal experience and skills development. The team leaders' job is both challenging and rewarding; certain management skills are required for effective management of the team, including communication, coordination, motivation, and flexibility. A good project manager shall have some understanding and training in operations management, not to mention that there are many ethical issues that exist in managing teams; together, these require the project manager to have well-rounded skills to perform the job satisfactorily. In this text, we will not cover the whole spectrum of system management responsibilities, but only focus on the technical aspects of systems design; for a more comprehensive review of project management, readers can refer to any operations management textbook such as Stevenson's *Operations Management* (2009). In the next section, we will introduce a work tool that is commonly used in systems engineering to manage the technical efforts of systems design: the systems engineering management plan, or SEMP.

10.2 Systems Engineering Management Plan (SEMP)

The systems engineering management plan (SEMP) is an important tool for systems engineers to manage and control systems design projects. According to INCOSE (2004), SEMP is a top-level plan for managing systems engineering design efforts and activities. It defines how a systems design project will be "organized, structured and conducted and how the total engineering process will be controlled to provide a systems that meets the stakeholder requirement." More specifically, the SEMP is a "document that identifies the plans and schedules that will be needed to perform the technical effort for the systems development. This document is used to tailor the various activities to the needs of the project/program and is used to control the systems development when completed and approved" (INCOSE).

Since systems engineering is a structured plan for systems design, SEMP can be considered as a "plan for the plan"; it defines and regulates design activities and efforts so the systems engineering process can be carried out in an efficient manner. In these sections, we will describe the general structure and functions of a typical SEMP document, based on INCOSE (2004). For any specific system, SEMP needs to be tailored to accommodate its special nature and needs.

The basic structure of a typical SEMP includes the following sections:

10.2.1 Cover Page

The cover page includes the title and contract information for the project; it should include "Systems Engineering Management Plan (SEMP)" as part of the title, and at a minimum, the cover page should also include a project number, the authors' names and their contact information, customer name, date and location, funding agency, and contract information if applicable.

10.2.2 Table of Contents

As a standard, a table of contents should be included to facilitate easy access within the document. Ideally, this table should be generated automatically by word processing software and should also include the list of figures and list of tables.

10.2.3 Scope of Project

The scope of the project should state the purpose of the system to be designed with a brief summary and the content of the system. For some complex systems, the level of the complexities and challenges involved may also be briefly described. Some scope sections may also need to state the environmental conditions in which the system will operate, and the organizational structure involved in systems design, including the composition of the technical teams and their responsibilities. The purpose of this section is to give an overview of the system and make readers aware of the fundamental players involved in the efforts, including the various types of stakeholders.

10.2.4 Applicable Documents

Some prefer to put this section at the end as part of the references of the document. This section lists all the documents that have been used as input to the project. These could include government documents, standards, specifications and codes, or any other documents needed, even internal planning documents. Usually, government documents are listed first, followed by the rest in the following format:

<u>Document Number</u>   <u>Title</u>

10.2.5 Systems Engineering Process

This section is the core component of the SEMP; it defines the major design processes that are needed to develop the system, serving as guidance for systems engineering teams. It should also cover the organization's policies and procedures for systems engineering activities; the main subsection should include systems engineering process planning, requirement analysis, functional analysis and allocation, synthesis, system analysis, and control.

10.2.5.1 Systems Engineering Process Planning

In this section, the planning effort of the systems design process is defined; this should include (but not be limited to)

> *Design data and major deliverables*: At each major design milestone, what is required to be delivered, such as design data, drawings, parts lists, specification documents, and baselines. For complex

systems, software may be required to produce these documents, such as Vitech's CORE. A particular format should be followed to develop these documents; this format needs to be specified in this section.

Design input: The source and rationale of the project is specified here; this might include the statement of work (SOW), request for proposal (RFP), documentation from the previous versions of the systems and major problems which need to be fixed, market surveys, and so on, to justify the need for the design of the system.

Technical objectives: The technical objectives specify the operational goals for the system to be designed; this serves as a high-level requirement source as well as supplementary material for design input.

Work breakdown structure: The work breakdown structure (WBS) is a hierarchical decomposition of the system design efforts into various levels of components in different phases. Usually constructed in a tree structure, WBS provides a framework to plan and control the work process needed for the design to complete the work in a timely and cost-efficient manner. It starts with the highest level of technical objectives, based on the outcomes of the design, rather than the work needed, to specify the outcomes in the sublevels, letting the outcomes drive the work. WBS is an important tool to map out the work efforts needed for the overall design of the project; it provides a basic structure for personnel and cost allocation, and has been a common practice technique for most complex projects. It has been adopted by government standards. For more details of WBS, readers can refer to INCOSE (2012) and MIL-STD-881C (DoD 2011).

Training: For complex systems, training may be an important issue; this section specifies planning for training, including training objectives, training performance measurement metrics, the proposed training facility, and support and major timelines/schedules for training.

Standards and procedures: In this section, all the standards and procedures that are related to the project should be listed. These standards and procedures are mainly internal and should be different from the applicable external documents. The standards and procedures of the organization should be followed, including those for human resources, accounting, quality control, and related policies.

Resource allocation: Resources include the personnel, capital, facilities, and time that are necessary for the successful completion of the project. According to the INCOSE SEMP template (INCOSE 2004), the resource application should include the resource requirements for the various phases of the design, and procedures for resource allocation.

The systems engineering process specifies the work authorization procedure and policies, and sometimes the verification planning, specifying the responsibilities of the verification personnel, and major verification plans and schedules, including the demonstrations and tests that may be required throughout the systems life cycle.

10.2.5.2 Requirement Analysis

This section defines the methods to be used for systems requirements analysis. This is the one of the most important design efforts in the conceptual design phases and even throughout the design life cycle. To obtain a good set of design requirements, these efforts need to be planned carefully, especially the methods and approaches that will be used to collect and analyze them. According to INCOSE (2004) at least 15 categories of requirements need to be covered in the requirements analysis section; they are: reliability and availability, maintainability, supportability and logistic control, sustainability, environmental compatibility, human engineering and human–system integration, safety and environmental impact, security, producibility, testing and evaluation, testability and diagnosis, computer resources, transportability, infrastructure support, and other engineering specialties required by the systems.

10.2.5.3 Functional Analysis/Functional Allocation

In this section, this refers to the scope and information needed in the functional analysis. Detailed information within the functional analysis, such as the level of function decomposition, functional requirements, risks, and function allocation to components, needs to be clearly defined in the SEMP for the consideration of the efforts required to complete the tasks.

10.2.5.4 Synthesis

In this section, the synthesis efforts are specified. Systems synthesis, as we have previously discussed, consists of a series of activities that brings the systems design concepts into actual systems. These activities translate the physical architecture of the systems, interfaces and prototyping, and proposed product and process solutions for the development. Possible topics include but are not limited to: trade studies, quality analysis, COTS items selection, simulation models (such as CAD, CATIA, Arena, etc.), performance prediction model, testing and evaluation model, and so on. The overall goal is to create a feasible plan to translate system requirements into detailed design parameters.

10.2.5.5 System Analysis and Control

This portion of the SEMP defines the tools and models to be used by designers to conduct system analysis and control the design process. Some of the tools include decision making using trade studies, cost-benefit analysis, risk assessment, configuration management, data and interface management, scheduling and planning, TPMs, technical reviews, supplier control, and selection and requirement management. Some of the topics have been covered in previous chapters; for a more detailed description of these items, please refer to INCOSE (2004, 2012). The purpose of this section is to define the criteria for each of the analyses, determine the methods and models for conducting the analysis, the basic procedures involved, and, most importantly, provide estimates of the resources required for completion of these analyses so that these efforts can be properly planned and controlled.

10.2.6 Transitioning Critical Technologies

According to INCOSE (2004), this section is described as the "key technologies for the program and their associated risks, including the activities and criteria for assessing and transitioning critical technologies form technology development and demonstration programs." Most of the technologies are developed in the scientific laboratory environment. Using the technology in a real-world system requires collaboration and partnerships between the government, academia, and industries. When a technology is ready to be transitioned to industry, the operational environment changes may bring out additional issues or risks that are not present in the originating laboratory. Issues such as operational environment and capacities, cost, residual use, purchasing and acquisition policies, and legal/political issues will bring risks and challenges to technology transition. A comprehensive plan is required for a smooth transition to lay out the foreseeable risks and identify the critical elements and management strategies, so that if unexpected events occur in the future, a management plan will be in place to handle them. This section will give a description of the candidate technology for the systems, their acquisition/purchasing procedures, possible risks involved, and alternatives that will be considered should any changes occur within the development.

10.2.7 Systems Integration Efforts

According to INCOSE (2004), systems integration defines how "various inputs into the Systems Engineering efforts will be integrated and how interdisciplinary teaming will be implemented to involve appropriate disciplines in a coordinated Systems Engineering effort." Systems integration is a process that brings together all components and subsystems into one system and ensures that they work and function as one. Nowadays, companies seldom produce everything within one organization, rather taking advantage of the global supply chain to find the most economical parts from various suppliers. To successfully integrate system components together, systems designers need to have a wide breadth of knowledge of systems components, such as hardware, software, and networks, as well as enterprise and government architecture and policies. General problem-solving skills are often required to handle the uncertain and challenging problems that may not be seen elsewhere, as every system is different. Common tasks in systems integration include (INCOSE 2004):

Team organization: Defines the multidisciplinary teams that will be formed for the project; key team players and their responsibilities shall be specified. A typical team composition should include management, engineers, accounting and finance, marketing, customers, suppliers, and subject matter experts. According to INCOSE (2004), these teams should have an organizational structure, and should be end-item oriented. The core controlling team is the "system team of teams," which includes the lead systems engineer, the technical director, and a representative from each of the end-item teams. The composition and organization of the teams integrate all the necessary expertise together to meet the project requirements, in terms of cost, time schedule, and performance objectives.

Technology verification: In this section, the team verification plan is developed for the candidate technology considered. The objective is to have the appropriate team composition to ensure the necessary expertise is covered for these technologies.

Process proofing: This defines the measures and efforts planned to ensure an appropriate process for the proposed system. It also defines the necessary team credentials for such process proofing and synergy with other ongoing concurrent projects within the organization.

Manufacturing of engineering test articles: This defines the team composition for the manufacturing of engineering test articles. Details include the definitions of the manufacturability and technology, and the process and development teams' relationships to ensure a suitable manufacturing process.

Development test and evaluation: This defines the teams that are responsible for development test and evaluation to achieve life cycle cost optimization. These tests and evaluations should cover the whole life cycle of system development and specify the levels and locations of the tests.

Implementation of software design for end items: This defines the teams who are responsible for developing the software end items. If any special software engineering efforts are needed, these should be defined clearly, regarding how to integrate them with the systems engineering efforts; if there are any foreseeable software development risks, they need to be described here, as well as the potential measures to minimize these risks.

Sustaining engineering and problem solution support: This defines the teams who are responsible for sustaining engineering efforts, including the levels of support and facility requirements, as well as the potential risks involved.

Other systems engineering implementation tasks: This defines the teams that are responsible for system implementation tasks, identifies the tasks that are related to system implementation and what efforts are needed to complete the tasks and coordinate them between different teams.

10.2.8 Additional Activities

This defines all other activities that are not included in the previous sections. These largely depend on the nature of the system, and describing these activities helps customers to understand the efforts planned. According to INCOSE (2004), these activities may include, but are not limited to: long lead items, engineering tools, design to cost, value engineering, system integration plans, compatibility with support, and other plans. For a more detailed description of these items, readers can refer to INCOSE (2004), section C-8.

10.2.9 Appendices

All the appendices are listed here, including support materials, background information, glossary, design documents (drawings, blueprints, checklists, etc.), and external links.

10.2.10 Systems Engineering Master Scheduling

A systems engineering master schedule (SEMS) is a top-level process control and measurement tool for systems engineers to plan, control, and monitor the progress of the systems design efforts and activities, to make sure these are carried out in a timely and efficient manner to ensure successful completion of the project. Using SEMS, the design efforts and processes, including the decision points and flow logic involved, are visible for in-process verification; the expected outcome, accomplishments, and factors are visible throughout the design process. Such visibility helps designers to control and monitor the progress and visualize the risk factors, which, in turn, helps to control these risks.

The SEMS is driven by technical events; there are four basic elements used to develop the SEMS:

Events: These include the major technical efforts, as well as the reviews and key demonstration points.

Required accomplishments for events: The final steps or outcomes for each of the technical events. Examples of such accomplishments are the completion of the physical design and the interface prototype, completed and documented reviews, and so on. These accomplishments are derived from the system specifications; they should be clearly defined yet allow flexibility in implementation.

Criteria for accomplishment: For each of the accomplishments defined above, criteria are specified that can be measured, such as "technical report completed," "reliability test accepted," and so on.

Timeline of events: For the sequence of events required, specify the time needed for each of them to plan the efforts to complete on time.

A typical SEMS development should start with identification of the major events, and then break these down into activities. Based on the process of these events, a sequence tree is developed to aggregate these events together; the input/output outcomes of these events are then specified, as well as the criteria for these outcomes. A decomposition of the SEMS to a detailed schedule is possible if multiple levels of activities are involved.

The SEMP is an essential part of the system design project; the sections listed here are quite general. Keep in mind that every design project is different, so the SEMP should be tailored to the needs and the nature of the system to be designed. As seen in the SEMP, system scheduling is one of the key components. In the next section, we will introduce the two most commonly used methods to control the system schedule: the critical path method (CPM) and the program evaluation review technique (PERT).

10.3 Systems Control Models

In this section, we introduce two of the most widely used techniques for planning, CPM and PERT. These methods are similar in nature but different in the types of data applied in planning. CPM is a deterministic approach with no randomness incorporated in its model, while PERT is considered a probabilistic model, as it includes random time data in the analysis.

10.3.1 Critical Path Method (CPM)

The critical path method, also sometimes called critical path planning, is commonly used in scheduling for systems planning. Based on the network structure of the work schedule, CPM can derive the minimum time needed for project completion by identifying the critical path of the project design activities, as well as the start and end times for all the activities involved in the project.

Before we get into the details of CPM models, let us define some of the elements that are essential for CPM modeling. First, CPM is a network model;

FIGURE 10.1
Event node.

FIGURE 10.2
Activity arrow for CPM.

the CPM network consists of a sequence of events and activities that are necessary for the completion of the project. An event is usually a milestone or certain phases in design process; for example, an event could be "requirement analysis completed," "prototype completed," or "functional model developed," and so on. In CPM, an event is represented by a circle, as seen in Figure 10.1.

The second element for CPM is the activity, which is represented by an arrow link from the start event to the end event, usually with time information on the arrow, as shown in Figure 10.2.

CPM represents the sequential relationships (predecessors and successors) of the design process using events, event nodes, and the corresponding activities. Completing the project means that all the activities have to be carried out and completed. Many people might confuse CPM with finding a short path for a network, as they are similar, but when finding the shortest path we do not have to visit every path, as long as we can get to the end point from the start point. In CPM, an event is not considered complete until *all* the activities prior to that event are completed. This is important to keep in mind as we introduce the algorithm.

Of particular interest in CPM is to find the critical path. A path is a sequence of events and activities from the start event to the end event. A critical path is the longest path (in terms of time) among them; the sum of the times on the critical path represents the shortest time required to complete the project. Any delay on the critical path is going to delay the whole project. This implies that, for all the events on the critical path, their earliest start time (t_{ES}) equals their latest start time (t_{LS}) or that there is no slack time for each critical event, that is to say, $s = t_{LS} - t_{ES} = 0$.

This gives us the indication of how to find the critical path; first, we start from the start node and work toward the end event, working out the earliest start time for each of the events; then, we work backward from the end event back to the start event, to work out the latest start time for each of the events. By calculating the slack time for each event

$$s = t_{LS} - t_{ES} \tag{10.1}$$

We can then identify the critical path by finding the events with zero slack time. Let us use an example to illustrate the procedure to find the critical path.

Example 10.1

Critical Path Method. Figure 10.3 shows a network diagram for a project. Find the critical path for this project.

Solution:

Step 1. Find the earliest start time for each event.

Event A: Obviously the earliest start time is $t_{ES}=0$.
Event B: In working out the earliest start time, we need to look at the immediate predecessors of that event; the earliest start time of the event is immediately after all the preceding events and activities complete. For event B, the immediate predecessor is event A. If we denote the earliest start time for B as $t_{ES}(B)$ and the activity time from A to B as d_{AB}, then for event B, $t_{ES}(B)=t_{ES}(A)+d_{AB}=0+3=3$.
Event C: Event C has more than one immediate predecessor; namely, A and B. The event has more than one immediate predecessor since it cannot start until all the previous events and activities are completed, so we take the maximum value of all the immediate predecessor events' earliest start times plus their corresponding activity times, or

$$t_{ES}(C)=\max\left\{\begin{array}{c}t_{ES}(A)+d_{AC}\\t_{ES}(B)+d_{BC}\end{array}\right\}=\max\left\{\begin{array}{c}0+5=5\\3+4=7*\end{array}\right\}=7$$

Event D: D has only one immediate predecessor, B, so

$$t_{ES}(D)=t_{ES}(B)+d_{BD}=3+6=9$$

Event E: E also has only one immediate predecessor, D, so

$$t_{ES}(E)=t_{ES}(D)+d_{DE}=9+5=14$$

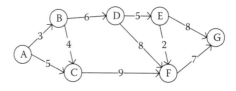

FIGURE 10.3
CPM network diagram.

Event F: F has three immediate predecessors, C, D, and E, so the earliest start time for F is

$$t_{ES}(F) = \max \begin{cases} t_{ES}(C) + d_{CF} \\ t_{ES}(D) + d_{DF} \\ t_{ES}(E) + d_{EF} \end{cases} = \max \begin{cases} 7 + 9 = 16 \\ 9 + 8 = 17 * \\ 14 + 2 = 16 \end{cases} = 7$$

Event G: G has two immediate predecessors, E and F:

$$t_{ES}(G) = \max \begin{cases} t_{ES}(E) + d_{EG} \\ t_{ES}(F) + d_{FG} \end{cases} = \max \begin{cases} 14 + 8 = 22 \\ 17 + 7 = 24 * \end{cases} = 24$$

We mark the earliest start time on the network, as shown in Figure 10.4.
Step 2. Find the latest start time for each event.

In finding the latest start time, we start from the end event, G, and work backward to the start point, A.

Event G: For the end event, the latest start point equals its earliest start point, so

$$t_{LS}(G) = 24$$

Event F: To work out the latest start time for an event, we need to look at its immediate successor in the network diagram. Event F has only successor, G, so its latest start time is

$$t_{LS}(F) = t_{LS}(G) - d_{FG} = 24 - 7 = 17$$

Event E: E has two immediate successors; namely, F and G. To work out the latest start time, we need first to calculate the latest possible start time for each of the successors by subtracting the

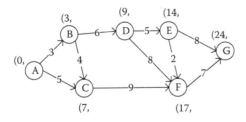

FIGURE 10.4
CPM network diagram with earliest start time.

activity time from the latest start time for that successor, and
then choosing the *minimum* value from the results of all the successors. The reason we choose the minimum value is to ensure
all the latest start times of the successors will not be affected.
We can obtain the latest start time for E as

$$t_{LS}(E) = \min\left\{\begin{array}{c}t_{LS}(G)-d_{EG}\\t_{LS}(F)-d_{EF}\end{array}\right\} = \min\left\{\begin{array}{c}24-8=16\\17-3=15*\end{array}\right\} = 15$$

Event D: D has two immediate successors, E and F, so the latest
start time for D is

$$t_{LS}(D) = \min\left\{\begin{array}{c}t_{LS}(E)-d_{DE}\\t_{LS}(F)-d_{DF}\end{array}\right\} = \min\left\{\begin{array}{c}15-5=10\\17-8=9*\end{array}\right\} = 9$$

Event C: C has only one successor, F, so its latest start time is

$$t_{LS}(C) = t_{LS}(F) - D_{CF} = 17 - 9 = 8$$

Event B: B has two immediate successors, C and D, so the latest
start time is

$$t_{LS}(B) = \min\left\{\begin{array}{c}t_{LS}(C)-d_{BC}\\t_{LS}(D)-d_{BD}\end{array}\right\} = \min\left\{\begin{array}{c}8-4=4\\9-6=3*\end{array}\right\} = 3$$

Event A: A is the start point, so its latest start time will be zero.
But, to reveal the critical path, it is a good idea to complete the
same procedure for A.

$$t_{LS}(A) = \min\left\{\begin{array}{c}t_{LS}(B)-d_{AB}\\t_{LS}(C)-d_{AC}\end{array}\right\} = \min\left\{\begin{array}{c}3-3=0*\\8-5=3\end{array}\right\} = 0$$

The earliest start (ES) and latest start (LS) results are shown in
Figure 10.5. The 3-tuple in the parentheses associated with each event
represents (ES, LS, slack). Slack is calculated by using Equation 10.1.
   From the algorithm above, trace back from G to A, identifying the
zero-slack events that are on the critical path (those with a * in the calculation), and mark them on the network diagram, as seen in Figure 10.6.
   The critical path is identified as A→B→D→F→G. The total time
required (or the minimum time needed) to complete the project is 24
time units.

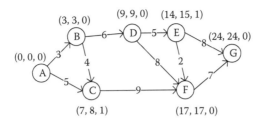

FIGURE 10.5
Earliest start time, latest start time, and slack for each of the events.

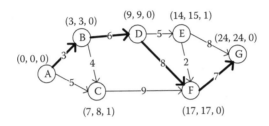

FIGURE 10.6
Critical path identified in bold lines.

We have described the CPM algorithm in detail. Tables 10.1 and 10.2 show how to use a spreadsheet to carry out the above algorithm. The earliest start time is illustrated in Table 10.1, and the latest start time is illustrated in Table 10.2.

An algorithm for CPM is easy to implement using a computer program. In such a program, it is more convenient to use numbers (1, 2, 3, ...) for the events rather than letters. The following shows a generic procedure for a CPM computer algorithm.

Step 1: Number all the events from start to end. Derive the matrix D to represent the CPM network diagram, with $D = [d_{ij}]$, $i = 1, 2, ..., n$; $j = 1, 2, ... n$; $d_{ij} > 0$ if there is a link between i and j; for the above example, the matrix would be

$$D = \begin{bmatrix} 0 & 3 & 5 & 0 & 0 & 0 & 0 \\ 0 & 0 & 4 & 6 & 0 & 0 & 0 \\ 0 & 0 & 0 & 0 & 0 & 9 & 0 \\ 0 & 0 & 0 & 0 & 5 & 8 & 0 \\ 0 & 0 & 0 & 0 & 0 & 2 & 8 \\ 0 & 0 & 0 & 0 & 0 & 0 & 7 \\ 0 & 0 & 0 & 0 & 0 & 0 & 0 \end{bmatrix}$$

TABLE 10.1

Earliest Start Time for Each Event

Event	Immediate Predecessor	$t_{ES}+d_{ij}$ for Each Predecessor	t_{ES}
A	–	–	0
B	A	$0+3=3$	3
C	B	$3+4=7^a$	7
	C	$0+5=5$	
D	B	$3+6=9$	9
E	D	$9+5=14$	14
F	C	$7+9=16$	
	D	$9+8=17^a$	17
	E	$14+2=16$	
G	E	$14+8=22$	
	F	$17+7=24^a$	24

ª Indicates the maximum value from the list.

TABLE 10.2

Latest Start Time for Each Event

Event	Immediate Successor	$t_{LS}-D_{ij}$ for Each Successor	t_{LS}
G	–	–	24
F	G	$24-7=17$	17
E	F	$17-2=15^a$	15
	G	$24-8=16$	
D	E	$15-5=10$	
	F	$17-9=9^a$	9
C	F	$17-9=8$	8
B	C	$8-4=4$	
	D	$9-6=3^a$	3
A	B	$3-3=0^a$	0
	C	$8-5=3$	

ª Indicates the minimum value from the list.

Step 2: Calculate the earliest start time for each node, as follows:
1. Let $E(1)=0$
2. For $j=2, 3, ..., n$, let

$$E(j) = \max_i \left(E(i)+d_{ij} \right)$$

Step 3: Calculate the latest start time for each node:
1. Let $L(n)=E(n)$

2. For $i = n-1, n-2, \ldots, 1$, let

$$L(i) = \min_{j} \left(L(j) - d_{ij} \right)$$

Step 4: Calculate the slack time for each node. The critical path is identified by finding the zero-slack time nodes.

10.3.2 Program Evaluation and Review Technique (PERT)

The program evaluation and review technique was first developed by the U.S. Navy's Special Project Office in the 1950s to support its Polaris nuclear submarine project. Its intention was to have a statistical technique to consider the uncertainty within the project and to measure and predict the program's progress. Its principle is very similar to that of the CPM method; they both use network models, and the algorithms are also similar. Unlike CPM, in which time is deterministic, PERT incorporates randomness into the time estimate of the activity; that is, for each of the activities between events, there are three pieces of time data being estimated:

1. Optimistic time (t_o): the time required to complete the activity under optimal conditions; this is the minimum time needed to complete the activity.
2. Most likely time (t_m): the amount of time that is the most probable to be taken to complete the activity.
3. Pessimistic time (t_p): the amount of time required to complete the activity under the worst-case scenario.

    In PERT analysis, we are interested in the expected (mean) time required to complete the activity t_e and its variance for the time (σ^2). Most of the PERT model uses the beta distribution as an assumption of the time; although there is no real theoretical justification for using this distribution, it has been argued that it provides great generality for different shapes of time distribution, whether symmetric or skewed (Figure 10.7).

The expected value t_e is given by Equation 10.2:

$$t_e = \frac{t_o + 4t_m + t_p}{6} \tag{10.2}$$

And the variance σ^2 is given by Equation 10.3:

$$\sigma^2 = \left[\frac{t_p - t_o}{6} \right]^2 \tag{10.3}$$

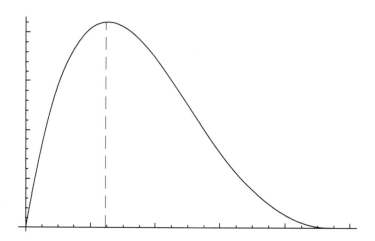

FIGURE 10.7
Illustration of beta distribution.

The basic steps for conducting a PERT analysis include the following:

1. Develop the network diagram for the project and identify the t_p, t_o, and t_m estimates for the activities;
2. Compute the expected (mean) time t_e and variance σ^2 for each of the activities, as shown in Equations 10.2 and 10.3.
3. Find the critical path using t_e.
4. Estimate the probability of project completion by a specified time or related estimate.

We will illustrate the PERT algorithm using Example 10.2.

Example 10.2

Figure 10.8 presents the information for a project, with the 3-tuple showing the t_o, t_m, and t_p of the time estimate (in days). (a) Find the critical path for the project; (b) determine the expected length of the project; and (c) calculate the probability that the project will be completed within 22 days or less.

Step 1: Based on the information given in Figure 10.8, compute the expected (mean) time t_e and variance σ^2, as shown in Table 10.3.
Step 2: Using CPM procedures, identify the critical path. The earliest start times are shown in Table 10.4 and the latest start times are shown in Table 10.5. Based on the above algorithm, the critical path can be easily obtained, as seen in Figure 10.9.

The second part of the question asks what the probability is that the project is completed in 22 days or less, or, what is $P(t \leq 22)$?

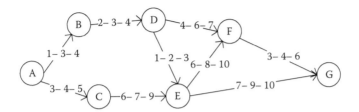

FIGURE 10.8
Network diagram for Example 10.2.

TABLE 10.3

Expected (Mean) Time t_e and Variance σ^2

Event	Immediate Successor	t_o	t_m	t_p	t_e	σ^2
A	B	1	3	4	2.83	0.25
A	C	3	4	5	4.00	0.11
B	D	2	3	4	3.00	0.11
C	E	6	7	9	7.17	0.25
D	E	1	2	3	2.00	0.11
D	F	4	6	7	5.83	0.25
E	F	6	8	10	8.00	0.44
E	G	7	9	10	8.83	0.25
F	G	3	4	6	4.17	0.25

TABLE 10.4

Earliest Start Time

Event	Immediate Predecessor	$t_{ES} + d_{ij}$ for Each Predecessor	t_{ES}
A	–	–	0
B	A	$0 + 2.83 = 2.83$	2.83
C	A	$0 + 4.00 = 4.00$	4.00
D	B	$2.83 + 3.00 = 5.83$	5.83
E	C	$4.00 + 7.17 = 11.17$[a]	11.17
	D	$5.83 + 2.00 = 7.83$	
F	D	$5.83 + 5.83 = 11.66$	19.17
	E	$11.17 + 8.00 = 19.17$[a]	
G	E	$11.17 + 8.83 = 20.00$	23.34
	F	$19.17 + 4.17 = 23.34$[a]	

[a] Indicates the smallest value for multiple successors.

TABLE 10.5

Latest Start Time

Event	Immediate Successor	$t_{LS} - D_{ij}$ for Each Successor	t_{LS}
G	–	–	23.34
F	G	$23.34 - 4.17 = 19.17$	19.17
E	F	$19.17 - 8.00 = 11.17^a$	11.17
	G	$23.34 - 8.83 = 14.51$	
D	E	$11.17 - 2.00 = 9.17^a$	9.17
	F	$19.17 - 5.83 = 13.34$	
C	E	$11.17 - 7.17 = 4.00$	4.00
B	D	$9.17 - 3.00 = 6.17$	6.17
A	B	$6.17 - 2.83 = 3.34$	0
	C	$4.00 - 4.00 = 0^a$	

[a] Indicates the smallest value for multiple successors.

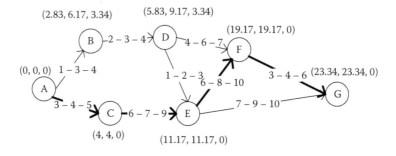

FIGURE 10.9
Critical path based on the expected time, $A \rightarrow C \rightarrow E \rightarrow F \rightarrow G$, with the total expected time to complete of 23.34.

In working out the probability, we assume that the overall completion time t follows the normal (Gaussian) distribution. One may ask why individual time follows the beta distribution while the overall time follows the Gaussian. According to the central limit theorem, the arithmetic sum of any sufficient large number of independent random variables will be approximately normally distributed, regardless of the distribution of those random variables. Recall that in previous chapters (such as in Section 5.3 on maintainability) we used the standard normal distribution to estimate the z-value. We use a similar procedure here.

The mean time for completion is $\mu = 23.34$ days and the standard deviation of the time equals the square root of the sum of activity variances on the critical path, or

$$\sigma = \sqrt{\sum \text{variances of critical path activities}} \qquad (10.4)$$

So, according to Equation 10.4, we can obtain the standard deviation by

$$\sigma = \sqrt{\sum (0.11 + 0.25 + 0.44 + 0.25)} \cong 1.0247$$

So, the z-value for 22 days can be obtained as

$$z = \frac{20 - \mu}{\sigma} = \frac{22 - 23.34}{1.0247} = -1.31$$

So, $P(t \le 22)$ is converted to $P(z \le -1.31)$

Using the normal distribution table in Appendix II, we find that the probability of $P(z \le -1.31) = 0.0951$, so the probability that the project is completed in 22 days or less is approximately 0.0951 or 9.51%. ∎

10.4 Summary

For large, complex systems design, a well-planned design process is essential for the success of the system development. This characteristic requires a team effort and also demands much of project management. In this chapter, major methods for systems engineering management and control were reviewed, including SEMP, CPM, and PERT. The SEMP is an organized and structured document, and lists practices to define and control technical efforts for systems development. It is used to make sure the systems design efforts are on track and to meet the time and cost limits. The major sections in the SEMP include the cover page, table of contents, scope of the project, applicable documents, systems engineering processes, transitioning critical technologies, systems integration efforts, other additional activities, and master scheduling. CPM is a very commonly used deterministic scheduling tool. Based on the work structure, CPM uses the network diagram to identify the earliest and latest start times for each event. The critical path can then be identified by finding the activities with zero slack time. PERT is a more general approach than CPM; it incorporates randomness into activity time, by using the beta distribution, a critical path can be obtained using the mean activity time, and the probability of completing the project within a certain window of time can be estimated using the mean and variance of the times of activities on the critical path.

PROBLEMS

1. Why are system management and control important in systems engineering processes?
2. What is SEMP? Describe the major sections included in SEMP.
3. Find out the critical path for the following project network.

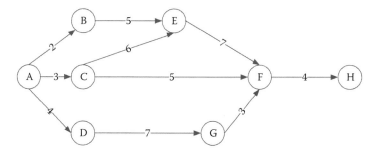

4. Find out the critical path for the following project network.

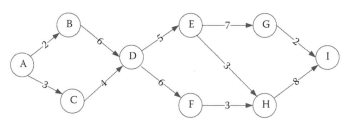

5. In the following project, find the critical path and calculate the probability of completing the project in 25 days.

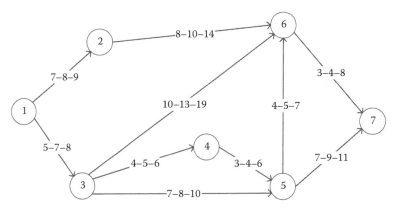

Bibliography

American Board for Engineering and Technology (ABET). n.d. http://www.abet.org.

Andre, T. S. and Schopper, A. W. 1997. *Human Factors Engineering in Systems Design.* Washington, DC: Defense Technical Information Center, Department of Defense.

Banks, J., Carson, J., Nelson, B. and Nicol, D. 2001. *Discrete-Event System Simulation.* Upper Saddle River, NJ: Prentice Hall.

Bentley, L. and Whitten, J. 2007. *System Analysis & Design for the Global Enterprise*, 7th edn. Boston: McGraw-Hill Education.

Blanchard, B. S. 1995. *Maintainability: A Key to Effective Serviceability and Maintenance Management.* New York: Wiley.

Blanchard, B. S. 2004. *Systems Engineering Management.* New York: Wiley.

Blanchard, B. S. and Fabrycky, W. J. 2006. *Systems Engineering and Analysis*, 4th edn. Upper Saddle River, NJ: Prentice Hall.

Boardman, A. E., Greenberg, D. H., Vining, A. R. and Weimer, D. L. 2001. *Cost-Benefit Analysis: Concepts and Practice*, 2nd edn. Upper Saddle River, NJ: Prentice Hall.

Boehm, B. 1988. A spiral model of software development and enhancement. *IEEE Computer*, 21(5), 61–72.

Booch, G., Rumbaugh, J., and Jacobson, I. 2005. *The Unified Modeling Language User Guide*, 2nd edn. Addison-Wesley Professional.

Buede, D. M. 2009. *The Engineering Design of Systems: Models and Methods.* Hoboken, NJ: Wiley.

Canada, J. R. and Sullivan, W. G. 1989. *Economic and Multiattribute Evaluation of Advanced Manufacturing Systems.* Englewood Cliffs, NJ: Prentice-Hall.

Chapanis, A. 1996. *Human Factors in Systems Engineering.* Hoboken, NJ: Wiley.

Chengalur, S. N., Hodgers, S. H. and Bernard, T. E. 2004. *Kodak's Ergonomics Design for People at Work.* Hoboken, NJ: Wiley.

Clark, D. W., Cramer, M. L., and Hoffman, M. S. 1986. Human factors and product development: Solutions for success. Workshop Notes, Retail System Division, NCR Corporation.

Clark, R. E., Feldon, D., Van Merrienboer, J. J. G., Yates, K., and Early, S. 2008. Cognitive task analysis. In J. M. Spector, M. D. Merrill, J. J. G. van Merrienboer, and M. P. Driscoll (Eds.), *Handbook of Research on Educational Communications and Technology*, 3rd edn. Mahwah, NJ: Lawrence Erlbaum Associates.

Cole, G. F. 1990. Market-driven customer manuals using QFD. *Proceedings of the AUTOFACT '90 Conference*, Detroit, MI, MS 90-03, pp. 31–35.

Defense Acquisition University. 2001. Systems engineering fundamentals. Fort Belvoir, VA: Defense Acquisition University Press. http://ocw.mit.edu/courses/aeronautics-and-astronautics/16-885j-aircraft-systems-engineering-fall-2005/readings/sefguide_01_01.pdf.

DoD (Department of Defense). 2011. MIL-STD-881C, Work breakdown structures for defense material items, October. DoD.

Ebeling, C. E. 1997. *An Introduction to Reliability and Maintainability Engineering.* Boston: McGraw-Hill.

Elsayed, A. E. 1996. *Reliability Engineering.* Addision Wesley Longman.

Elsayed, A. E. 2012. *Reliability Engineering*. New York: Wiley.

Evans, J. R., and Minieka, E. 1992. *Optimization Algorithm for Networks and Graphs*, 2nd edn. CRC Press.

Faulconbridge, R. I. and Ryan, M. J. 2003. *Managing Complex Technical Projects: A Systems Engineering Approach*. Norwood, MA: Artech House.

Fishburn, P. C. 1968. Utility theory. *Management Science*, 14(5), 335–378.

Fishburn, P. C. 1970. *Utility Theory for Decision Making*. Huntington, NY: Robert E. Krieger.

Forsberg, K., Mooz, H. and Cotterman, H. 2005, *Visualizing Project Management*, 3rd edn. New York: Wiley.

Friedental, S., Moore, A. and Steiner, R. 2009. *A Practical Guide to SysML*. Burlington: Elsevier.

Gotel, O. and Finkelstein, A. 1994. An analysis of the requirements traceability problem. *Proceedings of First International Conference on Requirements Engineering*, Colorado Springs, CO, pp. 94–101.

Hall, A. D. 1969. Three dimensional morphology of systems engineering. *IEEE Transactions on Systems Science and Cybernetics*, SSC 5(2), 156–160.

Hallam, C. R. 2001. An overview of systems engineering. MIT ESD.83 Research seminar in Engineering Systems. http://web.mit.edu/esd.83/www/notebook/syseng.doc

Hammond, W. 1999. *Space Transportation: A Systems Approach to Analysis and Design*. AIAA.

Henley, E. J. and Lynn, J. W. 1976. *Generic Techniques in System Reliability Assessment*. Leyden, The Netherlands: Nordhoff.

Honour, E. (n.d.). Systems engineering and complexity. Retrieved December 12, 2011. From http://sse.stevens.edu/fileadmin/cser/2006/papers/193-4-Honour-SE&Complexity.pdf

Hood, C., Wiedemann, S., Fichtinger, S. and Pautz, U. 2007. *Requirements Management: Interface between Requirements Development and All Other Engineering Processes*. Berlin: Springer.

Hughes, A. C. and Hughes, T. P. 2000. *Systems, Experts, and Computers: The Systems Approach in Management and Engineering, World War II and After*. Cambridge, MA: MIT Press.

International Council on Systems Engineering (INCOSE). 2004. Systems Engineering Handbook, Version 2A, June. San Diego, CA: INCOSE.

International Council on Systems Engineering (INCOSE). 2012. *Systems Engineering Handbook: A Guide for System Life Cycle Processes and Activities*, version 3.2.2. San Diego, CA: INCOSE. http://www.incose.org.

International Electrotechnical Commission. 2006. *Fault Tree Analysis*. Edition 2.0. International Electrotechnical Commission, IEC 61025. Retrieved January 17, 2010. From http://webstore.iec.ch/preview/info_iec61025%7Bed2.0%7Den_d.pdf.

International Organization for Standardization. 2008. ISO 9001 QMS requirement. ISO.

Jacobson, I., Christerson, M., Jonsson, P. and Övergaard, G. 1992. *Object-Oriented Software Engineering: A Use Case Driven Approach*. Reading, MA: Addison-Wesley.

Jacobson, I., Spence, I. and Bittner, K. 2011. *Use Case 2.0: The Guide to Succeeding with Use Cases*. ebook: http://www.ivarjacobson.com/Use_Case2.0_ebook/.

Justis, R. T. and Kreigsmann, B. 1979. The feasibility study as a tool for venture analysis. *Business Journal of Small Business Management*, 17(1), 35–42.

Laplace, P. S. 1902. *A Philosophical Essay on Probabilities*. New York: John Wiley and Sons, New York.

Laplante, P. 2009. *Requirements Engineering for Software and Systems*, 1st edn. Redmond, WA: CRC.

Law, A. M. and Kelton, W. D. 2000. *Simulation Modeling and Analysis*. New York: McGraw-Hill.

Liu, D. 2002. Web design using a quality function deployment methodology. PhD dissertation, University of Nebraska, Lincoln.

Liu, D. and Findlay, M. A. 2014. Assessment of resource scheduling changes on flight training effectiveness using discrete event simulation. *Human Factors and Ergonomics in Manufacturing and Service Industries*, 24, 226–240.

Liu, D., Bishu, R. and Lotfollah, N. 2005. Using analytical hierarchy process as a tool for assessing service quality. *International Journal of Industrial Engineering and Management Systems*, 4(2), 129–136.

Logan, P., Harvey, D. and Spencer, D. 2012. Documents are an essential part of model based systems engineering. *INCOSE International Symposium*, 22, 1899–1913.

Long, D. and Scott, Z. 2012. *A Primer for Model Based Systems Engineering*. Vitech Corporation. lulu.com.

Michels, J. M. 1965. Computer evaluation of the safety fault tree model. In *Proceedings System Safety Symposium*. Seattle, WA: University of Washington.

Nielsen, J. 1993. *Usability Engineering*. Boston: Academic Press.

Norman, R. 1996. *Object-Oriented Systems Analysis and Design*. Upper Saddle River, NJ: Prentice Hall.

Parnell, G. S., Driscoll, P. J., and Henderson, D. L. (Eds.) 2008. *Decision Making in Systems Engineering and Management*. Hoboken, NJ: Wiley-Interscience.

Project Management Institute. 2009. *A Guide to the Project Management Body of Knowledge*, 4th edn. Newtown Square, PA: Project Management Institute.

Ralph, P. and Wand, Y. 2009. A proposal for a formal definition of the design concept. In Lyytinen, K., Loucopoulos, P., Mylopoulos, J. and Robinson, W. (Eds.), *Design Requirements Engineering: A Ten-Year Perspective*, pp. 103–136. Berlin: Springer.

Rangone, A. 1996. An analytical hierarchy process framework for comparing the overall performance of manufacturing departments. *International Journal of Operations & Production Management*, 16(8), 104–119.

Revelle, J. B., Moran, J. W., and Cox, C. A. 1998. *The QFD Handbook*. New York: John Wiley.

Rogers, Y., Sharp, H. and Preece, J. 2011. *Interaction Design: Beyond Human-Computer Interaction*. Chichester, UK: Wiley.

Ross, S. M. 1997. *Introduction to Probability Models*. New York: Academic Press.

Roth, G. L. 2007. Decision making in systems engineering: The foundation. In Parnell, G. S., Driscoll, P. J. and Henderson, D. L. (Eds.), *Decision Making in Systems Engineering and Management*, pp. 236–246. Wright-Patterson AFB, OH: Future Systems Concepts & Integration Branch.

Royce, W. W. 1970. Managing the development of large software systems. *WESCON Technical Papers* 14(August), 1–9.

Saaty, T. L. 1980. *The Analytic Hierarchy Process*. New York: McGraw Hill.

Saaty, T. L. 2008. Decision making with the analytic hierarchy process. *International Journal of Services Sciences*, 1(1), 83–98.

Sage, A. P. 1992. *Systems Engineering*. New York: Wiley-Interscience.

Sage, A. P. and Rouse, W. B. 1999. *Handbook of Systems Engineering and Management*. New York: Wiley.

Shillito, M. L. 1994. *Advanced QFD: Linking Technology to Market and Customer Needs*. New York: John Wiley.

Smith, W. 2013. Essential model-based systems engineering: Applied and practical. *INCOSE International Symposium*, 23, 1634.

Son, H., 2004. *Operation Feasibility Analysis of Freeway Diversion at Urban Network.* Research Report. Charlottesville, VA: University of Virginia, Center for Transportation Studies.

Stamatelatos, M., Vesely, W., Dugan, J., Fragola, J., Minarick, J. and Raisback, J. 2002. *Fault Tree Handbook with Aerospace Applications.* National Aeronautics and Space Administration. Retrieved January 17, 2010. From http://www.hq.nasa.gov/office/codeq/doctree/fthb.pdf.

Stellman, A. and Greene, J. 2005. *Applied Software Project Management.* Sebastopol, CA: O'Reilly Media.

Stevenson, W. J. 2009. *Operations Management.* New York: McGraw-Hill-Irwin.

Stewart, J. 2011. *Single Variable Calculus,* 7th edn. Cengage Learning.

Tague, N. R. 2005. *Quality Toolbox.* Milwaukee, WI: ASQ Quality Press.

Telelogic. (2003). Getting it right for the first time: Writing good requirements, Telelogic User Manual, ERS60-007.

Tompkins, J. A., White, J. A. Bozer, Y. A. and Tanchoco, J. M. A. 2003. *Facility Planning,* 3rd edn. Hoboken, NJ: Wiley.

U.S. Department of Defense. 1966. MIL-HDBK-472 maintainability prediction.

U.S. Department of Defense. 1973. MIL-STD-471A maintainability verification/demonstration/evaluation.

U.S. Department of Defense. 1988. DOD-HDBK-791 maintainability design techniques.

U.S. Department of Defense. 1989a. MIL-STD-1472D, military standard: Human engineering, design criteria for military systems, equipment and facilities.

U.S. Department of Defense. 1989b. MIL-STD-470B maintainability program requirements for systems and equipment.

U.S. Department of Defense. 1998. Electronic reliability design handbook. MIL-HDBK-338B. Retrieved January 17, 2010. From http://www.sre.org/pubs/Mil-Hdbk-338B.pdf.

Vesely, W. E. 1970. A time-dependent methodology for fault tree evaluation. *Nuclear Engineering and Design*, 13(2), 337–360.

Vesely, W. E., Goldberg, F. F., Roberts, N. H. and Haasl, D. F. 1981. *Fault Tree Handbook.* Nuclear Regulatory Commission. NUREG–0492. Retrieved January 17, 2010. From http://www.nrc.gov/reading-rm/doc-collections/nuregs/staff/sr0492/sr0492.pdf.

Vitech Corporation. 2012. CORE guided tour. Retrieved August 2012. From http://www.vitechcorp.com/support/documentation.shtml.

Vitech Corporation. 2012. CORE system definition guide. August 2012. From http://www.vitechcorp.com/support/documentation/core/600/systemdefinitionguide.pdf.

von Neumann, J. and Morgenstern, O. 1944. *Theory of Games and Economic Behavior.* Princeton, NJ: Princeton University Press.

Weaver, W. 1963. *Lady Luck.* New York: Anchor Books.

Weigel, A. L. 2003. An overview of systems engineering knowledge domain. *ESD.83.* Research Seminar in Engineering Systems. Retrieved December 2011. From http://web.mit.edu/esd.83/www/notebook/sysengkd.pdf.

Wickens, C. D., Lee, J. D., Liu, Y. and Gordon-Becker, S. 2003. *An Introduction to Human Factors Engineering.* Upper Saddle River, NJ: Prentice Hall.

Winston, R. 1970. Managing the development of large software systems. *Proceedings of IEEE WESCON*, 26, 1–9.

Winston, W. L. 1994. *Operations Research: Application and Algorithms.* Belmont, CA: Duxbury Press.

Winston W. L. 2005. *Operations Research: Application and Algorithms.* Belmont, CA: Thompson.

Wymore, A. W. 1993. *Model Based Systems Engineering.* Boca Raton, FL: CRC.

Young, R. R. 2003. *The Requirement Engineering Handbook.* Boston, MA: Artech House (Print on Demand).

Appendix I: Introduction to Probability and Statistics

As Heraclitus said, "No man ever steps in the same river twice"; the world surrounding us is fast moving and full of randomness. Either the world is created that way, such that everything is random, or our understanding or knowledge is not sufficient to understand the laws behind the randomness. Regardless of this, we have to deal with the randomness and uncertainty that is represented in the random data we observe. The only tools we humans possess to deal with randomness are probability and statistics. Probability is a branch of theoretical mathematics to study the likeliness that a certain event will occur; statistics, on the other hand, is a branch of applied mathematics that measures the events occurring in the world through observation and presents the data collected. Both fields are important for the understanding of phenomena occurring in the world. Probability provides a fundamental foundation for data analysis in statistics, while statistics uses probability theory to guide the measurement and analysis of the random data observed in the world. Systems engineering lasts years, even decades; it involves many people and thus inevitably a great deal of randomness and uncertainty; for example, the number of defects appearing in the production batch, the market share of the developed system, or the errors occurring in the manufacturing process. We have seen many examples in the book's chapters. Understanding the basic concepts of probability and statistics is essential for the successful development and implementation of a complex system. It is expected that probability and statistics should be included in any systems engineering curriculum as a required course. We here include a brief review of the basic concepts of probability theory for readers to refresh their memory on these subjects.

I.1 Basic Probability Concepts

As mentioned above, probability studies the randomness of the universe. For example, if we toss a coin, when it lands, it could be showing either heads or tails. We cannot say for sure whether it will be heads or tails before it lands, as the result of tossing a coin is a typically random event that many people like to use to demonstrate probability events. If this coin is fair, we know the chances of a head or tail would be 50%. This simple example includes all the fundamental concepts of probability: experiment, event, sample, sample space, and probability.

I.1.1 Experiment, Sample, Sample Space and Universe/Population

I.1.1.1 Experiment

In probability theory, an *experiment* can be loosely defined as an orderly action procedure carried out for a certain scientific purpose, such as verifying the chances of certain things occurring. Examples of experiments include tossing a coin, throwing a die, drawing a card from a deck of cards, testing a rocket launch, predicting tomorrow's weather temperature, finding out the number of customers in a store, and so on.

I.1.1.2 Sample, Sample Space, and Event

The outcome of an experiment is called a *sample*. For example, when tossing a coin, the possible sample would be either heads or tails; when throwing a die, the sample would be 1, 2, 3, 4, 5, or 6 dots on the uppermost face. We denote a sample as ω, so for throwing a die, $\omega = 1, 2, 3, 4, 5,$ or 6. The set that includes all the possible sample results is called a sample space (Ω), for example, for tossing the dice, $\Omega = \{1, 2, 3, 4, 5, 6\}$. An *event* is a collection of samples resulting from a certain experiment; for example, if we toss a coin twice, the possible events for this experiment could be, $\{H,H\}$, $\{H,T\}$, $\{T,H\}$, or $\{T,T\}$. From this perspective, we can see events are the subsets of sample space, if we denote the event as A_i, then $A_i \subset \Omega$. If an outcome of an experiment $\omega \in A$, then we say that event A occurs. The set of all the events possible together is the largest possible sample space; we also call it the *universe*.

Set theory can be used for operations of events, such as intersection ($\cap$) or union ($\cup$) of the events. For example, for event A and B, $A \cup B$, $A \cap B$, $A - B = A \cap \bar{B}$ are all new events ($\bar{B} = \Omega - B$, the compliment of B). Some of the notations for events are:

$A = B$: A and B are the same events

$A \cup B$: Union of A and B, meaning at least one event occurs

$A \cap B$: Intersection of A and B, meaning both events occur

$A - B$: A occurs and B does not occur

All the computation rules of set operations will apply for events; for example,

$$A \cup B = B \cup A \tag{I.1}$$

$$A \cap B = B \cap A \tag{I.2}$$

$$A \cap (B \cap C) = (A \cap B) \cap C \qquad (I.3)$$

$$A \cup (B \cup C) = (A \cup B) \cup C \qquad (I.4)$$

$$A \cap (B \cap C) = (A \cup B) \cap (A \cup C) \qquad (I.5)$$

$$A \cap (B \cup C) = (A \cap B) \cup (A \cap C) \qquad (I.6)$$

I.1.1.3 Probability

Probability measures the likelihood of certain events occurring; that is, the probability of event A is denoted as $P(A)$. Probability has the following characteristics:

$$0 \le P(A) \le 1 \qquad (I.7)$$

$$P(\Omega) = 1 \qquad (I.8)$$

$$P(\varnothing) = 0 \qquad (I.9)$$

$$A \cup P(\varnothing) = 0 \qquad (I.10)$$

$$P(A + B) = P(A) + P(B) - P(A \cap B) \qquad (I.11)$$

I.1.1.3.1 Mutually Exclusive Events

Events A and B are said to be mutually exclusive if A and B cannot both occur at the same time. For example, if a coin is tossed, A (the event of a head showing) and B (the event of a tail showing) are two mutually exclusive events, since a head and a tail cannot both be showing at the same time; or $P(A \cap B) = 0$.

For mutually exclusive events, the probability of either event occurring equals the sum of the probabilities of the individual events occurring, or

$$P(A + B) = P(A) + P(B)$$

If this is a fair coin, meaning the probability of a head or tail showing is 50%, then $P(A + B) = P(A) + P(B) = 0.5 + 0.5 = 1$

I.1.1.3.2 Independent Events

Event A is said to be independent to Event B if the probability of A occurring does not depend upon the probability of Event B occurring; in other words, the occurrence of A has nothing to do with Event B or vice versa.

For example, if we toss a coin, the event of having a head on the first toss (Event A) is independent to the event of having a head on the second toss (Event B). If two events are independent to each other, then the probability of the occurrence of the two events are the products of the two individual probabilities. So, the probability of two heads showing in a row if we toss a fair coin twice is

$$P(A \cdot A) = P(A)P(A) = 0.5(0.5) = 0.25$$

I.1.1.3.2.1 Conditional Probability   Suppose that we throw a fair die twice in a row; we know that the probability of a sum of 2 for the two throws is 1/36 (i.e., throwing 1 on the first throw and 1 on the second throw, and these two throws are independent to each other). However, if we have observed the first throw result, the probability becomes a conditional probability since the first throw results will impact the probability of the sum. Suppose if we see the first throw is 1, then the probability having a sum of 2 for two throws is 1/6; however, if we observe that the first throw is 3, then the probability of having a sum of 2 is zero, since there is no way we can reach a total of 2 if the first throw exceeds 2. This example illustrates the concept of conditional probability. If we denote A and B as two events, the conditional probability of A occurring given that B has occurred is denoted by a conditional probability

$$P(A \mid B)$$

The conditional probability equals the probability of A and B both occurring divided by the probability of B occurring,

$$P(A \mid B) = \frac{P(AB)}{P(B)}$$

(I.12)

In conditional probability, Bayes' theorem (formula) is one of the most important and fundamental concepts that one should remember.

For two events A and B, the conditional probability $P(A|B)$ is obtained by

$$P(A \mid B) = \frac{P(B \mid A)P(A)}{P(B)}$$

(I.13)

Given that events B_1, B_2, ..., B_n are mutually exclusive to each other, and suppose that event A has occurred, we can now determine the probability of event of B_1, B_2, ..., B_n occurring by using a more general form of Bayes' theorem:

$$P(B_j \mid A) = \frac{P(A \mid B_j)P(B_j)}{\sum_{i=1}^{n} P(B_i)} \qquad (\text{I.14})$$

B_j is called the prior probability and $P(B_j|A)$ is the posterior probability.

I.2 Random Variables and Distribution Function

Using events to describe randomness is quite flexible; however, it is not suitable for numeric analysis. We need a more formal way to represent random events and outcomes of experiments. These quantities of random event outcomes, in the format of real values, are known as random variables.

A random variable is a variable whose value is subject to the outcome of random events. Each value is associated with a chance of such value occurring, which is the probability of the random variable. The function that describes the relations between the values of random variables and their associated probability is called the *probability distribution function*, sometimes called the *probability density function* (p.d.f), or *probability mass function*.

For example, if we let X be the random variable for the outcome of a throw of a fair die, then

$$X = 1,2,3,4,5, \text{or } 6$$

and their probabilities are specified as follows:

$$P(X = 1) = P(X = 2) = P(X = 3) = P(X = 4) = P(X = 5) = P(X = 6) = \frac{1}{6}$$

For another experiment of tossing two fair dice, if we define the random variable X as the sum of the values of two dice, then

$$X = \{2,3,4,5,6,7,8,9,10,11,12\}$$

And the probability distribution function for this random variable is

$$P(X = 2) = \frac{1}{36}$$

$$P(X = 3) = \frac{2}{36}$$

$$P(X = 4) = \frac{3}{36}$$

$$P(X = 5) = \frac{4}{36}$$

$$P(X = 6) = \frac{5}{36}$$

$$P(X = 7) = \frac{6}{36}$$

$$P(X = 8) = \frac{5}{36}$$

$$P(X = 9) = \frac{4}{36}$$

$$P(X = 10) = \frac{3}{36}$$

$$P(X = 11) = \frac{2}{36}$$

$$P(X = 12) = \frac{1}{36}$$

The reader can verify that the above values are mutually exclusive and the sum of all the probabilities equals 1, that is to say,

$$\sum_{i=2}^{12} P(X = i) = 1$$

Here, we formally define the probability distribution function.

The discrete probability function is defined as follows: If the values of the random variables are either finite or countable (a countable set means that you can assign 1-to-1 numbering from natural numbers, 1, 2, 3, …, to the elements of the set; a countable set may be finite or infinite).

We denote the discrete random variable as X, and its possible values as $\{x_1, x_2, x_3, …, x_n, …\}$, the p.d.f of X is denoted as

$$f(X = x_i) = P(X = x_i), i = 1, 2, 3, …$$

and

$$\sum_i f(x_i) = 1$$

The cumulative distribution function, denoted as $F(X)$, is defined as

$$F(a) = P(X \le a) = \sum_{all\ x_i \le a} P(x_i) \tag{I.15}$$

For example, if we toss a fair die, the cumulative probability $F(4)$ is obtained as

$$F(4) = \sum_{all\ x_i \le 4} P(x_i) = P(1) + P(2) + P(3) + P(4) = \frac{1}{6} + \frac{1}{6} + \frac{1}{6} + \frac{1}{6} = \frac{2}{3}$$

I.2.1 Continuous Random Variables

If the random variable X is a real value from a real value interval, then X is called a continuous random variable. The probability function of a continuous random variable is defined as

$$f(a) = P(x = a)$$

And for the cumulative probability function $F(a)$,

$$F(a) = P\{X \in (-\infty, a)\} = \int_{-\infty}^{a} f(x) dx \tag{I.16}$$

Interpreting this graphically, the cumulative probability of a continuous random variable is the area to the left of the value under the p.d.f.

I.2.1.1 Expected Value and Variance of the Random Variable

The expected value of random variable X is also sometimes called the mean value of X, or the first moment of X, denoted as $E[X]$, or sometimes μ.

If X is a discrete random variable with probability density function of $p(x)$, then the expected value of X is defined as

$$E[X] = \sum_{x:P(x)>0} xp(x) \tag{I.17}$$

We can see the expected value of a discrete random variable is the weighted sum of its value in the sample space multiplied by its corresponding probability.

If X is a continuous random variable, then its expected value is defined as

$$E[X] = \int_{-\infty}^{+\infty} xf(x)\,dx \tag{I.18}$$

Expectation of a function of a random variable:

If X is a discrete random variable with probability density function of $p(x)$, and suppose g is a real-value function defined on X, then the expected value of the composite function is

$$E[g(X)] = \sum_{x:P(x)>0} g(x)p(x) \tag{I.19}$$

If X is a continuous random variable with probability density function of $f(x)$, and suppose g is a real-value function defined on X, then the expected value of the composite function is

$$E[g(X)] = \int_{-\infty}^{+\infty} g(x)f(x)\,dx \tag{I.20}$$

The variance of the random variable X is another important measure of the variable, usually denoted as σ^2. The variance for a discrete random variable is defined as

$$\sigma^2 = Var(X) = E\left[\left(X - E[X]\right)^2\right] \tag{I.21}$$

So for a discrete random variable

$$\sigma^2 = Var(X) = E\left[(X - E[X])^2\right] = \sum_{x:P(x)>0} (x - E[X])^2 p(x) \qquad (\text{I.22})$$

For a continuous random variable,

$$\sigma^2 = Var(X) = E\left[(X - E[X])^2\right] = \int_{-\infty}^{+\infty} (x - E[X])^2 f(x) dx \qquad (\text{I.23})$$

In the next section, we will review some of the popular random variables and distribution functions.

I.3 Some Commonly Used Probability Functions

I.3.1 Discrete Random Variables

I.3.1.1 Bernoulli Random Variable

The Bernoulli random variable has the following distribution:
    For an experiment, there are possible two cases, success or failure (such as tossing a coin, if we call a head success and a tail failure). If we let $X = 1$ if a success occurs, and $X = 0$ if a failure occurs, then X is called a Bernoulli random variable, with a probability of p, $0 \le p \le 1$:

$$p(1) = p$$

$$p(0) = 1 - p$$

The expected value $E[X]$ for a Bernoulli random variable is

$$E[X] = p(1) + (1 - p)(0) = p$$

And the variance is

$$\sigma^2 = Var(X) = E\left[(X - E[X])^2\right] = p(1-p)^2 + (1-p)(0-p)^2$$

$$= p(1-p)^2 + (1-p)p^2$$

I.3.1.2 Binomial Random Variable

Suppose we perform the above Bernoulli experiment n times. If we denote X as the number of successes occurring in the n trials, then X is called a binomial random variable. The p.d.f of a binomial random variable is

$$p(i) = \binom{n}{i} p^i (1-p)^{n-i} \tag{I.24}$$

where

$$\binom{n}{i} = \frac{n!}{i!(n-i)!}$$

The expected value for a binomial random variable is

$$E[X] = np \tag{I.25}$$

and the variance of a binomial random variable is

$$\sigma^2 = Var(X) = np(1-p) \tag{I.26}$$

I.3.1.3 Poisson Random Variable

A random variable X is said to be a Poisson random variable if X takes a value from the nonnegative integers set, $\{0, 1, 2, 3, \ldots\}$, and the probability function is

$$p(i) = P[X = i] = e^{-\lambda} \frac{\lambda^i}{i!}, \quad i = 0, 1, 2, \ldots \tag{I.27}$$

The expected value of a Poisson distribution is

$$E[X] = \lambda \tag{I.28}$$

and the variance of a Poisson distribution is the same, that is

$$\sigma^2 = Var(X) = \lambda \tag{I.29}$$

I.3.2 Continuous Random Variables

I.3.2.1 Uniform Random Variable

The general uniform random variable defined on the interval [a,b] has the following p.d.f given by

$$f(x) = \begin{cases} \dfrac{1}{b-a}, & \text{if } a < x < b \\ 0, & \text{otherwise} \end{cases} \qquad (I.30)$$

Figure I.1 illustrates the p.d.f. for a uniform random variable.
The cumulative probability distribution for a uniform random variable is

$$f(x) = \begin{cases} 0, & \text{if } x < a \\ \dfrac{x-a}{b-a}, & \text{if } a < x < b \\ 1, & \text{if } x > b \end{cases} \qquad (I.31)$$

The expected value for the uniform distribution is

$$E[X] = \frac{1}{2}(a+b) \qquad (I.32)$$

and the variance for the uniform distribution is

$$Var(X) = \frac{1}{12}(b-a)^2 \qquad (I.33)$$

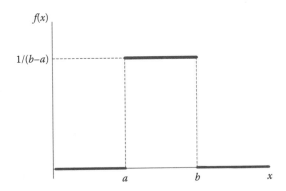

FIGURE I.1
Uniform distribution.

I.3.2.2 Exponential Random Variables

In the chapter on queuing theory, we saw that the exponential distribution was used widely. An exponential random variable with parameter λ $(\lambda > 0)$ has the following distribution function:

$$f(x) = \begin{cases} \lambda e^{-\lambda x}, & \text{if } x \geq 0 \\ 0, & \text{if } x < 0 \end{cases} \tag{I.34}$$

Figure I.2 illustrates the exponential function.
The cumulative distribution function F is

$$F(a) = \int_0^a \lambda e^{-\lambda x} dx = 1 - e^{-\lambda x}, \text{ for } a > 0 \tag{I.35}$$

The expected value of the exponential function is

$$E[X] = \frac{1}{\lambda} \tag{I.36}$$

and the variance of the exponential variable is

$$Var(x) = \frac{1}{\lambda^2} \tag{I.37}$$

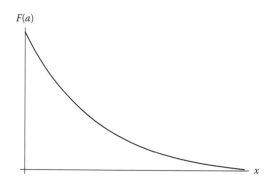

FIGURE I.2
Exponential distribution function.

I.3.2.3 Normal Random Variable

The probability distribution function for a normal random variable is

$$f(x) = \frac{1}{\sqrt{2\pi}\sigma} e^{-(x-\mu)^2/2\sigma^2} \tag{I.38}$$

The normal distribution is illustrated in Figure I.3.

As easily seen, the normal distribution is a symmetric bell-shaped curve about its mean value μ.

The expected value is

$$E[X] = \mu \tag{I.39}$$

The variance is

$$Var(X) = \sigma^2 \tag{I.40}$$

I.3.2.4 Lognormal Random Variable

X is said to be a lognormal random variable if its logarithmic $\log(X)$ is a normal random variable. The lognormal distribution has the following form with parameters μ and σ:

$$f(x) = \frac{1}{x\sigma\sqrt{2\pi}} e^{[-(\ln x - \mu)^2/2\sigma^2]}, \ x > 0 \tag{I.41}$$

When parameters μ and σ change, the shape of the lognormal function also changes. μ is called its scale parameter and σ is called its shape parameter. Figures I.4 and I.5 illustrate the lognormal function with different parameters.

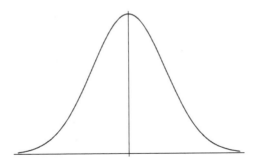

FIGURE I.3
Normal distribution.

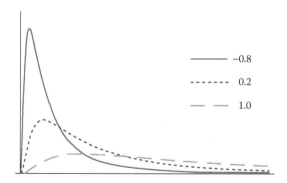

FIGURE I.4
Lognormal functions with different values for μ (σ = 1).

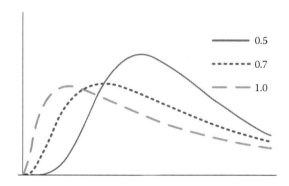

FIGURE I.5
Lognormal functions with different values for σ (μ = 0.2).

The expected value of the lognormal distribution is

$$E[X] = e^{\mu + \sigma^2/2} \tag{I.42}$$

and the variance of the lognormal distribution is

$$Var(x) = e^{(\sigma^2-1)}e^{(2\mu+\sigma^2)} \tag{I.43}$$

I.3.2.5 Weibull Random Variable

A Weibull random variable has the following distribution function:

$$f(x) = \frac{\alpha}{\beta^\alpha}x^{\alpha-1}e^{-(x/\beta)^\alpha}, \; x \geq 0 \tag{I.44}$$

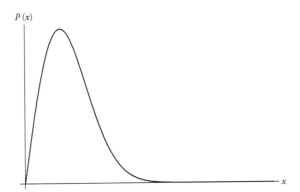

P (x)

x

FIGURE I.6
Weibull distribution.

where $\alpha > 0$, $\beta > 0$.

Figure I.6 illustrates the Weibull distribution. The Weibull distribution is a more general form as it has the following characteristics:

1. If $\alpha = 1$, the Weibull distribution becomes an exponential distribution with parameter β.
2. If $\alpha = 3.4$, then the Weibull distribution can approximate a normal distribution.

The mean for the Weibull distribution is

$$E[X] = \frac{\alpha}{\beta} \Gamma\left(\frac{1}{\alpha}\right) \tag{I.45}$$

and the variance is

$$Var(x) = \frac{\beta^2}{\alpha}\left\{2\Gamma\left(\frac{2}{\alpha}\right) - \frac{1}{\alpha}\left[\Gamma\left(\frac{1}{\alpha}\right)\right]^2\right\} \tag{I.46}$$

$\Gamma()$ is the gamma function, which is

$$\Gamma(y) = \int_0^\infty t^{y-1} e^{-t} dt \tag{I.47}$$

Appendix II: Cumulative Normal Distribution Table

Z	0.09	0.08	0.07	0.06	0.05	0.04	0.03	0.02	0.01	0.00
−3.5	0.00017	0.00017	0.00018	0.00019	0.00019	0.00020	0.00021	0.00022	0.00022	0.00023
−3.4	0.00024	0.00025	0.00026	0.00027	0.00028	0.00029	0.00030	0.00031	0.00032	0.00034
−3.3	0.00035	0.00036	0.00038	0.00039	0.00040	0.00042	0.00043	0.00045	0.00047	0.00048
−3.2	0.00050	0.00052	0.00054	0.00056	0.00058	0.00060	0.00062	0.00064	0.00066	0.00069
−3.1	0.00071	0.00074	0.00076	0.00079	0.00082	0.00084	0.00087	0.00090	0.00094	0.00097
−3.0	0.00100	0.00104	0.00107	0.00111	0.00114	0.00118	0.00122	0.00126	0.00131	0.00135
−2.9	0.00139	0.00144	0.00149	0.00154	0.00159	0.00164	0.00169	0.00175	0.00181	0.00187
−2.8	0.00193	0.00199	0.00205	0.00212	0.00219	0.00226	0.00233	0.00240	0.00248	0.00256
−2.7	0.00264	0.00272	0.00280	0.00289	0.00298	0.00307	0.00317	0.00326	0.00336	0.00347
−2.6	0.00357	0.00368	0.00379	0.00391	0.00402	0.00415	0.00427	0.00440	0.00453	0.00466
−2.5	0.00480	0.00494	0.00508	0.00523	0.00539	0.00554	0.00570	0.00587	0.00604	0.00621
−2.4	0.00639	0.00657	0.00676	0.00695	0.00714	0.00734	0.00755	0.00776	0.00798	0.00820
−2.3	0.00842	0.00866	0.00889	0.00914	0.00939	0.00964	0.00990	0.01017	0.01044	0.01072
−2.2	0.01101	0.01130	0.01160	0.01191	0.01222	0.01255	0.01287	0.01321	0.01355	0.01390
−2.1	0.01426	0.01463	0.01500	0.01539	0.01578	0.01618	0.01659	0.01700	0.01743	0.01786
−2.0	0.01831	0.01876	0.01923	0.01970	0.02018	0.02068	0.02118	0.02169	0.02222	0.02275
−1.9	0.02330	0.02385	0.02442	0.02500	0.02559	0.02619	0.02680	0.02743	0.02807	0.02872
−1.8	0.02938	0.03005	0.03074	0.03144	0.03216	0.03288	0.03362	0.03438	0.03515	0.03593
−1.7	0.03673	0.03754	0.03836	0.03920	0.04006	0.04093	0.04182	0.04272	0.04363	0.04457
−1.6	0.04551	0.04648	0.04746	0.04846	0.04947	0.05050	0.05155	0.05262	0.05370	0.05480
−1.5	0.05592	0.05705	0.05821	0.05938	0.06057	0.06178	0.06301	0.06426	0.06552	0.06681
−1.4	0.06811	0.06944	0.07078	0.07215	0.07353	0.07493	0.07636	0.07780	0.07927	0.08076
−1.3	0.08226	0.08379	0.08534	0.08691	0.08851	0.09012	0.09176	0.09342	0.09510	0.09680
−1.2	0.09853	0.10027	0.10204	0.10383	0.10565	0.10749	0.10935	0.11123	0.11314	0.11507
−1.1	0.11702	0.11900	0.12100	0.12302	0.12507	0.12714	0.12924	0.13136	0.13350	0.13567
−1.0	0.13786	0.14007	0.14231	0.14457	0.14686	0.14917	0.15151	0.15386	0.15625	0.15866
−0.9	0.16109	0.16354	0.16602	0.16853	0.17106	0.17361	0.17619	0.17879	0.18141	0.18406
−0.8	0.18673	0.18943	0.19215	0.19489	0.19766	0.20045	0.20327	0.20611	0.20897	0.21186
−0.7	0.21476	0.21770	0.22065	0.22363	0.22663	0.22965	0.23270	0.23576	0.23885	0.24196
−0.6	0.24510	0.24825	0.25143	0.25463	0.25785	0.26109	0.26435	0.26763	0.27093	0.27425
−0.5	0.27760	0.28096	0.28434	0.28774	0.29116	0.29460	0.29806	0.30153	0.30503	0.30854
−0.4	0.31207	0.31561	0.31918	0.32276	0.32636	0.32997	0.33360	0.33724	0.34090	0.34458
−0.3	0.34827	0.35197	0.35569	0.35942	0.36317	0.36693	0.37070	0.37448	0.37828	0.38209
−0.2	0.38591	0.38974	0.39358	0.39743	0.40129	0.40517	0.40905	0.41294	0.41683	0.42074
−0.1	0.42465	0.42858	0.43251	0.43644	0.44038	0.44433	0.44828	0.45224	0.45620	0.46017
0.0	0.46414	0.46812	0.47210	0.47608	0.48006	0.48405	0.48803	0.49202	0.49601	0.50000

Z	0.00	0.01	0.02	0.03	0.04	0.05	0.06	0.07	0.08	0.09
0.0	0.50000	0.50399	0.50798	0.51197	0.51595	0.51994	0.52392	0.52790	0.53188	0.53586
0.1	0.53983	0.54380	0.54776	0.55172	0.55567	0.55962	0.56356	0.56749	0.57142	0.57535
0.2	0.57926	0.58317	0.58706	0.59095	0.59483	0.59871	0.60257	0.60642	0.61026	0.61409
0.3	0.61791	0.62172	0.62552	0.62930	0.63307	0.63683	0.64058	0.64431	0.64803	0.65173
0.4	0.65542	0.65910	0.66276	0.66640	0.67003	0.67364	0.67724	0.68082	0.68439	0.68793
0.5	0.69146	0.69497	0.69847	0.70194	0.70540	0.70884	0.71226	0.71566	0.71904	0.72240
0.6	0.72575	0.72907	0.73237	0.73565	0.73891	0.74215	0.74537	0.74857	0.75175	0.75490
0.7	0.75804	0.76115	0.76424	0.76730	0.77035	0.77337	0.77637	0.77935	0.78230	0.78524
0.8	0.78814	0.79103	0.79389	0.79673	0.79955	0.80234	0.80511	0.80785	0.81057	0.81327
0.9	0.81594	0.81859	0.82121	0.82381	0.82639	0.82894	0.83147	0.83398	0.83646	0.83891
1.0	0.84134	0.84375	0.84614	0.84849	0.85083	0.85314	0.85543	0.85769	0.85993	0.86214
1.1	0.86433	0.86650	0.86864	0.87076	0.87286	0.87493	0.87698	0.87900	0.88100	0.88298
1.2	0.88493	0.88686	0.88877	0.89065	0.89251	0.89435	0.89617	0.89796	0.89973	0.90147
1.3	0.90320	0.90490	0.90658	0.90824	0.90988	0.91149	0.91309	0.91466	0.91621	0.91774
1.4	0.91924	0.92073	0.92220	0.92364	0.92507	0.92647	0.92785	0.92922	0.93056	0.93189
1.5	0.93319	0.93448	0.93574	0.93699	0.93822	0.93943	0.94062	0.94179	0.94295	0.94408
1.6	0.94520	0.94630	0.94738	0.94845	0.94950	0.95053	0.95154	0.95254	0.95352	0.95449
1.7	0.95543	0.95637	0.95728	0.95818	0.95907	0.95994	0.96080	0.96164	0.96246	0.96327
1.8	0.96407	0.96485	0.96562	0.96638	0.96712	0.96784	0.96856	0.96926	0.96995	0.97062
1.9	0.97128	0.97193	0.97257	0.97320	0.97381	0.97441	0.97500	0.97558	0.97615	0.97670
2.0	0.97725	0.97778	0.97831	0.97882	0.97932	0.97982	0.98030	0.98077	0.98124	0.98169
2.1	0.98214	0.98257	0.98300	0.98341	0.98382	0.98422	0.98461	0.98500	0.98537	0.98574
2.2	0.98610	0.98645	0.98679	0.98713	0.98745	0.98778	0.98809	0.98840	0.98870	0.98899
2.3	0.98928	0.98956	0.98983	0.99010	0.99036	0.99061	0.99086	0.99111	0.99134	0.99158
2.4	0.99180	0.99202	0.99224	0.99245	0.99266	0.99286	0.99305	0.99324	0.99343	0.99361
2.5	0.99379	0.99396	0.99413	0.99430	0.99446	0.99461	0.99477	0.99492	0.99506	0.99520
2.6	0.99534	0.99547	0.99560	0.99573	0.99585	0.99598	0.99609	0.99621	0.99632	0.99643
2.7	0.99653	0.99664	0.99674	0.99683	0.99693	0.99702	0.99711	0.99720	0.99728	0.99736
2.8	0.99744	0.99752	0.99760	0.99767	0.99774	0.99781	0.99788	0.99795	0.99801	0.99807
2.9	0.99813	0.99819	0.99825	0.99831	0.99836	0.99841	0.99846	0.99851	0.99856	0.99861
3.0	0.99865	0.99869	0.99874	0.99878	0.99882	0.99886	0.99889	0.99893	0.99896	0.99900
3.1	0.99903	0.99906	0.99910	0.99913	0.99916	0.99918	0.99921	0.99924	0.99926	0.99929
3.2	0.99931	0.99934	0.99936	0.99938	0.99940	0.99942	0.99944	0.99946	0.99948	0.99950
3.3	0.99952	0.99953	0.99955	0.99957	0.99958	0.99960	0.99961	0.99962	0.99964	0.99965
3.4	0.99966	0.99968	0.99969	0.99970	0.99971	0.99972	0.99973	0.99974	0.99975	0.99976
3.5	0.99977	0.99978	0.99978	0.99979	0.99980	0.99981	0.99981	0.99982	0.99983	0.99983

Appendix III: Interest Factor Tables

TABLE III.1

5% Interest Factors for Annual Compounding

	Single Payment		Equal-Payment Series			
	Future-Worth Factor	Present Worth Factor	Future-Worth Factor	Annual-Worth Factor	Present-Worth Factor	Annual-Worth Factor
	To Find F	To Find P	To Find F	To Find A	To Find P	To Find A
	Given P	Given F	Given A	Given F	Given A	Given P
n	$F/P,i,n$	$P/F,i,n$	$F/A,i,n$	$A/F,i,n$	$P/A,i,n$	$A/P,i,n$
1	1.0500	0.9524	1.0000	1.0000	0.9524	1.0500
2	1.1025	0.9070	2.0500	0.4878	1.8594	0.5378
3	1.1576	0.8638	3.1525	0.3172	2.7232	0.3672
4	1.2155	0.8227	4.3101	0.2320	3.5460	0.2820
5	1.2763	0.7835	5.5256	0.1810	4.3295	0.2310
6	1.3401	0.7462	6.8019	0.1470	5.0757	0.1970
7	1.4071	0.7107	8.1420	0.1228	5.7864	0.1728
8	1.4775	0.6768	9.5491	0.1047	6.4632	0.1547
9	1.5513	0.6446	11.0266	0.0907	7.1078	0.1407
10	1.6289	0.6139	12.5779	0.0795	7.7217	0.1295
11	1.7103	0.5847	14.2068	0.0704	8.3064	0.1204
12	1.7959	0.5568	15.9171	0.0628	8.8633	0.1128
13	1.8856	0.5303	17.7130	0.0565	9.3936	0.1065
14	1.9799	0.5051	19.5986	0.0510	9.8986	0.1010
15	2.0789	0.4810	21.5786	0.0463	10.3797	0.0963
16	2.1829	0.4581	23.6575	0.0423	10.8378	0.0923
17	2.2920	0.4363	25.8404	0.0387	11.2741	0.0887
18	2.4066	0.4155	28.1324	0.0355	11.6896	0.0855
19	2.5270	0.3957	30.5390	0.0327	12.0853	0.0827
20	2.6533	0.3769	33.0660	0.0302	12.4622	0.0802
21	2.7860	0.3589	35.7193	0.0280	12.8212	0.0780
22	2.9253	0.3418	38.5052	0.0260	13.1630	0.0760
23	3.0715	0.3256	41.4305	0.0241	13.4886	0.0741
24	3.2251	0.3101	44.5020	0.0225	13.7986	0.0725
25	3.3864	0.2953	47.7271	0.0210	14.0939	0.0710
26	3.5557	0.2812	51.1135	0.0196	14.3752	0.0696
27	3.7335	0.2678	54.6691	0.0183	14.6430	0.0683
28	3.9201	0.2551	58.4026	0.0171	14.8981	0.0671
29	4.1161	0.2429	62.3227	0.0160	15.1411	0.0660
30	4.3219	0.2314	66.4388	0.0151	15.3725	0.0651
31	4.5380	0.2204	70.7608	0.0141	15.5928	0.0641
32	4.7649	0.2099	75.2988	0.0133	15.8027	0.0633
33	5.0032	0.1999	80.0638	0.0125	16.0025	0.0625
34	5.2533	0.1904	85.0670	0.0118	16.1929	0.0618
35	5.5160	0.1813	90.3203	0.0111	16.3742	0.0611

TABLE III.2

6% Interest Factors for Annual Compounding

	Single Payment		Equal-Payment Series			
	Future-Worth Factor	Present-Worth Factor	Future-Worth Factor	Annual-Worth Factor	Present-Worth Factor	Annual-Worth Factor
	To Find F	To Find P	To Find F	To Find A	To Find P	To Find A
	Given P	Given F	Given A	Given F	Given A	Given P
n	$F/P,i,n$	$P/F,i,n$	$F/A,i,n$	$A/F,i,n$	$P/A,i,n$	$A/P,i,n$
1	1.0600	0.9434	1.0000	1.0000	0.9434	1.0600
2	1.1236	0.8900	2.0600	0.4854	1.8334	0.5454
3	1.1910	0.8396	3.1836	0.3141	2.6730	0.3741
4	1.2625	0.7921	4.3746	0.2286	3.4651	0.2886
5	1.3382	0.7473	5.6371	0.1774	4.2124	0.2374
6	1.4185	0.7050	6.9753	0.1434	4.9173	0.2034
7	1.5036	0.6651	8.3938	0.1191	5.5824	0.1791
8	1.5938	0.6274	9.8975	0.1010	6.2098	0.1610
9	1.6895	0.5919	11.4913	0.0870	6.8017	0.1470
10	1.7908	0.5584	13.1808	0.0759	7.3601	0.1359
11	1.8983	0.5268	14.9716	0.0668	7.8869	0.1268
12	2.0122	0.4970	16.8699	0.0593	8.3838	0.1193
13	2.1329	0.4688	18.8821	0.0530	8.8527	0.1130
14	2.2609	0.4423	21.0151	0.0476	9.2950	0.1076
15	2.3966	0.4173	23.2760	0.0430	9.7122	0.1030
16	2.5404	0.3936	25.6725	0.0390	10.1059	0.0990
17	2.6928	0.3714	28.2129	0.0354	10.4773	0.0954
18	2.8543	0.3503	30.9057	0.0324	10.8276	0.0924
19	3.0256	0.3305	33.7600	0.0296	11.1581	0.0896
20	3.2071	0.3118	36.7856	0.0272	11.4699	0.0872
21	3.3996	0.2942	39.9927	0.0250	11.7641	0.0850
22	3.6035	0.2775	43.3923	0.0230	12.0416	0.0830
23	3.8197	0.2618	46.9958	0.0213	12.3034	0.0813
24	4.0489	0.2470	50.8156	0.0197	12.5504	0.0797
25	4.2919	0.2330	54.8645	0.0182	12.7834	0.0782
26	4.5494	0.2198	59.1564	0.0169	13.0032	0.0769
27	4.8223	0.2074	63.7058	0.0157	13.2105	0.0757
28	5.1117	0.1956	68.5281	0.0146	13.4062	0.0746
29	5.4184	0.1846	73.6398	0.0136	13.5907	0.0736
30	5.7435	0.1741	79.0582	0.0126	13.7648	0.0726
31	6.0881	0.1643	84.8017	0.0118	13.9291	0.0718
32	6.4534	0.1550	90.8898	0.0110	14.0840	0.0710
33	6.8406	0.1462	97.3432	0.0103	14.2302	0.0703
34	7.2510	0.1379	104.1838	0.0096	14.3681	0.0696
35	7.6861	0.1301	111.4348	0.0090	14.4982	0.0690

TABLE III.3

7% Interest Factors for Annual Compounding

	Single Payment		Equal-Payment Series			
	Future-Worth Factor	Present-Worth Factor	Future-Worth Factor	Annual-Worth Factor	Present-Worth Factor	Annual-Worth Factor
	To Find F	*To Find P*	*To Find F*	*To Find A*	*To Find P*	*To Find A*
	Given P	*Given F*	*Given A*	*Given F*	*Given A*	*Given P*
n	*F/P,i,n*	*P/F,i,n*	*F/A,i,n*	*A/F,i,n*	*P/A,i,n*	*A/P,i,n*
1	1.0700	0.9346	1.0000	1.0000	0.9346	1.0700
2	1.1449	0.8734	2.0700	0.4831	1.8080	0.5531
3	1.2250	0.8163	3.2149	0.3111	2.6243	0.3811
4	1.3108	0.7629	4.4399	0.2252	3.3872	0.2952
5	1.4026	0.7130	5.7507	0.1739	4.1002	0.2439
6	1.5007	0.6663	7.1533	0.1398	4.7665	0.2098
7	1.6058	0.6227	8.6540	0.1156	5.3893	0.1856
8	1.7182	0.5820	10.2598	0.0975	5.9713	0.1675
9	1.8385	0.5439	11.9780	0.0835	6.5152	0.1535
10	1.9672	0.5083	13.8164	0.0724	7.0236	0.1424
11	2.1049	0.4751	15.7836	0.0634	7.4987	0.1334
12	2.2522	0.4440	17.8885	0.0559	7.9427	0.1259
13	2.4098	0.4150	20.1406	0.0497	8.3577	0.1197
14	2.5785	0.3878	22.5505	0.0443	8.7455	0.1143
15	2.7590	0.3624	25.1290	0.0398	9.1079	0.1098
16	2.9522	0.3387	27.8881	0.0359	9.4466	0.1059
17	3.1588	0.3166	30.8402	0.0324	9.7632	0.1024
18	3.3799	0.2959	33.9990	0.0294	10.0591	0.0994
19	3.6165	0.2765	37.3790	0.0268	10.3356	0.0968
20	3.8697	0.2584	40.9955	0.0244	10.5940	0.0944
21	4.1406	0.2415	44.8652	0.0223	10.8355	0.0923
22	4.4304	0.2257	49.0057	0.0204	11.0612	0.0904
23	4.7405	0.2109	53.4361	0.0187	11.2722	0.0887
24	5.0724	0.1971	58.1767	0.0172	11.4693	0.0872
25	5.4274	0.1842	63.2490	0.0158	11.6536	0.0858
26	5.8074	0.1722	68.6765	0.0146	11.8258	0.0846
27	6.2139	0.1609	74.4838	0.0134	11.9867	0.0834
28	6.6488	0.1504	80.6977	0.0124	12.1371	0.0824
29	7.1143	0.1406	87.3465	0.0114	12.2777	0.0814
30	7.6123	0.1314	94.4608	0.0106	12.4090	0.0806
31	8.1451	0.1228	102.0730	0.0098	12.5318	0.0798
32	8.7153	0.1147	110.2182	0.0091	12.6466	0.0791
33	9.3253	0.1072	118.9334	0.0084	12.7538	0.0784
34	9.9781	0.1002	128.2588	0.0078	12.8540	0.0778
35	10.6766	0.0937	138.2369	0.0072	12.9477	0.0772

TABLE III.4

8% Interest Factors for Annual Compounding

	Single Payment		Equal-Payment Series			
	Future-Worth Factor	Present-Worth Factor	Future-Worth Factor	Annual-Worth Factor	Present-Worth Factor	Annual-Worth Factor
	To Find F	*To Find P*	*To Find F*	*To Find A*	*To Find P*	*To Find A*
	Given P	*Given F*	*Given A*	*Given F*	*Given A*	*Given P*
n	$F/P,i,n$	$P/F,i,n$	$F/A,i,n$	$A/F,i,n$	$P/A,i,n$	$A/P,i,n$
1	1.0800	0.9259	1.0000	1.0000	0.9259	1.0800
2	1.1664	0.8573	2.0800	0.4808	1.7833	0.5608
3	1.2597	0.7938	3.2464	0.3080	2.5771	0.3880
4	1.3605	0.7350	4.5061	0.2219	3.3121	0.3019
5	1.4693	0.6806	5.8666	0.1705	3.9927	0.2505
6	1.5869	0.6302	7.3359	0.1363	4.6229	0.2163
7	1.7138	0.5835	8.9228	0.1121	5.2064	0.1921
8	1.8509	0.5403	10.6366	0.0940	5.7466	0.1740
9	1.9990	0.5002	12.4876	0.0801	6.2469	0.1601
10	2.1589	0.4632	14.4866	0.0690	6.7101	0.1490
11	2.3316	0.4289	16.6455	0.0601	7.1390	0.1401
12	2.5182	0.3971	18.9771	0.0527	7.5361	0.1327
13	2.7196	0.3677	21.4953	0.0465	7.9038	0.1265
14	2.9372	0.3405	24.2149	0.0413	8.2442	0.1213
15	3.1722	0.3152	27.1521	0.0368	8.5595	0.1168
16	3.4259	0.2919	30.3243	0.0330	8.8514	0.1130
17	3.7000	0.2703	33.7502	0.0296	9.1216	0.1096
18	3.9960	0.2502	37.4502	0.0267	9.3719	0.1067
19	4.3157	0.2317	41.4463	0.0241	9.6036	0.1041
20	4.6610	0.2145	45.7620	0.0219	9.8181	0.1019
21	5.0338	0.1987	50.4229	0.0198	10.0168	0.0998
22	5.4365	0.1839	55.4568	0.0180	10.2007	0.0980
23	5.8715	0.1703	60.8933	0.0164	10.3711	0.0964
24	6.3412	0.1577	66.7648	0.0150	10.5288	0.0950
25	6.8485	0.1460	73.1059	0.0137	10.6748	0.0937
26	7.3964	0.1352	79.9544	0.0125	10.8100	0.0925
27	7.9881	0.1252	87.3508	0.0114	10.9352	0.0914
28	8.6271	0.1159	95.3388	0.0105	11.0511	0.0905
29	9.3173	0.1073	103.9659	0.0096	11.1584	0.0896
30	10.0627	0.0994	113.2832	0.0088	11.2578	0.0888
31	10.8677	0.0920	123.3459	0.0081	11.3498	0.0881
32	11.7371	0.0852	134.2135	0.0075	11.4350	0.0875
33	12.6760	0.0789	145.9506	0.0069	11.5139	0.0869
34	13.6901	0.0730	158.6267	0.0063	11.5869	0.0863
35	14.7853	0.0676	172.3168	0.0058	11.6546	0.0858

TABLE III.5

9% Interest Factors for Annual Compounding

	Single Payment		Equal-Payment Series			
	Future-Worth Factor	Present-Worth Factor	Future-Worth Factor	Annual-Worth Factor	Present-Worth Factor	Annual-Worth Factor
	To Find F	*To Find P*	*To Find F*	*To Find A*	*To Find P*	*To Find A*
	Given P	*Given F*	*Given A*	*Given F*	*Given A*	*Given P*
n	$F/P,i,n$	$P/F,i,n$	$F/A,i,n$	$A/F,i,n$	$P/A,i,n$	$A/P,i,n$
1	1.0900	0.9174	1.0000	1.0000	0.9174	1.0900
2	1.1881	0.8417	2.0900	0.4785	1.7591	0.5685
3	1.2950	0.7722	3.2781	0.3051	2.5313	0.3951
4	1.4116	0.7084	4.5731	0.2187	3.2397	0.3087
5	1.5386	0.6499	5.9847	0.1671	3.8897	0.2571
6	1.6771	0.5963	7.5233	0.1329	4.4859	0.2229
7	1.8280	0.5470	9.2004	0.1087	5.0330	0.1987
8	1.9926	0.5019	11.0285	0.0907	5.5348	0.1807
9	2.1719	0.4604	13.0210	0.0768	5.9952	0.1668
10	2.3674	0.4224	15.1929	0.0658	6.4177	0.1558
11	2.5804	0.3875	17.5603	0.0569	6.8052	0.1469
12	2.8127	0.3555	20.1407	0.0497	7.1607	0.1397
13	3.0658	0.3262	22.9534	0.0436	7.4869	0.1336
14	3.3417	0.2992	26.0192	0.0384	7.7862	0.1284
15	3.6425	0.2745	29.3609	0.0341	8.0607	0.1241
16	3.9703	0.2519	33.0034	0.0303	8.3126	0.1203
17	4.3276	0.2311	36.9737	0.0270	8.5436	0.1170
18	4.7171	0.2120	41.3013	0.0242	8.7556	0.1142
19	5.1417	0.1945	46.0185	0.0217	8.9501	0.1117
20	5.6044	0.1784	51.1601	0.0195	9.1285	0.1095
21	6.1088	0.1637	56.7645	0.0176	9.2922	0.1076
22	6.6586	0.1502	62.8733	0.0159	9.4424	0.1059
23	7.2579	0.1378	69.5319	0.0144	9.5802	0.1044
24	7.9111	0.1264	76.7898	0.0130	9.7066	0.1030
25	8.6231	0.1160	84.7009	0.0118	9.8226	0.1018
26	9.3992	0.1064	93.3240	0.0107	9.9290	0.1007
27	10.2451	0.0976	102.7231	0.0097	10.0266	0.0997
28	11.1671	0.0895	112.9682	0.0089	10.1161	0.0989
29	12.1722	0.0822	124.1354	0.0081	10.1983	0.0981
30	13.2677	0.0754	136.3075	0.0073	10.2737	0.0973
31	14.4618	0.0691	149.5752	0.0067	10.3428	0.0967
32	15.7633	0.0634	164.0370	0.0061	10.4062	0.0961
33	17.1820	0.0582	179.8003	0.0056	10.4644	0.0956
34	18.7284	0.0534	196.9823	0.0051	10.5178	0.0951
35	20.4140	0.0490	215.7108	0.0046	10.5668	0.0946

TABLE III.6

10% Interest Factors for Annual Compounding

	Single Payment		Equal-Payment Series			
	Future-Worth Factor	Present-Worth Factor	Future-Worth Factor	Annual-Worth Factor	Present-Worth Factor	Annual-Worth Factor
	To Find F	*To Find P*	*To Find F*	*To Find A*	*To Find P*	*To Find A*
	Given P	*Given F*	*Given A*	*Given F*	*Given A*	*Given P*
n	$F/P,i,n$	$P/F,i,n$	$F/A,i,n$	$A/F,i,,n$	$P/A,i,n$	$A/P,i,n$
1	1.1000	0.9091	1.0000	1.0000	0.9091	1.1000
2	1.2100	0.8264	2.1000	0.4762	1.7355	0.5762
3	1.3310	0.7513	3.3100	0.3021	2.4869	0.4021
4	1.4641	0.6830	4.6410	0.2155	3.1699	0.3155
5	1.6105	0.6209	6.1051	0.1638	3.7908	0.2638
6	1.7716	0.5645	7.7156	0.1296	4.3553	0.2296
7	1.9487	0.5132	9.4872	0.1054	4.8684	0.2054
8	2.1436	0.4665	11.4359	0.0874	5.3349	0.1874
9	2.3579	0.4241	13.5795	0.0736	5.7590	0.1736
10	2.5937	0.3855	15.9374	0.0627	6.1446	0.1627
11	2.8531	0.3505	18.5312	0.0540	6.4951	0.1540
12	3.1384	0.3186	21.3843	0.0468	6.8137	0.1468
13	3.4523	0.2897	24.5227	0.0408	7.1034	0.1408
14	3.7975	0.2633	27.9750	0.0357	7.3667	0.1357
15	4.1772	0.2394	31.7725	0.0315	7.6061	0.1315
16	4.5950	0.2176	35.9497	0.0278	7.8237	0.1278
17	5.0545	0.1978	40.5447	0.0247	8.0216	0.1247
18	5.5599	0.1799	45.5992	0.0219	8.2014	0.1219
19	6.1159	0.1635	51.1591	0.0195	8.3649	0.1195
20	6.7275	0.1486	57.2750	0.0175	8.5136	0.1175
21	7.4002	0.1351	64.0025	0.0156	8.6487	0.1156
22	8.1403	0.1228	71.4027	0.0140	8.7715	0.1140
23	8.9543	0.1117	79.5430	0.0126	8.8832	0.1126
24	9.8497	0.1015	88.4973	0.0113	8.9847	0.1113
25	10.8347	0.0923	98.3471	0.0102	9.0770	0.1102
26	11.9182	0.0839	109.1818	0.0092	9.1609	0.1092
27	13.1100	0.0763	121.0999	0.0083	9.2372	0.1083
28	14.4210	0.0693	134.2099	0.0075	9.3066	0.1075
29	15.8631	0.0630	148.6309	0.0067	9.3696	0.1067
30	17.4494	0.0573	164.4940	0.0061	9.4269	0.1061
31	19.1943	0.0521	181.9434	0.0055	9.4790	0.1055
32	21.1138	0.0474	201.1378	0.0050	9.5264	0.1050
33	23.2252	0.0431	222.2515	0.0045	9.5694	0.1045
34	25.5477	0.0391	245.4767	0.0041	9.6086	0.1041
35	28.1024	0.0356	271.0244	0.0037	9.6442	0.1037

TABLE III.7

12% Interest Factors for Annual Compounding

	Single Payment		Equal-Payment Series			
	Future-Worth Factor	Present-Worth Factor	Future-Worth Factor	Annual-Worth Factor	Present-Worth Factor	Annual-Worth Factor
	To Find F	*To Find P*	*To Find F*	*To Find A*	*To Find P*	*To Find A*
	Given P	*Given F*	*Given A*	*Given F*	*Given A*	*Given P*
n	*F/P,i,n*	*P/F,i,n*	*F/A,i,n*	*A/F,i,n*	*P/A,i,n*	*A/P,i,n*
1	1.1200	0.8929	1.0000	1.0000	0.8929	1.1200
2	1.2544	0.7972	2.1200	0.4717	1.6901	0.5917
3	1.4049	0.7118	3.3744	0.2963	2.4018	0.4163
4	1.5735	0.6355	4.7793	0.2092	3.0373	0.3292
5	1.7623	0.5674	6.3528	0.1574	3.6048	0.2774
6	1.9738	0.5066	8.1152	0.1232	4.1114	0.2432
7	2.2107	0.4523	10.0890	0.0991	4.5638	0.2191
8	2.4760	0.4039	12.2997	0.0813	4.9676	0.2013
9	2.7731	0.3606	14.7757	0.0677	5.3282	0.1877
10	3.1058	0.3220	17.5487	0.0570	5.6502	0.1770
11	3.4785	0.2875	20.6546	0.0484	5.9377	0.1684
12	3.8960	0.2567	24.1331	0.0414	6.1944	0.1614
13	4.3635	0.2292	28.0291	0.0357	6.4235	0.1557
14	4.8871	0.2046	32.3926	0.0309	6.6282	0.1509
15	5.4736	0.1827	37.2797	0.0268	6.8109	0.1468
16	6.1304	0.1631	42.7533	0.0234	6.9740	0.1434
17	6.8660	0.1456	48.8837	0.0205	7.1196	0.1405
18	7.6900	0.1300	55.7497	0.0179	7.2497	0.1379
19	8.6128	0.1161	63.4397	0.0158	7.3658	0.1358
20	9.6463	0.1037	72.0524	0.0139	7.4694	0.1339
21	10.8038	0.0926	81.6987	0.0122	7.5620	0.1322
22	12.1003	0.0826	92.5026	0.0108	7.6446	0.1308
23	13.5523	0.0738	104.6029	0.0096	7.7184	0.1296
24	15.1786	0.0659	118.1552	0.0085	7.7843	0.1285
25	17.0001	0.0588	133.3339	0.0075	7.8431	0.1275
26	19.0401	0.0525	150.3339	0.0067	7.8957	0.1267
27	21.3249	0.0469	169.3740	0.0059	7.9426	0.1259
28	23.8839	0.0419	190.6989	0.0052	7.9844	0.1252
29	26.7499	0.0374	214.5828	0.0047	8.0218	0.1247
30	29.9599	0.0334	241.3327	0.0041	8.0552	0.1241
31	33.5551	0.0298	271.2926	0.0037	8.0850	0.1237
32	37.5817	0.0266	304.8477	0.0033	8.1116	0.1233
33	42.0915	0.0238	342.4294	0.0029	8.1354	0.1229
34	47.1425	0.0212	384.5210	0.0026	8.1566	0.1226
35	52.7996	0.0189	431.6635	0.0023	8.1755	0.1223

TABLE III.8

15% Interest Factors for Annual Compounding

	Single Payment		Equal-Payment Series			
	Future-Worth Factor	Present-Worth Factor	Future-Worth Factor	Annual-Worth Factor	Present-Worth Factor	Annual-Worth Factor
	To Find F	To Find P	To Find F	To Find A	To Find P	To Find A
	Given P	Given F	Given A	Given F	Given A	Given P
n	$F/P,i,n$	$P/F,i,n$	$F/A,i,n$	$A/F,i,n$	$P/A,i,n$	$A/P,i,n$
1	1.1500	0.8696	1.0000	1.0000	0.8696	1.1500
2	1.3225	0.7561	2.1500	0.4651	1.6257	0.6151
3	1.5209	0.6575	3.4725	0.2880	2.2832	0.4380
4	1.7490	0.5718	4.9934	0.2003	2.8550	0.3503
5	2.0114	0.4972	6.7424	0.1483	3.3522	0.2983
6	2.3131	0.4323	8.7537	0.1142	3.7845	0.2642
7	2.6600	0.3759	11.0668	0.0904	4.1604	0.2404
8	3.0590	0.3269	13.7268	0.0729	4.4873	0.2229
9	3.5179	0.2843	16.7858	0.0596	4.7716	0.2096
10	4.0456	0.2472	20.3037	0.0493	5.0188	0.1993
11	4.6524	0.2149	24.3493	0.0411	5.2337	0.1911
12	5.3503	0.1869	29.0017	0.0345	5.4206	0.1845
13	6.1528	0.1625	34.3519	0.0291	5.5831	0.1791
14	7.0757	0.1413	40.5047	0.0247	5.7245	0.1747
15	8.1371	0.1229	47.5804	0.0210	5.8474	0.1710
16	9.3576	0.1069	55.7175	0.0179	5.9542	0.1679
17	10.7613	0.0929	65.0751	0.0154	6.0472	0.1654
18	12.3755	0.0808	75.8364	0.0132	6.1280	0.1632
19	14.2318	0.0703	88.2118	0.0113	6.1982	0.1613
20	16.3665	0.0611	102.4436	0.0098	6.2593	0.1598
21	18.8215	0.0531	118.8101	0.0084	6.3125	0.1584
22	21.6447	0.0462	137.6316	0.0073	6.3587	0.1573
23	24.8915	0.0402	159.2764	0.0063	6.3988	0.1563
24	28.6252	0.0349	184.1678	0.0054	6.4338	0.1554
25	32.9190	0.0304	212.7930	0.0047	6.4641	0.1547
26	37.8568	0.0264	245.7120	0.0041	6.4906	0.1541
27	43.5353	0.0230	283.5688	0.0035	6.5135	0.1535
28	50.0656	0.0200	327.1041	0.0031	6.5335	0.1531
29	57.5755	0.0174	377.1697	0.0027	6.5509	0.1527
30	66.2118	0.0151	434.7451	0.0023	6.5660	0.1523
31	76.1435	0.0131	500.9569	0.0020	6.5791	0.1520
32	87.5651	0.0114	577.1005	0.0017	6.5905	0.1517
33	100.6998	0.0099	664.6655	0.0015	6.6005	0.1515
34	115.8048	0.0086	765.3654	0.0013	6.6091	0.1513
35	133.1755	0.0075	881.1702	0.0011	6.6166	0.1511

TABLE III.9

20% Interest Factors for Annual Compounding

	Single Payment		Equal-Payment Series			
	Future-Worth Factor	Present-Worth Factor	Future-Worth Factor	Annual-Worth Factor	Present-Worth Factor	Annual-Worth Factor
	To Find F	*To Find P*	*To Find F*	*To Find A*	*To Find P*	*To Find A*
	Given P	*Given F*	*Given A*	*Given F*	*Given A*	*Given P*
n	*F/P,i,n*	*P/F,i,n*	*F/A,i,n*	*A/F,i,n*	*P/A,i,n*	*A/P,i,n*
1	1.2000	0.8333	1.0000	1.0000	0.8333	1.2000
2	1.4400	0.6944	2.2000	0.4545	1.5278	0.6545
3	1.7280	0.5787	3.6400	0.2747	2.1065	0.4747
4	2.0736	0.4823	5.3680	0.1863	2.5887	0.3863
5	2.4883	0.4019	7.4416	0.1344	2.9906	0.3344
6	2.9860	0.3349	9.9299	0.1007	3.3255	0.3007
7	3.5832	0.2791	12.9159	0.0774	3.6046	0.2774
8	4.2998	0.2326	16.4991	0.0606	3.8372	0.2606
9	5.1598	0.1938	20.7989	0.0481	4.0310	0.2481
10	6.1917	0.1615	25.9587	0.0385	4.1925	0.2385
11	7.4301	0.1346	32.1504	0.0311	4.3271	0.2311
12	8.9161	0.1122	39.5805	0.0253	4.4392	0.2253
13	10.6993	0.0935	48.4966	0.0206	4.5327	0.2206
14	12.8392	0.0779	59.1959	0.0169	4.6106	0.2169
15	15.4070	0.0649	72.0351	0.0139	4.6755	0.2139
16	18.4884	0.0541	87.4421	0.0114	4.7296	0.2114
17	22.1861	0.0451	105.9306	0.0094	4.7746	0.2094
18	26.6233	0.0376	128.1167	0.0078	4.8122	0.2078
19	31.9480	0.0313	154.7400	0.0065	4.8435	0.2065
20	38.3376	0.0261	186.6880	0.0054	4.8696	0.2054
21	46.0051	0.0217	225.0256	0.0044	4.8913	0.2044
22	55.2061	0.0181	271.0307	0.0037	4.9094	0.2037
23	66.2474	0.0151	326.2369	0.0031	4.9245	0.2031
24	79.4968	0.0126	392.4842	0.0025	4.9371	0.2025
25	95.3962	0.0105	471.9811	0.0021	4.9476	0.2021
26	114.4755	0.0087	567.3773	0.0018	4.9563	0.2018
27	137.3706	0.0073	681.8528	0.0015	4.9636	0.2015
28	164.8447	0.0061	819.2233	0.0012	4.9697	0.2012
29	197.8136	0.0051	984.0680	0.0010	4.9747	0.2010
30	237.3763	0.0042	1181.8816	0.0008	4.9789	0.2008
31	284.8516	0.0035	1419.2579	0.0007	4.9824	0.2007
32	341.8219	0.0029	1704.1095	0.0006	4.9854	0.2006
33	410.1863	0.0024	2045.9314	0.0005	4.9878	0.2005
34	492.2235	0.0020	2456.1176	0.0004	4.9898	0.2004
35	590.6682	0.0017	2948.3411	0.0003	4.9915	0.2003

TABLE III.10

25% Interest Factors for Annual Compounding

	Single Payment		Equal-Payment Series			
	Future-Worth Factor	Present-Worth Factor	Future-Worth Factor	Annual-Worth Factor	Present-Worth Factor	Annual-Worth Factor
	To Find F	*To Find P*	*To Find F*	*To Find A*	*To Find P*	*To Find A*
	Given P	*Given F*	*Given A*	*Given F*	*Given A*	*Given P*
n	*F/P,i,n*	*P/F,i,n*	*F/A,i,n*	*A/F,i,n*	*P/A,i,n*	*A/P,i,n*
1	1.2500	0.8000	1.0000	1.0000	0.8000	1.2500
2	1.5625	0.6400	2.2500	0.4444	1.4400	0.6944
3	1.9531	0.5120	3.8125	0.2623	1.9520	0.5123
4	2.4414	0.4096	5.7656	0.1734	2.3616	0.4234
5	3.0518	0.3277	8.2070	0.1218	2.6893	0.3718
6	3.8147	0.2621	11.2588	0.0888	2.9514	0.3388
7	4.7684	0.2097	15.0735	0.0663	3.1611	0.3163
8	5.9605	0.1678	19.8419	0.0504	3.3289	0.3004
9	7.4506	0.1342	25.8023	0.0388	3.4631	0.2888
10	9.3132	0.1074	33.2529	0.0301	3.5705	0.2801
11	11.6415	0.0859	42.5661	0.0235	3.6564	0.2735
12	14.5519	0.0687	54.2077	0.0184	3.7251	0.2684
13	18.1899	0.0550	68.7596	0.0145	3.7801	0.2645
14	22.7374	0.0440	86.9495	0.0115	3.8241	0.2615
15	28.4217	0.0352	109.6868	0.0091	3.8593	0.2591
16	35.5271	0.0281	138.1085	0.0072	3.8874	0.2572
17	44.4089	0.0225	173.6357	0.0058	3.9099	0.2558
18	55.5112	0.0180	218.0446	0.0046	3.9279	0.2546
19	69.3889	0.0144	273.5558	0.0037	3.9424	0.2537
20	86.7362	0.0115	342.9447	0.0029	3.9539	0.2529
21	108.4202	0.0092	429.6809	0.0023	3.9631	0.2523
22	135.5253	0.0074	538.1011	0.0019	3.9705	0.2519
23	169.4066	0.0059	673.6264	0.0015	3.9764	0.2515
24	211.7582	0.0047	843.0329	0.0012	3.9811	0.2512
25	264.6978	0.0038	1054.7912	0.0009	3.9849	0.2509
26	330.8722	0.0030	1319.4890	0.0008	3.9879	0.2508
27	413.5903	0.0024	1650.3612	0.0006	3.9903	0.2506
28	516.9879	0.0019	2063.9515	0.0005	3.9923	0.2505
29	646.2349	0.0015	2580.9394	0.0004	3.9938	0.2504
30	807.7936	0.0012	3227.1743	0.0003	3.9950	0.2503
31	1009.7420	0.0010	4034.9678	0.0002	3.9960	0.2502
32	1262.1774	0.0008	5044.7098	0.0002	3.9968	0.2502
33	1577.7218	0.0006	6306.8872	0.0002	3.9975	0.2502
34	1972.1523	0.0005	7884.6091	0.0001	3.9980	0.2501
35	2465.1903	0.0004	9856.7613	0.0001	3.9984	0.2501

TABLE III.11

30% Interest Factors for Annual Compounding

	Single Payment		Equal-Payment Series			
	Future-Worth Factor	Present-Worth Factor	Future-Worth Factor	Annual-Worth Factor	Present-Worth Factor	Annual-Worth Factor
	To Find F	*To Find P*	*To Find F*	*To Find A*	*To Find P*	*To Find A*
	Given P	*Given F*	*Given A*	*Given F*	*Given A*	*Given P*
n	*F/P,i,n*	*P/F,i,n*	*F/A,i,n*	*A/F,i,n*	*P/A,i,n*	*A/P,i,n*
1	1.3000	0.7692	1.0000	1.0000	0.7692	1.3000
2	1.6900	0.5917	2.3000	0.4348	1.3609	0.7348
3	2.1970	0.4552	3.9900	0.2506	1.8161	0.5506
4	2.8561	0.3501	6.1870	0.1616	2.1662	0.4616
5	3.7129	0.2693	9.0431	0.1106	2.4356	0.4106
6	4.8268	0.2072	12.7560	0.0784	2.6427	0.3784
7	6.2749	0.1594	17.5828	0.0569	2.8021	0.3569
8	8.1573	0.1226	23.8577	0.0419	2.9247	0.3419
9	10.6045	0.0943	32.0150	0.0312	3.0190	0.3312
10	13.7858	0.0725	42.6195	0.0235	3.0915	0.3235
11	17.9216	0.0558	56.4053	0.0177	3.1473	0.3177
12	23.2981	0.0429	74.3270	0.0135	3.1903	0.3135
13	30.2875	0.0330	97.6250	0.0102	3.2233	0.3102
14	39.3738	0.0254	127.9125	0.0078	3.2487	0.3078
15	51.1859	0.0195	167.2863	0.0060	3.2682	0.3060
16	66.5417	0.0150	218.4722	0.0046	3.2832	0.3046
17	86.5042	0.0116	285.0139	0.0035	3.2948	0.3035
18	112.4554	0.0089	371.5180	0.0027	3.3037	0.3027
19	146.1920	0.0068	483.9734	0.0021	3.3105	0.3021
20	190.0496	0.0053	630.1655	0.0016	3.3158	0.3016
21	247.0645	0.0040	820.2151	0.0012	3.3198	0.3012
22	321.1839	0.0031	1067.2796	0.0009	3.3230	0.3009
23	417.5391	0.0024	1388.4635	0.0007	3.3254	0.3007
24	542.8008	0.0018	1806.0026	0.0006	3.3272	0.3006
25	705.6410	0.0014	2348.8033	0.0004	3.3286	0.3004
26	917.3333	0.0011	3054.4443	0.0003	3.3297	0.3003
27	1192.5333	0.0008	3971.7776	0.0003	3.3305	0.3003
28	1550.2933	0.0006	5164.3109	0.0002	3.3312	0.3002
29	2015.3813	0.0005	6714.6042	0.0001	3.3317	0.3001
30	2619.9956	0.0004	8729.9855	0.0001	3.3321	0.3001
31	3405.9943	0.0003	11349.9811	0.0001	3.3324	0.3001
32	4427.7926	0.0002	14755.9755	0.0001	3.3326	0.3001
33	5756.1304	0.0002	19183.7681	0.0001	3.3328	0.3001
34	7482.9696	0.0001	24939.8985	0.0000	3.3329	0.3000
35	9727.8604	0.0001	32422.8681	0.0000	3.3330	0.3000

Index

National Institute for
    Occupational Safety and
    Health (NOISH), 215
Natural systems, 6
Needs identification, for conceptual
    design, 37–38
Network, 144, 145
NOISH, *see* National Institute for
    Occupational Safety and
    Health (NOISH)
Nongovernment customers, systems
    engineering and, 22
"Nonparametric" statistics, *see* Chi-
    square testing
Normal random variable, 431

O

Open systems, 8–9
Operational concepts, 24
Operational feasibility, 39
Operational needs
    government agency and, 21–22
    nongovernment customers and, 22
    typical, 23
Operations Management (Stevenson),
    393
Operations research, 14
Optimistic criterion, *see* Maximax
    criterion
Optimization; *see also* System
    optimization models
    definition, 281, 282
    generalized theory of, 282
    methods and models, 282
OR-gate, FTA and, 190–191

P

Parallel (AND) construct, in CORE,
    130, 131
Parallel component structure, 171–172
Performance requirements, of
    system, 80
PERT, *see* Program Evaluation and
    Review Technique (PERT)
Physical models, 114
Physical systems, 7–8

Poisson process, 190, 325, 357, 365
    exponential time arrival and,
    325–328
    Poisson random variable, 428
Polaris nuclear submarine
    project, 407
Preliminary design, 51
    design reviews, 60–61
    design tools and analytical models,
    58–60
    functional analysis and function
    allocation, 55–58
Present value (*P*)
    annual value and, 373–376
    future value and, 370–371
Present value equivalence (*PE*) criterion,
    378–380
Preventive maintenance
    time (M_{pt}), 197
Proactive maintenance, 193
Probability
    characteristics, 421
    conditional, 422–423
    continuous random variables,
    425–427, 429–433
    definition, 419
    discrete random variables, 427–428
    experiment, 420
    independent events, 422
    mutually exclusive events, 421
    sample, sample space and event,
    420–421
Probability density function, *see*
    Probability distribution
    function
Probability distribution function,
    423–424
Probability mass function, 423–424
Process baseline (type
    D specification), 62
Product baseline (type C
    specification), 62
Program evaluation and review
    technique (PERT), 15, 407–411
    algorithm for, 408–411
    steps for conducting, 408
    time data, 407
ProModel, 59
Prototype-based simulation, 63